THE

ALPINE FLOWERS

OF BRITAIN AND EUROPE

Text and black & white illustrations by

Christopher Grey-Wilson

Colour illustrations by

Marjorie Blamey

Collins, St James's Place, London

William Collins Sons & Co Ltd
London · Glasgow · Sydney · Auckland
Toronto · Johannesburg

Dedicated to

The Alpine Garden Society

First published 1979

© 1979 Christopher Grey-Wilson (text, maps and drawings),
       Marjorie Blamey (colour paintings)

Paperback edition: 0 00–219288–8
Hardback edition: 0 00–219749–9

Filmset by Jolly & Barber Ltd, Rugby
Colour reproduction by Adroit Photo Litho, Birmingham
Printed by Wm. Collins Sons & Co Ltd, London & Glasgow

Contents

Introduction 7

Main Key 15

Conifer Families 22
Willow Family: Willows, Poplars 24-6
Birch Family: Birches, Alders 28
Hazel Family 28
Beech Family: Beech, Oaks 28
Dock Family: Docks, Bistorts 30
Pink Family: Catchflies, Campions, Soapworts, Pinks 32-6
Buttercup Family: Hellebores, Monkshoods, Delphinium, 38-54
        Anemones, Pasque Flowers, Clematis,
        Pheasant's-eyes, Buttercups, Meadow-rues,
        Columbines
Paeony Family 56
Barberry Family 56
Fumitory Family: Corydalis, Fumitories 56
Poppy Family: Poppies, Greater Celandine 58
Cress Family: Rockets, Woad, Cresses, Cuckoo Flower, 60-72
        Whitlow-grasses, Scurvy-grasses,
        Pennycress, Candytufts, Buckler Mustards,
        Pepperworts
Mignonette Family 72
Sundew Family 74
Stonecrop Family: Houseleeks, Stonecrops, Roseroots 74-8
Gooseberry Family 78
Grass of Parnassus Family 78
Saxifrage Family 80-88
Rose Family: Spiraea, Meadowsweets, Brambles, 90-108
        Raspberry, Roses, Agrimony, Avens,
        Cinquefoils, Strawberries, Lady's Mantles,
        Pears, Whitebeam, Hawthorns, Cherries
Pea Family: Laburnum, Brooms, Greenweeds, Gorse,
        Milk-vetches, Vetches, Peas, Vetchlings,
        Restharrows, Melilots, Medicks, Clovers, Trefoils,
        Kidney-vetches, Sainfoins
Wood-sorrel Family 128
Geranium Family: Cranesbills, Storksbills 130-32
Spurge Family 134
Flax Family 136
Milkwort Family 136
Mallow Family 138
Balsam Family 138
Oleaster Family 138
Daphne Family: Thymelaeas, Mezereons 138-40

Rockrose Family   140
St. John's Wort Family   142
Tamarix Family   142
Violet Family: Violets, Pansies   144-46
Willowherb Family   148
Dogwood Family   150
Ivy Family   150
Carrot Family: Masterworts, Eryngos, Hare's-ears,   150-58
         Lovages, Hogweeds
Wintergreen Family   160
Birdsnest Family   160
Diapensia Family   162
Crowberry Family   162
Heather Family   162-64
Primrose Family: Primroses, Rock-jasmines, Snowbells,   166-74
         Cyclamen, Pimpernels
Thrift Family   174
Phlox Family   174
Bindweed Family: Bindweed, Dodders   174
Bogbean Family   176
Gentian Family: Gentians, Centauries, Felworts   176-82
Periwinkle Family   182
Milkweed Family   184
Borage Family: Gromwells, Golden Drops, Honeyworts,   184-90
         Lungworts, Comfreys, Forget-me-nots,
         Hound's Tongue
Verbena Family   190
Mint Family: Germanders, Bugles, Hemp-nettles,   190-204
         Salvias, Deadnettles, Catmints, Woundworts,
         Ground Ivy, Self-heals, Calamints,
         Thymes, Mints, Lavender
Figwort Family: Toadflaxes, Mulleins, Foxgloves,   204-18
         Speedwells, Cow-wheats, Eyebrights,
         Bartsias, Louseworts, Rattles, Toothwort
Gloxinia Family   218
Globularia Family   220
Butterwort Family   220
Broomrape Family   222
Bedstraw Family   224
Honeysuckle Family: Honeysuckles, Elders   224-26
Moschatel Family   226
Valerian Family   226
Bellflower Family: Sheepsbits, Rampions, Bellflowers,   228-36
         Harebells
Scabious Family   238
Daisy Family: Golden Rod, Asters, Fleabanes,   240-62
         Catsfoots, Edelweiss, Cudweeds,
         Moon Daisies, Sneezeworts, Milfoils,
         Coltsfoots, Butterburs, Leopard's-banes,
         Thistles, Cornflowers, Knapweeds,
         Vipergrasses, Hawkbits, Hawkweeds

Lily Family: Aphyllanthes, Autumn Crocuses,   264-74
      Gageas, Onions, False Helleborines,
      Snakesheads, Tulips, Lilies, Squills, Hyacinths,
      May Lily, Solomon's Seals, Herb Paris
Yam Family: Black Bryony, Pyrenean Yam   274
Arum Family   274
Daffodil Family: Snowflakes, Snowdrop, Daffodils   276
Iris Family: Crocuses, Irises   278
Orchid Family   280-86

Black-and-white Plates
Nettle Family   288
Sandalwood Family   288
Mistletoe Family   288
Birthwort Family   288
Dock Family (contd. from p.30)   290
Pink Family (contd. from p.36)   292
Cress Family (contd. from p.72)   302
Spindle-tree Family   304
Maple Family   306
Box Family   308
Buckthorn Family   308
Water Starwort Family   310
Olive Family   310
Bedstraw Family (contd. from p.224)   312
Lime Tree Family   314
Nightshade Family   316
Figwort Family (contd. from p.218)   318
Plantain Family   322
Daisy Family (contd. from p.262)   324

Appendix 1   Additional species   335
Appendix 2   Key to Monocotyledons   341
Appendix 3   Key to trees and large shrubs   342
Appendix 4   Crucifer fruits   344
Appendix 5   *Dianthus* calyces   345
Appendix 6   Rock-jasmine *(Androsace)* leaves   346
Appendix 7   Monkshood *(Aconitum)* leaves   347
Appendix 8   Umbellifer fruits   348
Appendix 9   Saxifrage leaves   350
Appendix 10   Lousewort *(Pedicularis)* flowers   353
Appendix 11   Cornflower *(Centaurea)* flower-bracts   354

Glossary   355
Mountain Flowers in Britain   361
Conservation   363
Societies to join   364
Further reading   364
Index of English names   365
Index of Scientific names   373

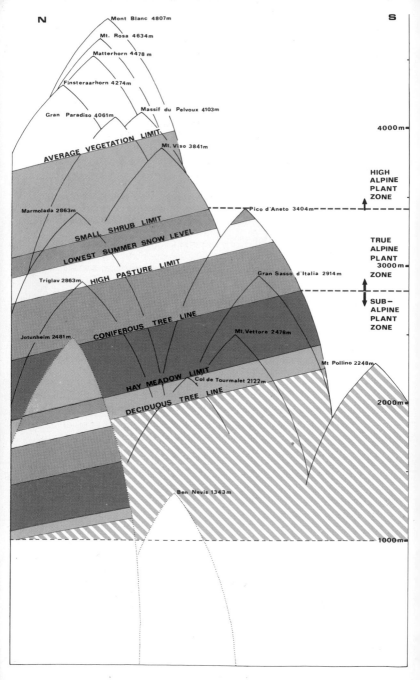

N   S

Mont Blanc 4807m

Mt. Rosa 4634m

Matterhorn 4478 m

Finsteraarhorn 4274m

Gran Paradiso 4061m

Massif du Pelvoux 4103m

4000m

Mt. Viso 3841m

AVERAGE VEGETATION LIMIT

HIGH
ALPINE
PLANT
ZONE

Marmolada 2863m

Pico d'Aneto 3404m

SMALL SHRUB LIMIT

LOWEST SUMMER SNOW LEVEL

TRUE
ALPINE
PLANT
ZONE

HIGH PASTURE LIMIT

Triglav 2863m

Gran Sasso d'Italia 2914m

3000m

SUB-
ALPINE
PLANT
ZONE

CONIFEROUS TREE LINE

Jotunheim 2481m

Mt.Vettore 2478m

Mt Pollino 2248m

HAY MEADOW LIMIT

Col de Tourmalet 2122m

DECIDUOUS TREE LINE

2000m

Ben Nevis 1343m

1000m

6

Introduction

What Are Alpine Plants?

The European mountains support many beautiful and often distinct plant species, variously adapted to the different habitats found in such regions.

Whereas many lowland species often reach up to alpine levels (above 1000m in this book) few true mountain plants come much below 1500m, at least in mainland Europe. In the extreme north and in the British Isles, typical mountain species can be found well below this level for reasons not entirely understood, but probably a combination of factors such as climate, exposure, soil types and the effects of the last ice ages has left pockets of mountain vegetation scattered in these areas.

Typical zones of vegetation at different altitudes can be seen in most mountain areas, as shown diagrammatically opposite. The zones often tend to merge gradually and may not all be present on a particular mountain, but the various tree and shrub lines can usually be easily identified. Various important mountain peaks are included to give an idea of the relative altitudes and zonings – these are not included for the British Isles. The further north one goes, the lower the altitude of the zones.

Sub-alpine zone This is the zone below the uppermost tree line and includes a rich flora of mixed lowland and mountain species. Meadows, woods, lanes and banks are often very floriferous with many large herbs, shrubs and trees. Any *cultivation* also occurs within or below this zone. At the uppermost limit the trees, generally pine, juniper or birch, become dwarf and stunted and often rather sparse.

True alpine plant zone Above the tree line where conditions are too rigorous to support tree growth, there is a poorer flora as regards numbers of species, but a much more highly specialised one. Dwarf and carpeting shrubs replace trees and short meadows support a rich growth of alpine plant species, often making rich splashes of colour when the spring arrives. If one walks higher still the meadows thin and the plants become even dwarfer and more highly adapted; ground or rock hugging species, often with cushion or tuft forming plants with small leaves and large brightly coloured flowers. In this high alpine zone plants are often widely scattered but are often the most rewarding to find and see.

Habitat is specially important for many mountain plants; many are highly specialised to a particular type. Occasionally rivers, screes or moraines may bring species down to much lower altitudes than they are normally found at. Exploring valleys which run down from a high mountain or a glacier will often reveal interesting plants.

The Text

All the native flowering plants of the Alps, Pyrenees and Apennines and

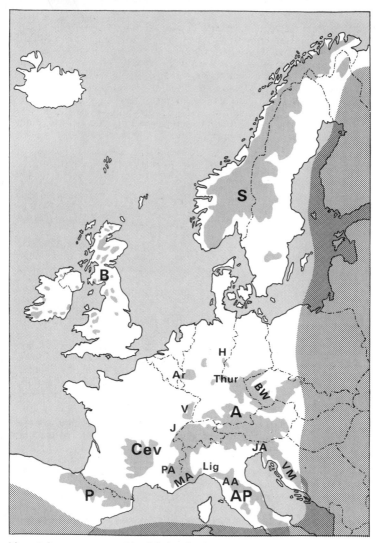

Mountain areas covered by this book, showing all land above 1000 m, and also lower areas in Northern Europe and Britain where mountain plants are found.

Main regions

A   Alps
AP Apennines
B   mountains of Britain and Ireland
Cev Cevennes and Massif Central
P   Pyrenees
S   mountains of Scandinavia

Subsidiary regions

Ar Ardennes
BW Böhmer Wald*
H   Harz Mountains*
J   Jura
JA Julian Alps
Lig Ligurian Apennines

MA Maritime Alps
PA Provence Alps
Th Thuringer Wald*
V   Vosges
VM Velebit Mountains**

*included in the text under nA (Northern Alps)
**included in the text under seA (south-eastern Alps)

the mountainous regions of Britain and Northern Europe that occur above one thousand metres (3281 ft) altitude are included in the text, with the exception of grasses, sedges and rushes. The non-flowering plants, which include ferns, horsetails and their allies, are omitted as also are introduced or alien species of flowering plants.

The less attractive and insignificant lowland species which may occasionally exceed 1000 m are also excluded to avoid overburdening the text with species which are not truly mountain plants. In Britain and Northern Europe a number of true alpine species are found well below 1000 m and this is taken into account in the text.

The **area chosen** is centred on the mountain masses of Central and Southwest Europe, but it also includes Northern Europe and the British Isles. In the southwest the whole of the Pyrenees, French and Spanish, and the Cevennes of Southern France are included. In Central Europe the whole of the Alps are covered, with the Bavarian Alps of Southern Germany as the most northerly point and to the west and south-west the Jura, the Dauphine, Maritime and Provence Alps. The eastern limit is the Niedere Hohen of Central Austria and the south-eastern the Karawanken Alps of South-east Austria and the Julian Alps of North-west Yugoslavia. The southward extension into Italy runs from the Ligurian Apennines to the Northern and Central Apennines, which floristically have much in common with the Alps (see map on pp.10-11). The Spanish mountains south of the Pyrenees have not been covered, i.e. the book will not be found to be comprehensive for them though the majority of their common plants are here covered, at least as regards the North of Spain.

Description. The text and illustrations complement each other and must be read in conjunction: descriptive details are generally confined to points which cannot easily be seen from the illustrations, such as height, whether the plant is evergreen, habitat, altitudinal range, distribution and flowering period. Important diagnostic characters are stressed in *italics*. A number of assumptions are made in the descriptions: unless otherwise stated plants are erect; and the flower measurements indicate diameter not length; and leaves are both toothed and stalked.

The characters described are, unless it has been absolutely unavoidable, ones that can be readily seen and which do not require the plant to be either picked or dug up.

Abbreviations used in the text

agg	— aggregate	lf, lvs	— leaf, leaves
ann	— annual	med	— medium
bien	— biennial	per	— perennial
fl, fls	— flower(s)	sp	— species
flg	— flowering	subsp	— subspecies
flheads	— flowerheads	var	— variety
fr	— fruit		

For other abbreviations used in the text, see below.

Family and generic descriptions. Notes on each **family** or **genus** are given in most cases, on the characters (only those useful for identification) common to its members. For example the number of

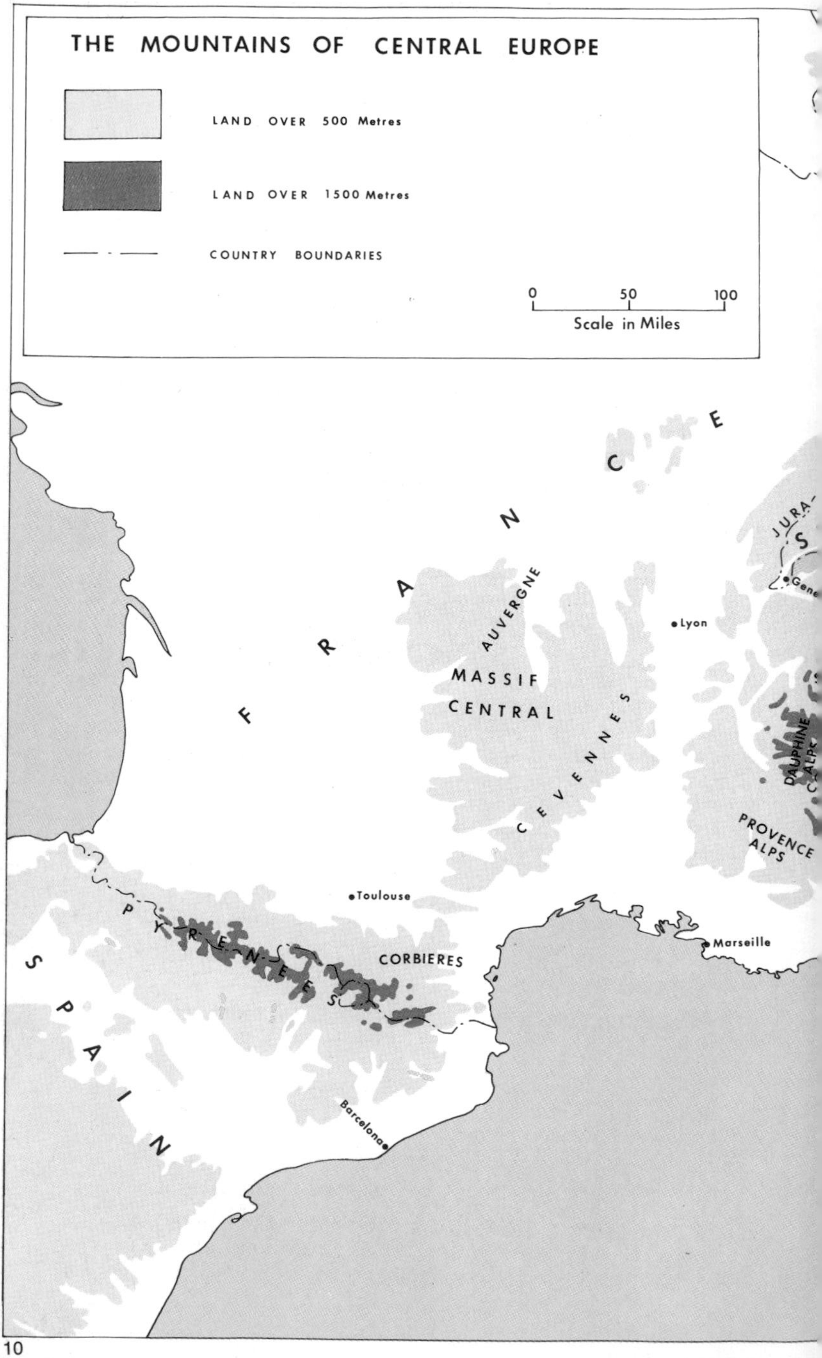

THE MOUNTAINS OF CENTRAL EUROPE

LAND OVER 500 Metres

LAND OVER 1500 Metres

— · — · — COUNTRY BOUNDARIES

0   50   100
Scale in Miles

FRANCE

JURA

AUVERGNE

MASSIF CENTRAL

CEVENNES

●Lyon

Gene●

DAUPHINE ALPS

PROVENCE ALPS

●Toulouse

PYRENEES

CORBIERES

●Marseille

SPAIN

Barcelona●

THURINGIAN MTNS.

CZECHOSLOVAKIA

WEST GERMANY

•Munich

A U S T R I A

•SES

BLACK FOREST

BAVARIAN ALPS

•Basel

Salzburg•

NIEDERE TAUERN

•Ingsbruck

HOHE TAUERN

NORIC ALPS

•Graz

TZERLAND

GLARUS ALPS

ÖTZTAI ALPS

BERNESE OBERLAND

LEPONTINE ALPS

RHAE...

ORTLER ALPS

CARNIC ALPS

KARAWANKEN

PENNINE ALPS

DOLOMITES

JULIAN ALPS

BERGAMASQUE ALPS

Y U G O S L A V I A

RAIAN ALPS

•Milan

•Verona

Venice•

•Trieste

VELEBIT MTNS.

LIGURIAN APENNINES

I

•TIME ALPS

APUAN ALPS

NORTHERN APENNINES

Rimini•

T

A

L

Y

CORSICA

CENTRAL APENNINES

•Rome

ABRUZZI MTNS.

SARDINIA

11

petals or sepals, or flower-shapes such as the composite flowers of the Daisy Family, the bell-flowers of the Campanulas or the pea-shaped flowers of the Pea Family. This saves much repetition, and the notes should be borne in mind when reading species descriptions.

Common names precede each species in the text. These follow, as far as it is possible, commonly used English names. Where no distinct common name exists we have derived one from the Latin epithet for the species, such as Swiss Rock-jasmine – *Androsace helvetica*. This has been done purely for convenience in order to avoid overburdening the reader with Latin names alone, although it is strictly against a botanist's best traditions.

Latin names follow as far as possible those used in *Flora Europaea* and this includes all the Dicotyledonous families and some genera of Monocotyledons. An + after the generic name indicates that the names used in the text under that particular genus do not follow *Flora Europaea*, because the accounts were not complete at the time of publication of this book. In such instances the names selected have been at the discretion of the author. Important or well known Latin synonyms are included in brackets after the main name.

Number Each plant in the text has a number that corresponds with its number on the opposite plate. Closely related plants are often placed under the same number with a subsidiary letter, 11a, 11b etc. In general the text descriptions of these subsidiary species, subspecies or varieties indicate only how it may differ from the main species, which should thus be referred to at the same time.

Aggregates Certain groups of plants are extremely difficult to distinguish except by detailed botanical examination. Such complex groups of species have been lumped together into aggregates (agg. in the text) to save confusion. This situation arises in certain genera in particular: *Alchemilla* and *Rubus* in the Rose Family, *Cirsium, Centaurea, Hieracium* and *Senecio* in the Daisy Family. To include illustrations of all the species involved in these complexes would have unnecessarily burdened the text.

Height is indicated as follows:

Tall	—	over 60cm (2 ft)
Medium	—	between 30–60cm (1–2 ft)
Short	—	between 10–30cm (4–12 in)
Low	—	0–10cm (0–4 in)

Status Annual, biennial or perennial status is shown in each case. Shrubs and trees are also indicated, together with a general guide to their average height (given in metres) and whether they are deciduous or evergreen.

Habit is often an important clue in the identification of alpine plants, thus some stress is made on the plant form – for instance tight or loose cushion habit, prostrate, creeping or tufted habit, and so on.

Hairiness or hairlessness is indicated in most cases. The presence of hairs on a particular part of a plant is often an important clue to identification.

Leaf-shape is another important and characteristic feature of many plants. See Glossary (p.355) for definition of terms such as *pinnate* and *palmate*.

Flower shape and colour Flower shape is usually indicated, such as bell, trumpet or saucer-shaped. Colour refers to petals, or to sepals when there are no petals present. Many plants show a remarkable variation in flower colour and many produce occasionally white or pale-coloured forms, but in such instances there are usually normal-coloured flowers close-by. The plants illustrated are generally the normal form and the one most likely to be spotted in the wild; but the colour range is also indicated in the text for each species.

Flowering time is indicated – April-June for instance. These periods give the most likely time to find a particular plant in flower, although this varies according to altitude and to location. This variation is particularly significant in mountainous regions: plants in the southern part of the range of a particular species flower earlier than those to the north and plants found at higher altitudes flower later than those from lower altitudes. Besides this, plants growing by snow patches may be considerably delayed in flowering: this is often true of certain plants, buttercups, crocuses and snowbells for instance.

One of the great joys of mountain and alpine plants is that they may be found in flower for a considerable period depending both on where they are found and at what altitude. Plants found in seed at a low altitude may often be found in flower by simply climbing further up the mountain slope.

Rarity is generally difficult to estimate. Certain plants occur extremely locally due to habitat conditions, but at the same time are to be found over a rather wide area. These cannot be considered as rarities. On the other hand certain species are to be found in a very limited area only and within this they may or may not be frequent: it is the infrequently occurring ones that can be considered rare. Such plants are indicated by an asterisk * in the text. If a plant is rare in only part of its range then this is also shown in many instances.

Distribution of each species described in the text is given. A system of coding has been devised to indicate various areas within the scope of this book. These are as follows:

A	— Alps		Lig	— Ligurian Apennines
AA	— Apuan Alps		MA	— Maritime Alps
Ap	— Apennines		P	— Pyrenees (including the
B	— Great Britain			Corbières)
Cev	— Cevennes (including		PA	— Provence Alps
	the Auvergne & other		S	— Scandinavia
	areas of the Massif		T	— Throughout the area
	Central over 1000 m)			
J	— Jura			
JA	— Julian Alps			

As well as these, prefixes in *small lettering* are added to facilitate subdivision of these areas where it is necessary

n —	north	w —	west
s —	south	c —	central
e —	east		

Thus 'eP, swA, Lig' indicates that the plant in question is to be found in the eastern Pyrenees, the south-west Alps and the Ligurian Apennines.

Occasionally it happens that a plant is found only in one small area of, for instance, the Alps. In such instances the name of the area is shown in full after the code (eA – Styrian).

Habitat The places where plants are to be found growing is often an important clue to their identity. Some plants occur in a variety of different habitats, (e.g. meadows, open woodland and rocky places) while others are very specific (e.g. limestone rocks). Knowing the rock or soil type may help to separate closely related plants, especially in genera such as the Gentians, Primroses, Rock-jasmines and Saxifrages.

Altitude is in metres and a conversion table to feet is given on page 384. Where a single figure is given (to 2900m) it implies that the plant in question is found between 1000 and 2900m, and although it may also occur below 1000m it is the upper altitudinal limit that is the more critical. Where a particular species occurs over 1000m then the altitudinal range is usually presented (1600-3400m). The altitudinal range is often somewhat speculative for it is dependant on existing records and it should be treated as an indicator rather than a hard and fast rule.

Identification

This book has been designed so as to be easily carried. Please *take the book to the plant* to avoid picking or spoiling specimens. Picking prevents a flower from setting seed and reproducing itself. Besides, identification is often much easier on the spot. It is also a good idea to carry a small notebook and pencil: a quick sketch of flowers, leaves and fruits can be a very useful record as well as a help in checking identifications.

The following points should be particularly considered:

- Height and nature of plant – whether shrubby, annual, perennial, herbaceous or cushion forming and so on.
- Shape and arrangement of leaves – whether stalked or toothed, hairy or not.
- Flower shape and colour – the number of sepals and petals, stamens and styles, whether the petals are joined together or not and to what degree.
- arrangement of flowers – solitary, in small tight clusters, in branched clusters, whether upright or drooping.
- Shape and colour of fruits – whether fleshy or dry, whether they split on ripening and if so, how many divisions.
- Habitat – whether they are growing in woodland, meadows, marshy areas or by streams and so on. If a rock plant, type of rock?
- Abundance – whether common or few.
- Locality – often important in determining a species or subspecies finally.
- Altitude – to within 50m if possible using maps as a guide.

Remember that often two similar-looking species occur in two quite different regions or habitats, for instance one in the eastern Pyrenees and one in the Alps.

The Main Key

Trees and large shrubs: see p.342   Herbs and undershrubs
see below

Leaves with parallel veins, linear or strap-shaped, sometimes arrow-shaped; flowers with usually 6, sometimes 3, petals; stamens 6 or 3, never more than the number of petals – MONOCOTYLEDONS see p. 341

Leaves with 'net veins', usually broad, rarely linear; flowers with 4, 5 or more petals or petal lobes, rarely 2, 3 or 6; stamens rarely 3 or 6, often more than the number of petals – DICOTYLEDONS see below

INDIVIDUAL FLOWERS LARGE OR CONSPICUOUS

Two petals
Alpine Enchanter's Night-shade **148**

Three petals
Ranunculus hyperboreus, **50**
Asarabacca **288**

Four petals – separated to the base or almost so

Erect Clematis **16**, Crucifers **60-72**, Southern Woodruff **224**, Alpine Bastard Toadflax **288**, Pearlworts **292**, Woodruff, Bedstraws **312**

Traveller's Joy **46**, Swiss Bedstraw etc **224**

Crucifers **60-72, 304**, Lady's Bedstraw etc **224**

Barrenwort **56**

Traveller's Joy **46**, Spindletree **304**, Mistletoe **288**

Blue Arabis **64**, Lomatogonium **182**

Blue Woodruff **224**, Cevennes Rockcress **302**

Crucifers **60-72**, Woodruffs **224**

Crucifers **60-72**, Pyrenean Woodruff **224**, Reddish Bedstraw **314**

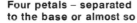

Baneberry **40**, Crucifers **60-72**, Poppies **58**

Rosebay Willowherb **148**

Candytufts **70**

Speedwells **208, 320**

Cranberry **162**

Alpine Clematis **16**

Four petals – joined together, often forming a tube
Apenninè Gentianella **180**, Lesser Centaury **182**

 Spurge Laurel **140**

  Gentians etc **178-82**

 Mezereon etc **140**

  Thymelaeas **178**

  Five petals – separated to the base or almost so

 Stonecrops **76-8**, Saxifrages **80-8**, Sandworts **298-300**

 Marsh Cinquefoil **98**

 Stonecrops **76-8**, Chickweed Wintergreen **174**, Bastard Toadflaxes **288**

  Stonecrops **76-8**

  Marsh Felwort **182**

 Flowers small or medium (to 15mm across): plants on pp. **30, 40, 48, 50, 74, 80-90, 98-102, 128, 162, 226**

 Flowers large (15mm or more across): plants on pp. **38, 42, 44, 48, 50, 78, 80-90, 98-102, 106, 136, 140**

 Flowers small or medium (to 15mm across): plants on pp. **34, 96-102, 128, 140, 142, 174**

 Flowers large (15mm or more across): plants on pp. **38, 40, 42, 44, 48, 50, 96-102, 136, 140, 142**

 Shrubby Rockrose **140**

 Mousetail **46**, Moschatel **226**, Alpine Spindle-tree **304**

 Love-in-a-mist **38**, Hepatica **44**, Meadow Cranesbill **130**, Flaxes **136**, Marsh Felwort etc **182**, Ramonda **218**

 Plants on pp. **34, 78, 90, 104, 106, 130-2, 136, 164, 300**

 Paeony **56**, Arctic Bramble **90**, Cranesbills etc **130-2**

 Campions etc **32-4**, Buttercups **48-52**, Roses **92-4**, Cinquefoils **98-102**, Dwarf Mallow **138**

 Spoon-leaved Soapwort **34**, Buttercups **48-52**, Alpine Avens **96**

 Campions etc **32-4**, Buttercups **48-52**, Roses **92-4**, Pink Cinquefoil **102**, Cranesbills **130-2**, Mallows **138**

  Lychnis etc **30**, Cranesbills **130-2**, Common Mallow **138**

 Catchflies **32**, Mouse-ears **294-6**

 Catchflies **32**

 Stitchworts **292**

 Alpine Catchfly **32**

16

Ragged Robin **30**

 Pinks **34-6**

 Sowbread **172**

 Black Nightshade etc **316**

 Small Pasque Flower **44**

 Wintergreens **160**, Cassiopes **162**

 Barberry **56**

 Wintergreens **160**

 Dwarf Alpenrose **164**

 Hellebores **38**

 Violet and Pansies **144-6**

Five petals – joined together, often forming a tube
 Rock-jasmines **170-2**, Forget-me-nots **188**

 Gromwells **184**

 Vitaliana **168**

 Forget-me-nots etc **188-90**

 Rock-jasmines **170-2**, Shrubby Gromwell **184**

 Gromwells **184**, alkanet **188**

 Gentians, Centauries **176-82**

 Brookweed etc **174**, Swallow-wort **184**

 Creeping Jenny etc **174**, Swallow-wort **184**

 Jacob's Ladder **174**, Edrai-anthus **230**, Bellflowers **232-6**

 Mucizonia **74**, Chaffweed etc **174**

 Hen-and-chickens House-leeks **76**

 Bogbean etc **176**

 Gentians **176-8**

 Viper's Bugloss **186**

 Wulfenia etc **208**

 Rhododendron etc **162-4**

 Primroses etc **166-8**

 Primroses **166-8**, Water Violet **172**

 Lesser Periwinkle **182**

 Bearberry etc **164**, Twin-flower etc **224**, Bellflowers **232-6**

 Yellow Birdsnest **160**

 Henbane **316**

 Blue-berried Honeysuckle **224**, Yellow Bellflower **234**

 Heather **164**, Bellflowers **232-6**

 Cowberry etc **162-4**

 Heaths **164**, Snowbells **172**

 Deadly Nightshade etc **316**

 Snowbells **172**

 Bilberry **162**, Marsh Andromeda **164**

 Tozzia **210**

 Figworts **318**

 Lungworts **186**, Hound's Tongue etc **190**, Valerians **226-8**

 Valerians **226-8**

 Golden Drops, Comfreys etc **184-8**

 Lesser Honeywort **186**

Six petals or more
also Key to Monocotyledons, p. **341**)

  Stonecrops **76-8**

 Stonecrops **76-8**, Crowberry **162**

Mousetail **46**

   Houseleeks **74**

 Anemones **42**, Hepatica **44**, Mountain Avens **96**, Chickweed Wintergreen **174**

 Callianthemum **38**

 Winter Aconite **38**, Pheasant's-eyes **46**, Buttercups **48-50**, Avens **96**, Yellow-wort **182**

  Wood Anemones **42**, Hepatica **44**

 Pheasant's-eyes **46**, Paeony **56**

 Globeflower **38**

 Yellow Anemone **42**, Alpine Pasque Flower **44**

  Pasque Flowers **44**

 Hen-and-chickens Houseleek **74**

 Gentians **178**

Flowers with a distinct spur or spurs

   Pansies and Violets **144-6**

 Wolfsbanes **40**

 Larkspurs **42**

 Fumitories etc **56**

 Touch-me-not **138**

 Toadflaxes **204**, Butterworts **220**

 Lesser Bladderwort **220**

     Valerians **226-8**

  Columbines **54**

One-lipped flowers

  Germanders **190**, Bugles **192**

  Honeysuckles **224**

 Two-lipped flowers

 Fumitories etc **56**

Milkworts **136**

Labiates **190-204**, Figwort Family **204-218**, Broomrapes **222**

INDIVIDUAL FLOWERS SMALL, IN CLUSTERS OR FLOWERHEADS (see also Onions, p. 266)

Globular heads, tightly packed (see also Onions, p. 266)

 Willows **24-6**, Salad Burnet **96**, Lady's Mantles **102-4**, Dodders **174**, Moschatel etc **226**

 Thrifts etc **174**, Globularias **220**, Blue Woodruff **224**, Rampions etc **228-30**, Devils-bit Scabious **238**

 Great Burnet **96**

Rampions **228-30**

 Thrifts etc **174**, Woodruffs **224, 312**

In long spikes

 Bistorts **30**, Mignonettes etc **72**, Italian Burnet **96**, Docks **290**

Oblong heads, tightly packed

Willows **24-6**

 Plantains **322**

In drooping spikes, catkin-like

 Poplars **26**, Hazel **28**

Oaks **28**, Nettles **288**

 Hacquetia **150**, Hare's-ears **156**

In umbrella-like clusters

Umbellifers, **150-8**

 Sanicle etc **150**

  Scabiouses **238**

In daisy or dandelion-like heads

Composites **240-62**

With conspicuous bunches of stamens

 Meadow-rues **52**

In thistle-like heads

 Eryngos **152**, Thistles **250-6**

Pine Family Pinaceae

Trees with needle leaves. Fruit a woody cone. Widely planted.

1 SILVER FIR *Abies alba.* Evergreen pyramidal tree to 50 m. Branches regularly whorled. Lvs single, grooved, whitish beneath, dark green above, leaving an *oval scar*. Cones erect, 10-20 cm, with *triangular bracts* between the scales. Woods and forests, to 2100 m. Fls Apr-May. T(B, S).

2 NORWAY SPRUCE *Picea abies.* Evergreen pyramidal tree to 60 m. Branches regularly whorled. Lvs single, 4-sided, pointed, grass-green, falling to leave a *peg-like projection.* Cones pendant, cylindric, 10-18 cm, *no bracts* protruding. Forests, to 2200 m. Fls Apr-June. T(B). **2a**, p.335.

3 LARCH *Larix decidua.* Deciduous pyramidal tree to 35 m. Twigs yellowish, hairless. Lvs single and tufted, slender pointed, pale green. Male fls yellow, female pink. Cones erect, *egg-shaped* 2-3·5 cm, *no bracts* protruding. Open forests and rocky slopes, to 2500 m. Fls Mar-June. A, Ap (B, S).

4 AUSTRIAN or BLACK PINE *Pinus nigra.* Evergreen pyramidal tree to 50 m. *Bark* grey-black. Lvs in pairs, straight, deep green. Cones pendant, pointed, 5-8 cm, yellowish-green. Forests, to 1800 m. Fls May-June. eA, n, cAp(T). **4a** *P.n. salzmannii* smaller tree. Lvs curved, bluish-green. Cones 4-6 cm. P, Cev. **4b Scots Pine** *P. sylvestris* is a smaller tree than 4, dome shaped, *bark* reddish, flaking. Lvs greyish, twisting. Cones egg-shaped, 3-6 cm, brown. Forests, moors and heaths, to 2300 m. Fls May-June. T.

5 MOUNTAIN PINE *Pinus mugo.* Evergreen *shrub* to 3·5 m, spreading, rather contorted. Bark greyish-brown scaly. Lvs paired, slightly twisted, deep green. Cones erect or slightly pendant, egg-shaped, 2-5 cm, shining. Rocky and stony places, peaty bogs and screes, to 2700 m. Fls May-June. A, n, cAp. **5a**, p.335 *Pinus uncinata* similar but a small tree to 15 m. *Cones* narrower, 5-7 cm. P, w, cA.

6 AROLLA PINE *Pinus cembra.* Evergreen pyramidal or irregular tree to 25 m, bark reddish-grey. Lvs in groups of 5, slender, blue-green. Cones erect, dark *violet-brown.* Open woods and stony places, to 2600 m. Fls May-June. A.

Cypress Family Cupressaceae

7 COMMON JUNIPER *Juniperus communis.* Evergreen shrub or small tree to 6 m. Lvs in groups of 3, spine-tipped, greyish. Fls yellow, male and female separate plants. Fr berry-like, green then blue-black when ripe. Rocky, stony places and moors, to 1500 m. Fls June-July. T. **7a** *J.c. hemisphaerica* is a rounded *shrub.* Lvs broader with a *white band* down the centre. To 2500 m. Fls May-June. A. **7b Dwarf Juniper** *J.c. nana* (= *J. nana*) is *prostrate,* lvs suddenly narrowed to a short point. To 3600 m. June-July. T.

8 FRENCH ALPINE JUNIPER *Juniperus thurifera.* Evergreen pyramidal tree to 20 m. Lvs scale-like in opposite pairs, overlapping, *not spiny,* deep green. Cone globose, 7-8 mm, green then dark purple when ripe. Rocky and stony places, to 1400 m. Fls Apr-June. swA-French. **8a**, p.335.

Yew Family Taxaceae

9 YEW *Taxus baccata.* Evergreen pyramidal tree or shrub to 20 m. Lvs single, in 2-rows, dark green. Fls yellowish, male and female on separate plants. Fr in a succulent *reddish-pink cup.* Woods and rocks slopes, usually on lime, to 1800 m. Fls. Mar-Apr. T.

Ephedra Family Ephedraceae

10 EPHEDRA *Ephedra distachya.* Low, twiggy, pale green shrub to 0·5 m. Lvs tiny, in opposite pairs sheathing the curved twigs. Fls tiny, greenish, male and female on separate plants. Fr globose 6-7 mm with *fleshy red scales.* Sandy and dry places, stream banks, to 1100 m. Fls May-June. P, A, Ap. **10a** *E.d. helvetica* has erect, deep green, twigs. swA.

Willow Family Salicaceae

Deciduous trees and shrubs with alternate lvs. Bud scales solitary (Willows) or several overlapping (Poplars). Stipules often present, small or large. Fls in upright catkins in Willows or drooping catkins in Poplars, petalless, male and female on separate plants, often appearing before, or with the young lvs; male with prominent yellow or orange stamens; female green. Fr a small capsule containing many woolly-white seeds.

1 NETTED WILLOW *Salix reticulata.* Prostrate mat-forming undershrub. Lvs rounded or oval, 10-30mm, *untoothed,* long-stalked, shiny green above, whitish and conspicuously *net-veined beneath.* Catkins long-stalked; with lvs. Damp rocks and screes, 1200-2500m. June-Aug. B, S, P, A. **1a Least Willow** *S. herbacea* □ has smaller, *slightly toothed,* lvs, 6-20mm, shiny green *beneath;* catkins short-stalked. Damp meadows and rocks, 1200-2800m. T. **1b Polar Willow** *S. polaris* like 1a but lvs *untoothed* and broadly elliptical; fr hairy. To 1700m. July. S.

2 RETUSE-LEAVED WILLOW *Salix retusa.* Prostrate mat-forming undershrub. Lvs oblong, *often notched,* 8-20mm, untoothed, short-stalked, shiny green above and beneath, hairless, veins *inconspicuous.* Catkins with lvs, short-stalked. Damp meadows and rocks, 1200-3000m. July-Aug. A, P, Ap. **2a** *S. serpyllifolia* is *more compact* and tightly pressed to the ground; lvs 4-10mm, overlapping. c, e, se A.

3 FINELY-TOOTHED WILLOW *Salix breviserrata.* Prostrate or dwarf undershrub. Branches crooked. Lvs oval, finely toothed, 10-20mm, shiny green above and beneath, *hairy on margins.* Catkins stalked, dark-purplish; with lvs. Damp meadows streamsides and moraines, 1700-3000m. June-July. P, A. **3a** *S. alpina* is always prostrate with *untoothed* lvs. To 2500m. eA. **3b** *S. myrsinites* like 3 but *dead lvs* persisting until end of following season. To 1750m. May-June. B, S.

4 SILKY WILLOW *Salix glaucosericea.* Knotted shrub 1-2m. Lvs lance-shaped, broadest above middle, 55-75mm, *silkily hairy,* pale green above and bluish-green beneath. Catkins long; with lvs. Damp and stony places, 1700-2550m. June-July. A. **4a** *S. glauca* has broader lvs with *tangled hairs.* To 1750m. S.

5 ALPINE WILLOW *Salix hegetschweileri.* Shrub 0·5-3m tall. Lvs oval, broadest above middle, *hairless,* slightly toothed, blunt ended, shiny green above, bluish-green beneath. Catkins short-stalked. Wet stony places, streamsides, 1500-2200m. May-June. A. **5a** *S. phylicifolia* has lvs *tapered* at both ends. To 1750m. B, S. **5b** *S. bicolor* like 5 but lvs *silkily-hairy* when young. P*.

6 PYRENEAN WILLOW *Salix pyrenaica.* Low shrub to 0·5m. Twigs reddish-brown. Lvs oval-elliptical, 10-30mm, *rounded* at base, untoothed, slightly hairy, always hairy-margined. Catkins long and loose; with lvs. Damp meadows and rocks, to 2500m. July. P.

7 APUAN WILLOW *Salix crataegifolia.* Prostrate or upright shrub to 1m. Twigs *dark purple.* Lvs broadly-elliptical, 65-115mm, finely toothed, bright green above, silkily-hairy beneath at first. Rocky places, on limestone, to 1800m. May-June. AA.

8 AUSTRIAN WILLOW *Salix mielichhoferi.* Shrub to 2m. Twigs hairy at first, blackish to brownish-green. Lvs lance-shaped or oval, scarcely toothed, dull green, conspicuously *net-veined* beneath. Catkins short-stalked; with lvs. Rocky places to 2100m. June-July. eA.

9 HAIRLESS WILLOW *Salix glabra.* Erect *hairless* shrub to 1·5m. Twigs dark brown. Lvs broadly elliptical or oval, toothed, *waxy green* above and greyish beneath. Catkins long-stalked; with lvs. Damp places 1300 to 2100m. June-July c, e, se A.

1a Least Willow

Willow Family *(contd.)*

1 LAGGER'S WILLOW *Salix laggeri.* Shrub 1-3m tall. Twigs knotted, brown or blackish, white-felted when young. Lvs narrow-elliptical to oblong, untoothed, deep green above, downy-white beneath, stipules arrow-shaped. Catkins large. Rocky places and banks to 2000m. May-June. A. **1a** *S. appendiculata* has *grey-brown*, hairy twigs and broader lvs, sometimes toothed, with conspicuous net-veins beneath, to 2300m. Apr-June. A, Ap.

2 MOUNTAIN WILLOW *Salix arbuscula.* Shrub to 2m. Twigs *hairless*, ridged under the bark. Lvs elliptical-lance-shaped, 5-40mm, *pointed,* toothed, shiny green above, greyish often hairless beneath, stipules inconspicuous. Catkins short-stalked, scales with rusty hairs. Damp meadows, stony places and moraines, to 1300m. May-June. G,S. **2a** *S. foetida* has smaller, deeply toothed lvs, with conspicuous *white glands.* On acid rocks, 1700-2800m. June-July. cP,w,cA. **2b** *S. waldsteiniana* has larger lvs than 2, scarcely toothed; catkins *long-stalked.* On limestone rocks, 1700-2800m. eA.

3 LARGE-STIPULED WILLOW *Salix hastata.* Variable shrub to 1·5m. Twigs greenish or brownish, shiny, hairless. Lvs broadly oval or elliptical, untoothed or finely toothed, dull pale green, hairless. *Stipules large,* oval, not persistent. Catkins large, long-stalked with long white hairs. Wet meadows, rocky places and stream banks, to 2500m. May-Aug. S, P*, A.

4 SWISS WILLOW *Salix helvetica.* Shrub 0·5-2m. Twigs thin, grey-brown becoming chestnut brown. Lvs oval, broadest above middle, 15-40mm, *untoothed,* shiny greenish above, *white-felted* beneath. Catkins large, short-stalked. Wet meadows and rocky places, 1700-3000m. June-July. A. **4a Lapland Willow** *S. lapponum* has lvs *greyish hairy* above and beneath, rather crowded at tips of twigs; catkins stalkless. Wet heaths, streamsides and stony places, to 2600m. May-July. B, S, c, eP.

5 BLUE-LEAVED WILLOW *Salix caesia.* Prostrate or upright shrub to 1m, *hairless.* Twigs brown, dull when young. Lvs elliptical or oval, 10-15mm, untoothed, dull bluish-green above and beneath. Catkins reddish-brown with violet anthers. Damp meadows and rocks, streamsides, 1700-2500m. June-July. A.

6 WOOLLY WILLOW *Salix lanata.* Variable shrub to 3m. Twigs *thickly felted.* Lvs broad-oval, 10-25mm, untoothed, *yellowish-hairy* at first, then grey-hairy becoming almost hairless. Catkins yellow-hairy. Damp and stony places, to 1750m. May-July. B, S. **6a** *S. glandulifera* has lvs broadest above the middle with a *glandular margin.* S.

A number of larger Willows reach above 1500m, though they are more characteristically lowland species.

7 ASPEN *Populus tremula.* Spreading tree to 20m, suckering freely. Bark smooth, greyish-brown. Buds slightly sticky. Lvs *rounded,* blunt-toothed, green, soon hairless, trembling in breeze on long thin stalks. Catkins drooping, reddish or purplish, *before lvs.* Damp woods and heaths to 2100m. Mar-May. T.

8 BLACK POPLAR *Populus nigra.* Spreading tree to 30m. Bark rugged, blackish. Buds sticky. Lvs *ace of spades,* pointed, toothed, deep green, hairless, long-stalked. Catkins drooping, before lvs. Moist places, often by river banks, to 1800m. Mar-Apr. A, Ap, but widely planted. **8a White Poplar** *P. alba* □ has 3-5-lobed lvs which are *white-downy* beneath. Damp woods, occasionally above 1200m, but widely planted.

8a White Poplar

Birch Family Betulaceae

Deciduous trees or shrubs with alternate lvs. Male and female fls *catkins,* separate but on the same tree.

1 SILVER BIRCH *Betula pendula.* Narrow erect tree to 30 m. Bark *silvery-white,* papery, peeling, brown and fissured below. Twigs hairless. Lvs diamond-shaped, teeth of different sizes. Catkins yellowish, male pendant, female shorter, erect. Woods, heaths and sandy places, to 2000 m. Mar-May. T. **1a Downy Birch** *B. pubescens* subsp. *carpatica* is shorter, often a shrub, *bark* greyish or brownish, not fissured below; lvs uniformly toothed. P, A, nAp.

2 DWARF BIRCH *Betula nana.* Dwarf shrub to 1 m, sometimes almost prostrate. Twigs hairless. *Lvs* rounded with deep, blunt teeth, downy when young. Catkins as 1 but much smaller, before the lvs. Bogs and moors, to 2200 m. Mar-May. B, S, nA*.

3 GREEN ALDER *Alnus viridis.* Dense shrub to 4 m. Twigs hairless. Lvs elliptical or rounded, double-toothed, bright green. Catkins yellowish, with the lvs, male pendant, female brownish, like miniature fir-cones. Meadows, woods and rocky places, 1500-2300 m. Apr-May. A, nAp. **3a Alder** *A. glutinosa* □ like 3 but often a small tree to 20 m. Bark dark brown, rugged; lvs *roundish,* blunt-tipped; *catkins* before lvs. Damp places, riverbanks, to 1800 m. Feb-Mar. T. **3b Grey Alder** *A. incana* is similar to 3 but taller, young twigs hairy, lvs oval, more pointed, *greyish below.* Drier places, to 1600 m. S, A, nAp.

Hazel Family Corylaceae

4 HAZEL *Corylus avellana.* Deciduous shrub to 6 m. Bark smooth brown. Lvs oval or roundish, pointed, toothed, downy. Male fls in hanging catkins, pale yellow; female tiny erect, bud-like with red styles; before lvs. Fr a nut partly enclosed in a leafy husk. Woods and scrubby areas, to 1800 m. Jan-Mar. T.

Beech Family Fagaceae

Containing the well-known Beech and Oak. Male and female fls separate but on the same tree. Widely planted.

5 BEECH *Fagus sylvatica.* Spreading deciduous tree to 30 m. Bark smooth grey. *Buds* scaly, pointed, red-brown. Lvs oval, pointed, veins prominent at edge, silky hairy below when young. Fls greenish, male in hanging *tassels,* female erect, with young lvs. Fr a pyramidal brown nut ('mast') enclosed in a bristly husk. Woods, often as pure stands, to 1900 m. Apr-May. T.

6 SESSILE or DURMAST OAK *Quercus petraea* (= *Q. sessiliflora*). Deciduous tree to 35 m. Bark grey, finely fissured. Twigs *hairless.* Lvs oval-lobed, tapered at the base, long-stalked. Catkins pale greenish-yellow, male long, female in short clusters. Acorns *scarcely stalked,* cup hairless. Woods, to 1800 m. Apr-May. T, notS. **6a Pyrenean Oak** *Q. pyrenaica* is a smaller tree with *downy* lvs and stems. Woods and rocky slopes, to 1400 m. P, MA, LA. **6b Pedunculate Oak** *Q. robur* has oblong lvs, the basal lobes overlapping the *short stalk;* twigs and lvs almost hairless; acorns *long-stalked.* Woods, often as pure stands, to 1400 m. T.

7 DOWNY or WHITE OAK *Quercus pubescens.* Deciduous tree or shrub to 20 m. Bark dark grey, rugged. Twigs thickly downy. Lvs oblong-lance-shaped, 6-12 cm long, lobed, stalked, downy below when young. Catkins pale greenish-yellow, male long, female in short clusters. Acorn cups downy. Woods, to 1500 m. A, Ap. **7a** *Q.p. palensis* is smaller, often *shrub-like,* the lvs *only 4-7 cm long.* P.

3 Green Alder

3a Alder

3b Grey Alder

Dock Family Polygonaceae

A large family; many are weeds but a few are genuine alpine species. Lvs alternate, untoothed, with a sheath (or ochrea) at the base encircling the stem. Fls small, 3-6 parted, no sepals, persisting and enclosing fr. Fr a tiny nut, often winged and flattened. **For Docks and Sorrels** (*Rumex* spp.) **see p. 290.**

1 KNOTGRASS *Polygonum aviculare.* Low, erect or prostrate, hairless ann. Lvs lance-shaped or oval, larger on main stem. Ochrea *silvery-transparent.* Fls greenish, pink or white, 1-6 together at base of upper lvs. Stony and bare ground, to 2300 m. June-Oct. T. **1a Iceland Purslane** *Koenigia islandica* is tiny, often reddish, with *broadly-elliptical* lvs; fls greenish, 3-parted. Damp muddy and bare ground, to 1500 m. B, S.

2 BISTORT *Polygonum bistorta.* Short/tall, almost hairless, per, patch-forming. Lvs narrow *oval-triangular,* the stalks winged in upper part. Fls bright pink in dense oblong clusters 10-15 mm broad. Damp grassy places, to 2500 m. June-Oct. B, P, A, Ap.

3 ALPINE BISTORT *Polygonum viviparum.* Low/med hairless, tufted, per. Lvs oblong to narrow-lance-shaped, tapered at base, margin *rolled under.* Fls pale pink or white in slender spikes 5-10 mm broad; lower part of spike with small brownish-purple *bulbils.* Grassy and rocky places, to 2300 m. June-Aug. T.

4 ALPINE KNOTGRASS *Polygonum alpinum.* Short/med hairless per. Lvs oblong-lance-shaped, tapered at both ends. Fls white or pink, in thin *branched spikes.* Damp meadows and rocky places, to 2200 m. July-Aug. P, A, Ap.

5 MOUNTAIN SORREL *Oxyria digyna.* Short hairless, tufted, per. Lvs mostly basal, *kidney-shaped,* long-stalked. Fls tiny, greenish, 4-petalled, in branched spikes. Fr a drooping winged nut. Damp rocks and streamsides, often on granite, to 3500 m. July-Aug. T.

Pink Family Caryophyllaceae

Lvs in opposite pairs, rarely toothed. Fls with 4-5 petals and sepals; sepals separate or joined into a tube; stamens 8-10. Fr a dry capsule. **For more white-flowered species in this family, see pp. 292–300.**

6 PURPLE LYCHNIS *Lychnis flos-jovis.* Short/tall hairy per. Lvs lance to spoon-shaped, pointed, the upper unstalked. Fls purplish or scarlet, rarely white, 14-18 mm, in dense heads; petals notched. Meadows, rocks and screes, to 2000 m. June-July. w, c, sA.

7 RAGGED ROBIN *Lychnis flos-cuculi.* Med/tall roughly-hairy per. Lvs oval-spoon-shaped, the upper pointed and unstalked. Fls bright pink, 30-40 mm, in loose branched clusters; petals *4-lobed,* ragged. Damp meadows and marshy places, to 2500 m. May-June. T.

8 STICKY CATCHFLY *Lychnis viscaria.* Short/tall tufted per, *sticky* below upper lf junctions. Lvs lance-shaped, mostly basal, hairless. Fls red or purplish, 20 mm, in *'whorled' closters;* petals, notched. Dry meadows and rocky places, to 1800 m. May-June. T, rare in sw.

9 ALPINE LYCHNIS *Lychnis alpina.* Low/short tufted, hairless per. Lvs linear or narrow spoon-shaped, mostly basal. Fls pale purple, rarely white, 8-12 mm, in compact heads; petals notched. Meadows, rocky and stony places, to 2000-3100 m. June-Aug. T, rare in Ap.

10 CORN COCKLE *Agrostemma githago.* Tall, greyish-hairy ann. Lvs narrow lance-shaped, pointed. Fls dull purple with a whitish centre, 30-50 mm, petals slightly notched; sepals forming a tube with 5 narrow lobes projecting beyond petals. Cornfield weed usually, to 2000 m. May-Aug. T.

11 PETROCOPTIS *Petrocoptis pyrenaica.* Low/short loosely tufted, per; thin stemmed. Lvs thin, oval-lance-shaped, green. Fls white or very pale purplish, 10-15 mm; petals slightly notched. Calyx whitish, 5-8 mm long. Rocky places and banks, 1300-2800 m. May-Aug. wP. **11a** *P. hispanica* has thicker bluish-green lvs. wcP; Jaca Region. **11b** p. 335.

Pink Family (contd.)

CATCHFLIES and CAMPIONS have showy fls with 5 separate, often notched, long-clawed petals; sepals joined into a tube with 5 teeth. Styles 3, protruding (except 1).

1 NORTHERN CATCHFLY Silene wahlbergella (= Melandrium apetalum). Short, unbranched, slightly hairy per. Lvs narrow-oblong. Fls solitary, 14-18mm, petals reddish-purple, *completely surrounded* by the *inflated*, whitish sepal-tube. Damp meadows and stony places, to 1900m. June-July. S. **1a**, p. 335.

2 NOTTINGHAM CATCHFLY Silene nutans. Variable med unbranched, hairy per, sticky above. Lvs oblong-spoon-shaped, stalked, the upper narrower and unstalked. Fls half drooping, petals white above, pink or greenish beneath, deeply cleft, *rolled back;* sepal-tube narrow, 9-12mm, hairy. Meadows, stony places and banks, to 2200m. May-Aug. T. **2a**, p. 335.

3 SPANISH CATCHFLY Silene otites. Variable short/med bien or per, stickily-hairy at base. Lvs oblong-spoon-shaped, the lower long-stalked. Fls small, 3-5mm, *greenish-yellow*, in whorled clusters; styles and stamens on different plants. Dry, often sandy, places, to 2000m. June-Sept. B, P, A, Ap.

4 LARGE-FLOWERED CATCHFLY Silene elisabetha. Low/short tufted per. Lvs mostly in basal rosettes, lance-shaped, hairless or slightly hairy along margins. Fls large, reddish-purple or pink, 25-35mm, usually *solitary*; petals *notched and toothed*. Limestone rocks and screes, 1500-2500m. July-Aug. sA-Italian.

5 HEART-LEAVED CATCHFLY Silene cordifolia. Short tufted hairy per. Lvs *oval-heart-shaped*. Fls white or pink, 10-15mm, in clusters of 1-4; petals deeply notched. Rocks and screes, 1200-2400m. July-Aug. MA. **5a**, p. 335.

6 BLADDER CAMPION Silene vulgaris. Med branched, greyish, often hairless per. Lvs oval to linear, often wavy-edged, 10-25mm broad. Fls white, 16-18mm, petals deeply notched; sepal-tube inflated, *bladder-like*. Meadows, rocks and banks, to 3100m. May-Sept. T. **6a**, **6b**, **6c**, p. 335.

7 VALAIS CATCHFLY Silene vallesia. Low/short, stickily-hairy, mat-forming per. Lvs oblong-lance-shaped to linear. Fls 1-3 clustered, 14-16mm, petals pale pink above, reddish beneath, deeply cleft, *curling in at tip*. Rocks and screes, to 2100m. July-Aug. w, swA, AA, cAP.

8 NARROW-LEAVED CATCHFLY Silene campanula. Low/short slender, hairless per. Lvs *linear*, pointed. Fls solitary or 2, petals white above, reddish-purple beneath, notched; sepal-tube 7-8mm, hairless. Damp limestone rocks, to 2200m. July-Aug. MA.

9 TUFTED CATCHFLY Silene saxifraga. Rather like 8 but *sticky* hairy below and forming *rounded tufts*. Fls whitish or greenish above, greenish or reddish beneath; sepal-tube 8-13mm, hairless. Limestone rocks and screes, to 2400m. May-Aug. P, A, Ap.

10 PYRENEAN CATCHFLY Silene borderi. Low *mat-forming* per. Lvs mostly basal, narrow-spoon-shaped, hairy margined, covered in *raised dots*, upper lvs linear. Fls pink in clusters of 1-4; petals deeply cleft; sepal-tube 8-10mm, hairy. Rocks, 2000-2200m. Aug. c, eP*.

11. MOSS CAMPION Silene acaulis. Low per forming *moss*-like cushions, hairless, bright green. Lvs tiny, linear. Fls solitary, short-stalked, pink, 6-10mm; petals notched. Damp rocks and screes, short turf, to 3700m. June-Aug. T. **11a** S.a. longiscapa has fls on *longer stalks*, 2-6mm. T.

12 ROCK CAMPION Silene rupestris. Short, hairless, branched, greyish per. Lvs elliptical, broadest towards tip. Fls small, white or pink, in *large clusters;* petals notched; sepal-tube 4-6mm, hairless. Rocks and screes, often on acid rocks, to 2900m. June-Sept. S, P, A, nAp.

13 ALPINE CATCHFLY Silene alpestris (= Heliosperma alpestre). Short, thin-stemmed, branched, almost hairless per, sticky above. Lvs oblong-lance-shaped to linear. Fls white, 8-10mm; petals *4-6-toothed*; sepal-tube 5-7mm, short-hairy. Limestone rocks, to 2500m. June-Aug. e, seA. **13a**, **13b**, p. 335.

Pink Family (contd.)

1 ALPINE GYPSOPHILA Gypsophila repens. Sprawling hairless per. Lvs bluish-green, narrow lance-shaped. Flstems upright from prostrate stems. Fls 8-10mm, white, pale pink or lilac, petals longer than sepals, notched. Rocky, stony and grassy places and banks, on limestone, to 2900m. May'Sept. P,A,Ap.

SOAPWORTS Saponaria. Tufted or sprawling perennials with solitary or clustered fls. Calyx cylindrical with 5 short teeth. Petals with a long narrow claw. Styles 2.

2 SPOON-LEAVED SOAPWORT Saponaria bellidifolia. Short/med tufted, generally hairless, per; fl stems upright, unbranched. Lvs spoon-shaped, stalked, the upper narrower, few. Fls 7-8mm, yellow in tight clusters, petals notched; stamen stalks yellow. Rocks and pastures, to 2000m. June-July. P,A,–except Austria and Switzerland,Ap.

3 YELLOW SOAPWORT Saponaria lutea. Low hairy cushion per; fl stems upright, unbranched. Basal lvs narrow lance-shaped, stem lvs linear, few. Fls 8mm, yellow, in tight clusters, petals blunt-tipped; stamen stalks violet-black, protruding. Limestone rocks, 1500-2600m. July-Aug. sw,cA.

4 TUFTED SOAPWORT Saponaria caespitosa. Low/short, scarcely hairy cushion per. Basal lvs narrow lance-shaped, stem lvs smaller, few. Fls 8-14mm, purplish, in small clusters, petals round-tipped. Rocks and screes, to 2100m. July-Aug. cP*.

5 DWARF SOAPWORT Saponaria pumilio (= Silene pumilio). Low almost hairless, cushion per, stems short, Ifless. Lvs linear, tufted. Fls solitary, 14-18mm, rose-red rarely white, petals broad, notched. Meadows on acid soils, 1900-2600m. Aug-Sept. eA, Italy and Austria.

6 ROCK SOAPWORT Saponaria ocymoides. Low/short sprawling, hairy per. Lvs oval-lance to spoon-shaped. Fls 6-10mm, pink or purplish, in branched clusters, petals blunt. Grassy, rocky or stony places, to 2000m. Mar-Oct. P,A,Ap.

7 TUNIC FLOWER Petrorhagia saxifraga. Short/med generally hairless per. Lvs linear, pointed, upright. Fls 5-8mm, white or pink, solitary or in a loose branched head, petals notched. Dry stony and sandy places, to 1300m. June-Aug. T.–except S,B*. **7a**, p.335.

PINKS Dianthus. Tufted perennials with stiff stems and greyish or grey-green linear lvs. Fls solitary or clustered. Calyx tubular, surrounded at base by several paired epicalyx-scales. Petals long-clawed, broad at top, serrated or deeply cut along margin. Styles 2, often protruding. (See also p.354)

8 PAINTED PINK Dianthus furcatus. Low/med densely-tufted, hairless per. Lvs linear, flat, soft. Fls 1-3, pink or whitish, 10-20mm, petals serrated; epicalyx-scales reaching halfway up the calyx. Dry meadows, rocky and stony places, to 2300m. June-Aug. P,MA,s,swA.

9 SÉQUIER'S PINK Dianthus seguieri. Med, loosely tufted, hairless per. Lvs green, narrow lance-shaped, 1-2mm broad. Fls pink or purplish with a ring of dark spots near the centre, 14-20mm, throat hairy, solitary or 2-4 clustered; epicalyx-scales as long as the calyx. Meadows and stony places, to 1600m. June-Sept. nP,sw,sA. **9a** D.s. italicus has larger fls and broader lvs, 4-6mm. AA.

10 SWEET WILLIAM Dianthus barbatus. Short/med, almost hairless per. Lvs lance-shaped. Fls purple, pink or reddish, often spotted, 20-34mm, in dense flat-topped clusters; epicalyx-scales green, pointed. Meadows and woodland clearings, to 2500m. June-Aug. P,s,eA,Ap. Often cultivated. **10a**, p.335.

11 FRINGED PINK Dianthus monspessulanus. Short/med, loosely tufted per. Lvs green linear, 1-3mm broad, thin and flexible. Fls pink or white, 20-30mm, in lax branched clusters of 2-5, petals deeply fringed, fragrant; epicalyx-scales half as long as the calyx. Meadows, stony places and woods, to 2000m. May-Aug. P,A-except e,Ap. **11a** D.m. marsicus is shorter with larger, usually solitary, fls, 30-40mm. cAp,Abruzzi. **11b** D.m. sternbergii (= D. sternbergii) like 11a but lvs and stems bluish-green. e,seA. **11c** D. superbus like 11 but petals fringed over half-way, the central part oblong. To 2400m. nP,A,nAp.

Pink Family (contd.)

1 ALPINE PINK *Dianthus alpinus.* Low/short, tufted, hairless per. Lvs *narrow-oblong,* the lowest 3-5mm broad, blunt, deep glossy-green. Fls purplish-red with *white spots,* 30-36mm, solitary; petals serrate-edged. Limestone rocks, to 2500m. June-Aug. e,seA.

2 GLACIER PINK *Dianthus glacialis.* Low, tufted, almost hairless per. Lvs linear, 1-2mm broad, fleshy. Fls purple-red, 15-18mm, solitary, *surrounded* by the lvs; petals serrate-edged. Meadows and stony places, on acid rocks, 1900-2900m. July-Aug. c,eA.

3 THREE-VEINED PINK *Dianthus pavonius* (= *D. neglectus*). Low, tufted, hairless per. Lvs linear grassy, pointed, *3-veined.* Fls pinkish-purple, 20-24mm, usually solitary; epicalyx-scales *as long* as the calyx. Meadows and stony places, 1200-3000m. July-Aug. swA, eA*.

4 CHEDDAR PINK *Dianthus gratianopolitanus* (= *D. caesius*). Low/short hairless, *bluish-green,* tufted per. Lvs linear, 1-2mm broad, more or less flat. Fls pink or purplish, 15-30mm, usually solitary, fragrant; calyx brownish-purple, the epicalyx-scales *very short.* Meadows and stony places, to 2200m. May-July. B–Cheddar Gorge, A*–except se.

5 WOOD PINK *Dianthus sylvestris.* Variable low/short per; stems branched usually, except in high alpine forms. Lvs *green,* linear, grassy, the basal only 1mm broad and usually *recurved.* Fls pink, 15-25mm, rarely slightly fragrant, solitary or two together; epicalyx-scales short. Meadows and stony places, 1400-2800m. July-Aug. nP, Cev, Jura, A, Ap.

6 SHORT PINK *Dianthus subacaulis.* Low/short neat, densely tufted, hairless per. Lower lvs narrow lance-shaped, blunt, 1mm broad, the stem lvs shorter and *pressed against* the stem. Fls *small* pale pink, 6-12mm, solitary; epicalyx-scales short. Meadows, rocky and stony places, to 1700m. June-Aug. P. **6a** *D. pungens* has narrower, 0·5mm, and longer lvs and darker fls. eP.

7 PYRENEAN PINK *Dianthus pyrenaicus.* Short/med, loosely tufted per, with trailing, slightly woody, stems. Lvs narrow lance-shaped, green, *pungent.* Fls small, pale pink, 6-8mm, *in branched clusters;* epicalyx-scales short. Meadows and rocky places, to 1500m. P. **7a**, p.335.

8 COMMON PINK *Dianthus plumarius.* Short/med bluish-green, loosely tufted per. Lvs linear, 1mm broad, pointed. Fls white to bright pink, 24-36mm, usually solitary, fragrant, petals *deeply and narrowly lobed;* epicalyx-scales very short. Meadows and rocky places, usually limestone, to 2000m. Apr-Aug. cs,eA–Italy and Austria. **8a** *D. serotinus* is more slender with cream-coloured fls in groups of 2-5. eA.

9 MAIDEN PINK *Dianthus deltoides.* Short/med, loosely tufted, bluish-green per, stems *rough-hairy.* Lvs linear, often broadest above the middle, edges rough-hairy. Fls deep or pale pink (or rarely white) with a *darker band* near the centre and white spots, 15mm; epicalyx-scales half as long as the calyx. Dry grassy places and open woods, to 2000m. June-Oct. T.

10 DEPTFORD PINK *Dianthus armeria.* Short/med stiff-hairy, green ann/bien. Lvs oblong, blunt, flat and thin, the upper linear. Fls small, pink or reddish, 10-12mm, in *branched* clusters; epicalyx-scales *as long* as the hairy sepal-tube. Dry sandy and waste places, to 1250m. June-Aug. T–except S.

11 TALL PINK *Dianthus pontederae.* Short/tall rather slender per. Lvs linear, 2-4mm broad, flat, pointed. Fls purple, 6-10mm, in *dense clusters;* epicalyx-scales brown, pointed, half as long as the 10-13mm long calyx. Rocky and grassy places, to 1400m. s,seA. **11a** *D. giganteus* is more robust with larger fls, 10-16mm; calyx 17-20mm long. cAp–Monte Morrone.

12 CARTHUSIAN PINK *Dianthus carthusianorum.* Variable low/med hairless, tufted per. Lvs linear, pointed, 1-5mm broad. Fls deep pink to purple, 20-30mm, in *dense clusters;* calyx and epicalyx-scales *purplish brown,* scales half as long as the calyx. Dry grassy and stony places, open woods, to 2500m. May-Aug. P, A, Ap.

Buttercup Family Ranunculaceae

Alpine species generally perennial, occasionally annuals or woody climbers. Fls with numerous stamens and normally five, but sometimes more, petals or petal-like sepals. Fls sometimes with nectary spurs or honey lvs, which secrete nectar. Fr a collection of dry achenes or follicles, rarely a berry.

1 STINKING HELLEBORE *Helleborus foetidus.* Med/tall foetid per. Lvs all on the stem, hand-like with 7-11 narrow lance-shaped, toothed, segments, stalked, uppermost lvs undivided. Fl lantern-shaped, 1-3 cm, petalless, sepals yellowish-green, edged purple, in clusters. Woods and scrub on calcareous soils, to 1600 m. Jan-Apr. B, P, wA, Ap.

2 GREEN HELLEBORE *Helleborus viridis.* Short/med per with *two root lvs* that do not overwinter. Lvs hand-like with 7-13 narrow-elliptical lobes, toothed; stem lvs smaller, stalkless. Fls broad open-cups, 4-5 cm, sepals apple-green, scentless, overtopped by lvs. Woods, scrub and rocky places, usually on limestone, to 1600 m. Mar-Apr. B, P, A. **2a** *H. dumetorum* has smaller, scented fls *held above* the lvs. Mar-May. e, seA.

3 CHRISTMAS ROSE *Helleborus niger.* Low clump-forming per with *over-wintering* basal lvs. Lvs hand-like with 7-9 elliptical toothed segments. Fls large, saucer-shaped, 3-10 cm, white with yellow anthers. Woods and scrub on lime-stone, to 1900 m. Jan-Apr. swA*, c, eA, Ap.

4 WINTER ACONITE *Eranthis hyemalis.* Low hairless per, rootstock a small tuber. Lvs palmately lobed, all from roots and *appearing* as fls fade. Fls solitary cups, 2-3 cm, with 5-7 yellow sepals, surrounded by a *ruff* of green leafy-bracts. Damp woods, scrub and banks, to 1500 m. Jan-Mar. w, swA, *but* often naturalised else-where.

5 CALLIANTHEMUM *Callianthemum anemonoides.* Low hairless per. Basal lvs 2-pinnate, segments oblong, pale green, stalked; stem lvs similar, stalkless. Fls 3-3·5 cm, white with an orange central ring, sepals 5, *petals* 5-20. Fr 5mm long. Open coniferous woods on calcareous soils, to 2100 m. Mar-May. neA. **5a** *C. kerneranum* is *smaller* with fls 2-5 cm across. Rocky limestone slopes, to 1500 m. May-July. sA—Italy only. **5b** *C. coriandrifolium* is similar to 5 but petals *broader* and fr *achenes* only 3mm long. Turf and stony places on neutral or acid soils, often by melting snow, 1800-3000 m. July-Aug. P*, A.

6 LOVE-IN-A-MIST *Nigella arvensis.* Short upright, branched, ann. Lvs grey-green, *feathery* with many thread-like segments. Fls 2-3 cm, long stalked with 5 blue 'clawed' petals, often green-veined. Weed of cornfields and waste places. June-July. P, A, Ap, (B).

7 GLOBEFLOWER *Trollius europaeus.* Short/med hairless per. Basal lvs palmate, deeply cut, stalked; upper lvs smaller, stalkless. Fls large 3-5 cm, almost *spherical,* with up to ten yellow sepals *curving* in at the top. Damp meadows and open woods, to 2800 m. May-Aug. T.

1 Stinking Hellebore, fr   **2** Green Hellebore, fr.

1

2

3

7

6

5

4

Buttercup Family (contd.)

1 RUE-LEAVED ISOPYRUM *Isopyrum thalictroides*. Low short slender, hairless per. Basal lvs stalked, trifoliate, lflets 3-lobed, grey-green; stem lvs stalkless. Fls saucer-shaped, 10-20 mm, with five oblong white sepals. Damp shady woods, to 1200 m – rarely higher. Mar-May. P*,A,n,c,Ap.

2 BANEBERRY *Actaea spicata*. Medium hairless rhizomatous per, strong smelling. Lvs, *'umbellifer-like'*, 2-pinnate or 2-trifoliate, dark green above, paler below, stalked; upper lvs smaller. Fls in small oblong clusters, with 4 small petals and 4 long stalked stamens, *all* white. Fr a large *shiny black berry*. Damp woods, to 1900 m. May-July. Poisonous. B*,S,P,A.

3 MARSH MARIGOLD *Caltha palustris*. Short/med hairless per, much dwarfed at high altitudes, often creeping. Lvs large, *heart-shaped,* toothed, dark green, shiny. Fls open cups, 1·5-5 cm, with five golden yellow sepals, *no* petals. Fr pod-like clusters, conspicuous. Marshes, bogs and wet pastures, to 2500 m. Mar-July. T.

MONKSHOODS *Aconitum*. Perennials with stout leafy stems and tuberous brown rootstocks. Fls with five petal-like sepals, the upper one forming a hood or helmet, often spur-like. Fr consisting of 2-5 follicles as in Columbine and Delphinium. Very Poisonous (see also p. 347).

4 WOLFSBANE *Aconitum vulparia*. Tall hairless per. Lvs palmately cut to middle, segments 4-6, deep green. Fls pale yellow, in branched spike-like racemes; *helmet* long and narrow. Meadows, woods and stony places, to 2400 m. June-Aug. P,A,Ap. **4a** *A. lamarckii* has pale green *lvs* with 7-8 segments and denser, many flowered racemes. July-Aug. P,sA,Ap. **4b Northern Wolfsbane** *A. septentrionale* □ has hairy violet fls, the hood thicker at the base than 4. To 1300 m. S.

5 YELLOW MONKSHOOD *Aconitum anthora*. Med/tall hairless per. Lvs palmately cut to middle, segments narrow, lobed and pointed. Fls yellowish, *sometimes* blue, few in a loose cluster; helmet rounded, as broad as high. Dry meadows and rocky places often on limestone, to 2200 m. July-Sept. P,A,Ap.

6 VARIEGATED MONKSHOOD *Aconitum variegatum*. Med/tall hairless per. Lvs palmately cut to the middle, segment deeply lobed and toothed. Fls blue streaked with white, rarely all white, in loose racemes; helmet oblong, twice as high as broad. Meadows, woods and clearing to 2000 m. July-Sept. A,n,cAp. **6a** *A. paniculatum* □ has stickily-hairy fl stems and violet or mauve fls in a loose branched cluster; helmet as broad as high. Damp meadows and woods, to 2400 m. A,n,cAp.

7 COMMON MONKSHOOD *Aconitum napellus*. Med/tall hairless per. Lvs palmately cut almost to the middle, not crowded below the fls. Fls violet, deep blue or red/dish-violet, in dense branched spike-like racemes; helmet rounded, as broad as high. Damp meadows and woods, to 2500 m. July-Sept. B,P,A,not Italy or Yugoslavia. **7a** *A. tauricum* has broad segmented lvs *crowded* below the *fls* and racemes usually *unbranched*. e,seA. **7b** *A. compactum* is smaller than 7a, the lvs with *linear segments*. P,w,cA.

4b
Northern
Wolfsbane

6a
Aconitum
paniculatum

1

2

3

4

5

6

7

Buttercup Family *(contd.)*

LARKSPURS *Delphinium.* Fls in spike-like racemes; with 5 petal-like sepals, the upper with a long narrow spur pointing backwards. Fr of 3-5 follicles. Poisonous.

1 MOUNTAIN LARKSPUR *Delphinium montanum.* Short/med velvety-hairy per, stems erect. Lvs *long-stalked,* palmately cut almost to the middle, lobes oblong, toothed. Fls pale blue, 12-20 mm, sepals *narrow* oblong. Fr *hairy.* Meadows and stony places to 2000 m. June-Aug. P. **1a** *D. dubium* is taller with *short* lf stalks and dark blue fls, 16-23 mm. swA, not Switzerland.

2 ALPINE LARKSPUR *Delphinium elatum.* Rather similar to 1 but stems with *straight* not curved hairs. Fls deep or dirty blue or bluish-violet, 14-17 mm, sepals *broad* oval or rounded; spur longer than sepals. Fr *usually* hairless. Meadows and stony places, to 2000 m. June-Aug. A. **2a** *D.e. helveticum* has a spur *equal* in length to the sepals. wA. **2b** *D.e. austriacum* like 2a but fls *larger,* 17-21 mm. ceA.

ANEMONES Perennials with basal lvs and a whorl of short stalked or stalkless lvs on the stem below the fls. Fls solitary or clustered, cup or saucer shaped; sepals 5 or more, 'petal-like'. Fr a cluster of achenes *not* feathered.

3 WOOD ANEMONE *Anemone nemorosa.* Low hairless per. Stem lvs 3, *stalked,* palmately-lobed to the middle; basal lvs similar, appearing after the fls. Fls solitary white, often flushed pink, 20-35 mm, with 6-7 sepals. Woods, to 1800 m. Mar-May. T. **3a** *A. trifolia* similar to 3 but stem lvs *trifoliate.* Woods and rocky places, to 1900 m. May-July. s, eA, not Switzerland, Ap.

4 YELLOW ANEMONE *Anemone ranunculoides.* Short hairy per. Stem lvs like 3 but *almost* stalkless. Fls solitary or two, *yellow,* 15-20 mm, with 5-8 petals. Woods, to 1500 m. Mar-May. T, except B.

5 NARCISSUS-FLOWERED ANEMONE *Anemone narcissiflora.* Low/med hairy per. Lvs palmately-lobed to the middle. Fls white, often pink flushed below, 20-30 mm in *umbels* of 3-8, with 5-6 sepals. Meadows, usually on limestone, 1500-2600 m. June-July. P, A, Ap.

6 SNOWDROP WINDFLOWER *Anemone sylvestris.* Short/med hairy per. Lvs palmately-lobed almost to the middle. Fls solitary white, 30-70 mm, with *only* 5 sepals, hairy beneath. Dry woods on limestone, to 1200 m. Apr-June. A.

7 MONTE BALDO ANEMONE *Anemone baldensis.* Low hairy per. Basal lvs trifoliate, lflets each 3-lobed and toothed; stem lvs similar but smaller. Fls solitary, white, sometimes bluish outside, 25-40 mm, *with* 8-10 pointed sepals. Rocky places and screes, 1800-3000 m. July-Aug. A, often local.

8 BLUE WOOD ANEMONE *Anemone apennina.* Low hairy per. Stem lvs 3, trifoliate, toothed; basal lvs similar, appearing after the fls. Fls solitary *blue,* rarely white, 25-35 mm, with 8-14 oblong sepals; anthers pale yellow or white. Open woods and scrub to 1200 m. Mar-Apr. c, s, Ap.

3 Wood Anemones

1

2

3

4

5

6

8

7

43

Buttercup Family *(contd.)*

PASQUE FLOWERS *Pulsatilla.* Tufted hairy perennials usually with ferny lvs and large solitary, upright or nodding fls, stalked with a ruff of feathery leafy bracts below; sepals usually five or six, petal-like, hairy outside. Fr clustered, with long silky plumes, often persisting into late summer. Poisonous.

1 ALPINE PASQUE FLOWER *Pulsatilla alpina.* Med hairy per. Lvs 2-pinnate, hairy above; basal and stem lvs *stalked.* Fls large, white often flushed bluish-purple outside, upright cups, 4-6cm. Meadows over limestone, to 2700m. May-July. P,A,Ap. **1a** *P.a. apiifolia* ☐ has *pale yellow* fls. Meadows generally over acid rocks, to 2700m. P,A,Ap.

2 WHITE PASQUE FLOWER *Pulsatilla alba.* Low hairy per. Lvs 2-pinnate, more or less *hairless* above; basal and stem lvs *stalked.* Fls large, white, sometimes bluish-flushed, upright cups, 2·5-4·5cm. Meadows, generally over acid rocks, to 2200m. May-July. A.

3 SPRING PASQUE FLOWER *Pulsatilla vernalis.* Low/short hairy per. Lvs evergreen, pinnate, *much less* divided than 1-2, lflets oblong, toothed; stem lvs stalkless, linear. Fls white, flushed pink, violet or blue on the outside, deep cups, 4-6cm, at first drooping, later upright. Meadows and stony places, often by melting snow, 1300-3600m. Apr-July. S,P,A.

4 SMALL PASQUE FLOWER *Pulsatilla pratensis.* Low/short hairy per. Lvs 2-pinnate, feathery; stem lvs stalkless, narrowly-elliptic, pointed. Fls dark purple, sometimes pale violet or greenish-yellow, greyish hairy outside, 3-4cm, *always* drooping; sepals with recurved tips. Meadows, to 2100m. Apr-May. eA, not Switzerland or Italy. **4a** *P. montana* is similar but with bluish to dark violet fls, sepals *not recurved* and opening more. Meadows and open woods, to 2150m. Apr-May. eP,wA,not France and probably not wAustria. **4b** *P. rubrum* ☐ has dark red-brown, purplish-brown or reddish black fls. cev.

5 COMMON PASQUE FLOWER *Pulsatilla vulgaris.* Low hairy per. Lvs 2-pinnate, feathery, covered with long hairs at first, later almost hairless; stem lvs stalkless, linear. Fls large, dark to pale purple, bell-shaped 5·5-8·5cm, erect at first, then drooping, anthers bright yellow. Meadows, often on limestone, to 1200m. May-June. B,S,A. **5a** *P. halleri* similar to 5 but mature lvs and stems densely woolly-hairy. To 3000m. June-July. sw,cA. **5b** *P.h. styriaca* like 5a but with *longer* lf blades, 5-11cm instead of 3-7cm. seA–Steiermark.

6 HEPATICA *Hepatica nobilis* (= *H. triloba, Anemone hepatica*). Low evergreen, slightly hairy per. Lvs 3-lobed, heart shaped at base, stalked, green above, purplish below. Fls solitary saucers, 15-25mm, purple, bluish-violet or pink, rarely white, with 6-9 sepals and 3 sepal-like bracts. Woods, scrub, rocky and grassy places, often on limestone, to 2200m. Mar-Apr. S,P,A,Ap.

4b *Pulsatilla rubrum*

Buttercup Family *(contd.)*

CLEMATIS Perennials or woody climbers with opposite, usually 1-2-pinnate lvs. Fls with 4 petal-like sepals and numerous stamens. Fr clustered, with long hairy plumes giving 'old man's beard' appearance.

1 TRAVELLER'S JOY *Clematis vitalba.* Deciduous scrambling or climbing woody perennial, sometimes with very long stems. *Lvs* pinnate, lflets toothed, often with twisting 'tendril-like' stalks. Fls greenish-white, fragrant, 15-20mm with many conspicuous stamens, in large clusters. Fr □ in dense *greyish clusters,* typical Old Man's Beard. Woods, scrub and hedges, to 2100m. June-Sept. T – except S.

2 ERECT CLEMATIS *Clematis recta.* Tall erect per, *not* woody. Lvs pinnate, lflets *untoothed oval.* Fls white, upright, 15-20mm, in terminal branched clusters. Open woods and dry hills, rarely above 1100m. May-June. P, A, Ap.

3 SIMPLE-LEAVED CLEMATIS *Clematis integrifolia.* Med/tall erect per, stems usually unbranched. Lvs oval, pointed, untoothed, *not* pinnate. Fls nodding purple bells, 3-5cm, solitary or 2-3. Meadows, rarely over 1000m. June-Aug. eA.

4 ALPINE CLEMATIS *Clematis* (= *Atragene) alpina.* Deciduous climbing or scrambling per. *Lvs* 2-pinnate with twisting 'tendril-like' stalks. Fls large, *solitary, nodding,* 2·5-4cm, violet or purplish, with white, 'petal-like' staminodes inside. Rocky mountain woods and meadows, 1900-2900m. June-July. S, A*.

5 PYRENEAN PHEASANT'S-EYE *Adonis pyrenaica.* Short/med per, stems pale green *not* scaly at base. Lvs 2-3-pinnate, feathery. Fls large, 40-60mm, with *twelve* or more shiny golden-yellow petals; sepals *hairless.* Rocky and stony places and screes, to 2400m. June-July. P, MA. **5a** *A. vernalis* is similar but stems *scaly* at base and with larger fls; sepals hairy. Dry meadows and rocks, rarely much above 1200m. Apr-May. S, P, A, Ap.

6 APENNINE PHEASANT'S-EYE *Adonis distorta.* Low per with *curved* stems and lf-segments; stems *not* scaly at base. Lvs 2-3-pinnate, basal ones *long* stalked. Fls *smaller* than 5 and 5a, 30-45mm, petals yellow, shiny, sepals hairy. Rocky and stony places usually on limestone, 2000-2900m. cAp.

7 PHEASANT'S-EYE *Adonis annua.* Low/short hairless ann. Lvs 3-pinnate, feathery, the lower unstalked. Fls open cups, 15-25mm, with 5-8 scarlet petals; anthers numerous black or purplish; sepals *hairless.* Fr achenes 3·5-5mm long. Cornfields and waste places, often on calcareous soils, to 1500m. May-Aug. P, sA, Ap, (B). **7a** *A. aestivalis* has red fls and achenes 3·5-5mm long; lower lvs *stalked.* P, A, Ap. **7b** *A. flammea* is similar to 7 but fls 20-30mm, sepals *hairy*; achenes with a *black* beak. P, A, Ap.

8 MOUSETAIL *Myosurus minimus.* Low/short hairless ann. Lvs linear, in a basal tuft, rather fleshy. Fls small solitary, long stalked, with 5-7 greenish yellow petals and sepals. Fr head greatly elongated 'mouse-tail' like. Damp cultivated and bare ground, rarely above 1400m. Mar-May. T.

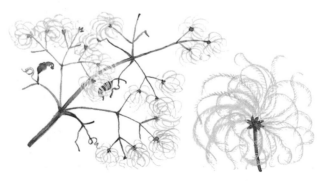

1 Traveller's Joy, fr.

Buttercup Family *(contd.)*

BUTTERCUPS *Ranunculus.* Fls open cups, shiny yellow or white, rarely pinkish, with *both* sepals and petals, distinguishing them from Anemones, Globe flowers and Marsh Marigolds. Fr a cluster of dry seeds (achenes).

1 MULTIFLOWERED BUTTERCUP *Ranunculus polyanthemos.* Short/med hairy per, much branched. Basal lvs *rounded* in outline, cut to middle into 5 lobes, lobes *narrow,* toothed, stem lvs similar but smaller. Fls golden-yellow, 18-25 mm, sepals erect. Meadows and grassy places, to 1250 m. May-Oct. S, c, s, eA, nAp. **1a** *R.p. polyanthemoides* is taller, the basal lvs only 3-lobed, the central lobe very long. A.

2 WOOD BUTTERCUP *Ranunculus nemorosus.* Short/med hairy per, branched. Basal lvs *pentagon* in outline, 3-lobed, lobes *broad* toothed; stem lvs scarcely toothed. Fls golden-yellow, 15-20 mm, sepals *erect.* Meadows and grassy places, to 2000 m. Apr-Oct. T – except B. **2a** *R.n. serpens* similar but bien, shorter stems *rooting* at the nodes. P, Jura, A. **2b** *R.n. polyanthemophyllus* is taller than 2, the basal lvs *rounded* in outlined, the central lobe often short-stalked. A. **2c Creeping Buttercup** *R. repens* ☐ rather like 2a but creeping with *rooting runners;* basal lvs *triangular* in outline with a stalked central lobe. To 2500 m. T.

3 WOOLLY BUTTERCUP *Ranunculus lanuginosus.* Med very hairy per. Basal lvs rounded in outline, 3-lobed, lobes broad, *less* deeply cut than 1, 2; stem lvs similar but smaller. Fls orange-yellow, 20-30 mm, sepals erect. Damp meadows and woods, to 1600 m. May-Aug. A, Ap. **3a Meadow Buttercup** *R. acris* ☐ has 3-7 lobed lvs, the end lobe *unstalked,* lobes narrowed at base; fls smaller golden yellow. To 2500 m. Apr-Oct. T.

4-8 are very similar and often confused. Fls solitary or two to three clustered; sepals erect.

4 GOUAN'S BUTTERCUP *Ranunculus gouanii.* Low/short hairy per. Basal lvs 3-5-lobed, lobes oval, broadest above middle, toothed; stem lvs with *untoothed* linear lobes, *clasping* stem at base. Fls golden-yellow, 20-40 mm, sepals *densely* hairy. Meadows and rocky places, to 2800 m. May-Aug. P. **4a** *R. ruscinonensis* is smaller, stem lvs *not* clasping. eP.

5 CARINTHIAN BUTTERCUP *Ranunculus carinthiacus.* Low/short per with *hairless* lvs. Basal lvs 3-5-lobed, lobes narrow, oblong or oval, toothed; stem lvs similar, *not* clasping. Fls golden-yellow, 8-22 mm, sepals *scarcely* hairy. Meadows, woods and screes, to 2800 m. May-Aug. P, A, Jura.

6 MOUNTAIN BUTTERCUP *Ranunculus montanus.* Short per with hairy *or* hairless lvs. Basal lvs 3-5-lobed, lobes oval, toothed; stem lvs narrower, *half-clasping* stem. Fls golden-yellow, 10-30 mm, sepals with short hairs. Achene with *slender* hooked beak. Meadows, woods, screes and snow patches, to 2800 m. May-Aug. A, Jura. **6a** *R. venetus* has densely hairy lvs and a *stout* beak to the achenes. seA.

7 GRENIER'S BUTTERCUP *Ranunculus grenieranus.* Low/short per with *very hairy* lvs. Basal lvs 3-5-lobed, lobes oval, toothed; stem lvs with linear or lance-shaped lobes *broadest* near the base. Fls golden-yellow, 8-22 mm, sepals hairy. Achenes with a *long* hooked beak. Meadows and screes, to 2800 m. May-Aug. A. **7a** *R. oreophilus* like 7; the stem lvs linear; achene with a *short* beak. A, Ap.

8 HOOKED BUTTERCUP *Ranunculus aduncus.* Similar to 6. Lvs *always* hairy, lower *clasping* stem at their base. Fls golden-yellow, 12-25 mm, sepals with *long* hairs. Meadows, to 2800 m. June-July. swA, not Switzerland.

2c Creeping Buttercup   **3a** Meadow Buttercup

Buttercup Family *(contd.)*

1 BULBOUS BUTTERCUP *Ranunculus bulbosus* agg. Short/med very hairy per, stem base *swollen.* Basal and lower lvs 3-lobed, lobes oval, toothed, central one *stalked;* upper lvs with linear-lance-shaped lobes. Fls golden-yellow, 20-30mm. Sepal hairy, *turned down.* Fl stalks furrowed. Meadows and grassy places, to 2500m. Apr-July. T. **1a** *R. sardous* is *annual,* hairier, the stem scarcely swollen; fls smaller, *pale yellow.* T.

2 PYGMY BUTTERCUP *Ranunculus pygmaeus.* Tiny low, scarcely hairy, per. Basal lvs *kidney-shaped* in outline with 3-5 blunt lobes. Fls yellow, 5-10mm. Short turf and by snow patches, 1800-2800m. July-Aug. S,cA*. **2a** *R. hyperboreus* is *creeping* with 5-lobed lvs and *3-petalled fls.* To 2100m. S. **2b** *R. nivalis* □ is taller than 2, with *larger* stem lvs and fls 12-15mm. Short grassy places and by snow patches, to 1550m. July. S.

3 THORE'S BUTTERCUP *Ranunculus thora.* Short hairless per. Lvs waxy, lowest *large, kidney-shaped,* toothed, unstalked, upper much smaller. Fls yellow, 10-20mm, solitary or two to five clustered; sepals hairless. Meadows and rocky places on limestone, to 2200m. May-July. cP,A,Jura,Ap.

4 HYBRID BUTTERCUP *Ranunculus hybridus.* Similar to 3 but with *two* large lower lvs, kidney-shaped, toothed and lobed towards the top. Fls yellow, 12-25mm, solitary or two to three. Achenes short beaked. Stony places on limestone, to 2500m. June-July. e,seA. **4a** *R. brevifolius* is smaller with several large lower lvs and achenes *long* beaked. cAp.

5 ALPINE BUTTERCUP *Ranunculus alpestris.* Low hairless, tufted, per. Lvs *shiny-green,* basal rounded in outline, *5-lobed,* long stalked; stem lvs with 3 linear lobes, stalkless. Fls white, 20mm, two or three clustered. Sepals hairless. Damp meadows, stony places and snow patches, 1300-3000m. June-Oct. P,A,Ap. **5a** *R. traunfellneri* □ is smaller, basal lvs *matt-green,* 3-lobed; fls 15mm, *solitary.* seA.

6 CRENATE BUTTERCUP *Ranunculus crenatus.* Low hairless per. Basal lvs *round-heart-shaped,* 3-lobed at top, long stalked; stem lvs linear-lance-shaped. Fls white, 20-25mm, solitary or two, petals notched or un-notched. Damp rocks and screes, 1700-2400m. June-July. eA, Ap*. **6a** *R. bilobus* has *distinctly veined* lower lvs and *more* notched petals. Limestone rocks, to 2000m. sA–Italian.

7 ACONITE-LEAVED BUTTERCUP *Ranunculus aconitifolius.* Short/med slightly hairy per. Lvs 3-5-lobed, lobes oblong-oval, central one cut to the middle; upper lvs stalkless. Fls white, 10-20mm, in *branched* clusters; *sepals* reddish-purple, falling as fls open. Meadows and woods, to 2600m. May-Aug. P,A,Ap. **7a** *R. platanifolius* □ is larger, the lvs 5-7-lobed, central lobes *not* cut to middle. Scrub and dry meadows, to 1600m. May-Aug. T.

8 SÉQUIER'S BUTTERCUP *Ranunculus seguieri.* Short downy per. Basal lvs long stalked, 3-5-lobed, each lobe with *angular-pointed* segments; stem lvs similar but smaller. Fls white, 20-25mm, solitary or two, petals slightly *notched.* Achenes downy. Damp meadows, stony places and screes on limestone, 1800-2400m. June-July. A,Ap.

9 GOLDILOCKS BUTTERCUP *Ranunculus auricomus.* Short, slightly hairy per. Lower lvs only slightly lobed. Fls few, yellow, petals 5 *often distorted* and of differing sizes, sometimes absent; sepals purple tinged. Woods and hedges, to 2100m. Apr-May. T.

Basal leaves

7 Aconite-leaved Buttercup   **7a** *R. platanifolius*

1

2

2b

5a

3

4

5

6

7

8

9

Buttercup Family *(contd.)*

1 GLACIER CROWFOOT *Ranunculus glacialis.* Low/short hairless per. Lvs thick, *3-lobed,* lobes short stalked, with angular teeth; upper lvs linear. Fls white *becoming* pink or purplish tinted, 25-40 mm, solitary or two to three; *sepals* with purple-brown hairs. Rocky debris, moraines and screes, often near snow patches, on acid rocks, 2300-4250 m. July-Oct. S, P, A.

2 LESSER SPEARWORT *Ranunculus flammula.* Variable short/med hairless per, erect or creeping, often rooting at lf junctions, but *without* runners. Lvs shiny-green, lance shaped, slightly toothed, the lower stalked. Fls yellow, 7-20 mm, in branched clusters. Wet meadows and stream banks, to 2000 m. June-Oct. T.
2a Creeping Spearwort *R. reptans* □ is more slender with *runners* and rooting at all lf junctions, the lvs narrow, spoon-shaped or elliptic; fls 5 mm. B, S, A*.

3 GREATER SPEARWORT *Ranunculus lingua.* Med/tall hairless per with *long runners.* Lvs oblong-lance-shaped, toothed, stalked or unstalked. Fls golden-yellow, 30-50 mm, stalks *not* furrowed. Wet meadows and marshes to 1200 m. June-Sept. T.

4 PYRENEAN BUTTERCUP *Ranunculus pyrenaeus.* Low/short more or less hairless per. Lvs bluish-green, linear to lance-shaped, *untoothed,* stalkless. Fls white, 10-20 mm, solitary or two to three; *sepals* whitish, hairless. Damp meadows and slopes on limestone, 1700-2800 m. May-July. P, A.

5 PARNASSUS-LEAVED BUTTERCUP *Ranunculus parnassifolius.* Low hairy or hairless per. Lvs shiny-green. *oval-heart-shaped,* or broad lance-shaped, ribbed, untoothed. Fls white often pink or red tinged, 20-25 mm, solitary or up to 5; sepals hairy. Rocks and moraines on limestone, often by snow patches, 1900-2900 m. July-Aug. P, A.

6 AMPLEXICAULE BUTTERCUP *Ranunculus amplexicaulis.* Low/short hairless per. Basal lvs oval-lance-shaped, pointed, stalked; stem lvs stalkless, *clasping* stem at base. Fls white, 20 mm; sepals greenish, hairless, soon *falling.* Meadows, to 2500 m. June-July. P.

MEADOW-RUES *Thalictrum.* Perennials with 2-4-pinnate, ferny, lvs and racemes or clusters of feathery fls. Fls with four tiny petals, no sepals and a bunch of conspicuous, coloured, stamens. Fr a cluster of dry, one seeded, achenes.

7 SMALL MEADOW-RUE *Thalictrum simplex.* Short/tall hairless per. Basal lvs 2-3-pinnate, lflets oval to linear, toothed or not. Fls yellow, *drooping* at first but becoming erect, in short-branched clusters. Meadows, to 2000 m. July-Aug. T – except B. (See p.54).

8 LESSER MEADOW-RUE *Thalictrum minus.* Variable med/tall hairless per. Basal lvs 2-4-pinnate, lflets rounded or oval, *toothed* in upper half. Fls yellowish, *drooping* at first but becoming erect, in *long*-branched clusters. Rocky and dry grassy places, to 2850 m. May-July. T. (See p. 54).

9 STINKING MEADOW-RUE *Thalictrum foetidum.* Short slightly hairy, *foetid,* per. Lvs *ash-grey,* 3-4-pinnate. Fls yellow, *drooping,* in long-branched clusters. Rocks and stony places, usually on limestone, 1500-2400 m. June-Aug. eP*, A. (See p.54).

2a Creeping Spearwort

Buttercup Family (contd.)

1 GREAT MEADOW-RUE *Thalictrum aquilegifolium*. Med/tall hairy per. Lvs 2-3-pinnate, lflets oval, *broadest* near the top, toothed. Fls with pale greenish-white petals, and lilac or whitish stamens, in larger branched clusters. Damp woods and meadows, to 2500 m. May-July. T, (B).

2 LARGE-FRUITED MEADOW-RUE *Thalictrum macrocarpum*. Med hairless per with lvs like 1. Fls with greenish petals and yellowish stamens, few in branched clusters. Fr achenes large, 8-10 mm, *long-beaked*. Damp limestone rocks, to 2200 m. June-Sept. w, cP*. **2a** *T. tuberosum* is tuberous rooted, with *white* petals and small achenes. Dry rocky places, to 2000 m. P.

3 ALPINE MEADOW-RUE *Thalictrum alpinum*. Short/low, rather insignificant, hairless per, *unbranched*. Lvs 2-pinnate. Fls with purple petals, stamens *violet* with yellow anthers. Damp meadows, stony and rocky places, 1900-2900 m. June-July. S, P, A*.

COLUMBINES *Aquilegia*. Tufted perennials, with usually branched stems and 2-trifoliate lvs. Fls drooping, with 5 coloured sepals and petals, the petals with long backward-pointing spurs. Fr a cluster of follicles like Delphinium. Poisonous.

4 COMMON COLUMBINE *Aquilegia vulgaris*. Med/tall per, stems usually hairy, lflets dull green, hairy *beneath*. Fls violet-blue or purplish, rarely white, 3-5 cm, spurs *hooked;* stamens yellow, *scarcely* protruding. Woods, meadows and rocky places, usually on limestone, to 2000 m. May-July. T−except Ap.

5 DARK COLUMBINE *Aquilegia atrata*. Med/tall per, stems hairy. Lflets hairless. Fls *dark* purple-violet, 3-4 cm, spurs hooked; stamens yellow, *protruding* well beyond petals. Woodland clearings and rocky places on limestone, to 2000 m. May-July. A, Ap. **5a** *A. nigricans* has *larger* fls, 5-6 cm and lflets *hairy* beneath. eA, not Italy.

6 EINSEL'S COLUMBINE *Aquilegia einseleana*. Short/med. slightly hairy per. Lflets hairless or with a few hairs above. Fls violet-blue or purplish, 2·5-3·5 cm, spurs *straight;* stamens *not* protruding. Grassy and stony places and woods, on limestone, to 1800 m. June-July. A, excluding W. Switzerland and France. **6a** *A. thalictrifolia* has *stickily hairy* stems and lvs and larger fls. To 1600 m. sA−Italian.

7 BERTOLONI'S COLUMBINE *Aquilegia bertolonii*. Short per, stems sometimes unbranched, stickily hairy in upper half. Fls blue-violet or dark blue, 2·5-3·5 cm, sepals *downward* pointing; spurs straight or curved. Woods and rocky places, to 1700 m. June-July. swA−French and Italian, MA, Lig.

8 PYRENEAN COLUMBINE *Aquilegia pyrenaica*. Short, more or less hairless per, stems usually branched. Fls bright blue or lilac, 3-5 cm, spurs *long, slender* and curved. Rocky and stony places, 1800-2500 m. July-Aug. P. **8a** *A. aragonensis* is shorter with *densely hairy* stems and lvs; lvs l-trifoliate; fls deep blue with short spurs. July-Aug. P−Spanish.

9 ALPINE COLUMBINE *Aquilegia alpina*. Short/tall hairy per. Fls *large,* bright blue, 5-8 cm, spurs long, straight or curved. Meadows, woods and rocky places, to 2600 m. July-Aug. A−mainly in the west, Ap.

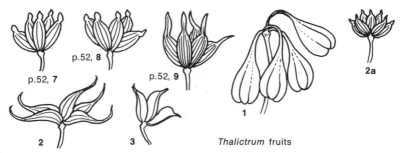

p.52, 8

p.52, 7

p.52, 9

2a

1

2

3

Thalictrum fruits

Paeony Family Paeoniaceae

1 PAEONY *Paeonia officinalis.* Med robust, clump-forming per. Basal lvs large, shiny green above, 2-3-trifoliate. Fls solitary, very large open cups, shiny red or pinkish, 7-13cm, petals 5-8; stamens many, pale yellow. Meadows and woods, to 1700m. May-June. sA, Ap.

Barberry Family Berberidaceae

Fls in short racemes with four to six petals, each with a honey petal (nectar producing) within.

2 BARRENWORT *Epimedium alpinum.* Short tufted per. Lvs 2-trifoliate, lflets oval or heart-shaped, *spiny-margined.* Fls dull red with four pale yellow honey petals, 9-13mm. Damp woods and shady places, to 1300m. Mar-May. se, eA, n, cAp.

3 BARBERRY *Berberis vulgaris.* Deciduous shrub to 4m, stems with *3-pointed spines.* Lvs oval, spiny-margined. Fls yellow, 6-8mm, in drooping racemes. Fr a bright red, *oblong,* berry, edible. Rocky hillslopes, generally on limestone, to 2300m. Apr-June. T.

Fumitory Family Fumariaceae

Often included in the Poppy Family *Papaveraceae.* Thin stemmed annuals or perennials, hairless, with 2-pinnate lvs. Fls in short spike-like racemes, two-lipped, spurred. Corydalis has oblong fr capsules, fumitory rounded ones.

4 YELLOW CORYDALIS *Corydalis lutea.* Short hairless per with leafy branched, stems. Lvs green above, greyish beneath. Bracts tiny oblong, *untoothed.* Fls golden-yellow, 12-20mm, spur short. Rocks, walls and shady places, to 1700m. May-Oct. c, eA, not Austria, (B). **4a** *C. acaulis* has lvs *greyish* above and beneath and fls *white* with a yellowish-green apex. seA-Jugoslavia.

5 BULBOUS CORYDALIS *Corydalis bulbosa.* Low/short, tuberous-rooted, per, stems *without* a scale below the lowest lf. Bracts large, *untoothed.* Fls dull purple, 18-30mm, spur downcurved. Woods, hedgerows and cultivated land to 2000m. Mar-May. P, A, Ap, (B). **5a** *C. intermedia* has stems with an *oval scale* below the lowest lf and fls purple, rarely white, 10-15mm, in racemes of 2-8. Woods and pastures, generally on limestone, to 2000m. Mar-Apr. S, A, n, cAp. **5b** *C. pumila* is similar to 5a but with *toothed bracts.* S, neA–Austria only.

6 SOLID-TUBERED CORYDALIS *Corydalis solida.* Short, tuberous rooted, per with an *oval scale* below the lowest lf. Bracts large, *toothed.* Fls purplish, 15-25mm, in racemes of 10-20, spurs slightly curved. Woods, hedgerows and cultivated ground, to 2000m. Mar-May. T, (B).

7 SARCOCAPNOS *Sarcocapnos enneaphylla.* Low much branched, *cushion forming,* per. Lvs pinnate, lflets heart-shaped. Fls long-stemmed, white or yellowish with a purple tip, 8-10mm, spurs short. Shady limestone and basic rocks, to 1200m. Apr-June. eP.

8 COMMON FUMITORY *Fumaria officinalis* agg. Weak bluish-green scrambling ann. Lf-segments flattened. Fls purplish-pink, darker at tip, 7-9mm, in *long-stalked* racemes. Bracts *more* than half the length of the fl stalk. Waste and stony places and cultivated land to 1500m. Apr-Oct. T. **8a** *F. schleicheri* has *shorter* bracts and deep pink fls, 5-6mm in racemes of 12 or more. To 1700m. T. **8b** *F. vaillantii* is similar to 8a but the racemes *short-stalked* with 6-12 fls; fls pale pink *tipped* with blackish-red. To 2100m. May-Oct. T.

1 Paeony, fr.

57

Poppy Family Papaveraceae

Annuals or perennials with 1-2-pinnate lvs and large, solitary fls with four, rather crumpled-looking, silky petals. Sepals two, falling as fls open. Fr a capsule with pores at the top.

1 COMMON or CORN POPPY *Papaver rhoeas.* Med/tall rather bristly ann. Lvs 1-2-pinnate. Fls deep scarlet, sometimes with a dark centre, 6-8·5cm. Fr almost round, *hairless.* Cultivated land, banks and waste places, to 1800m. May-July. T. **1a Long-Headed Poppy** *P. dubium* □ has *smaller* pale scarlet fls, 3-5·5cm, and fr *oblong* in outline. T.

2 PRICKLY POPPY *Papaver argemone.* Short/med bristly ann. Lvs 1-2-pinnate. Fls pale scarlet, often with a black centre, 3·5-4·5cm, petals not overlapping. Fr oblong-club-shaped in outline, *bristly.* Fields and waste places, to 1700m. May-July. T. **2a** *P. hybridum* □ has almost globular-shaped fr covered by many yellowish bristles; fls crimson. T.

3 PYRENEAN POPPY *Papaver suaveolens.* Low bristly tufted per. Lvs pinnate, *scarcely* bristly. Fls yellow or red, 20-30mm. Fr oval in outline, bristly. Rocky places, screes and moraines on limestone, 1800-2500m. July-Aug. P– Spanish. **3a** *P.s. endressii* has *bristly* 2-pinnate lvs.

4 RHAETIAN POPPY *Papaver rhaeticum.* Short tufted bristly per. Lvs 1-2-pinnate; lflets *not* opposite, oval-lance-shaped, blunt. Fls golden-yellow, sometimes red or white, 4-5cm. Limestone rocks, screes and moraines or river gravels, 1500-3050m. July-Aug. eP,sw,eA. **4a** *P. sendtneri* is smaller with sharply pointed lflets and white fls, 3-4cm, 2000-2700m. c,eA,cAp.

5 ALPINE POPPY *Papaver burseri.* Short tufted per, almost hairless. Lvs 1-2-pinnate; lflets usually *opposite,* lance-shaped or linear, sharply pointed. Fls white, 3-4cm. Limestone rocks, screes, moraines and river gravels, 1200-2000m. July-Aug. nA*. **5a** *P. kerneri* has *yellow* fls. seA.

6 ARCTIC POPPY *Papaver radicatum* Variable low/short tufted hairy per; juice white or yellow. Lvs pinnately-lobed. Fls pale to deep yellow, sometimes pinkish, 2-4cm. Fr oblong, *softly hairy.* Screes and gravels, to 1850m. June-Aug. S.

7 GREATER CELANDINE *Chelidonium majus.* Med/tall slightly hairy per, branched. Lvs greyish, alternate, pinnate, end lflet 3-lobed. Fls bright yellow, 1·5-2cm, in *umbel-like* clusters of 3-8. Fr a long narrow capsule. Banks, hedges and waste places, to 1700m. May-Oct. T.

8 WELSH POPPY *Meconopsis cambrica.* Med tufted, slightly hairy per. Lvs pinnate, basal ones long-stalked, but stem lvs short-stalked. Fls solitary, yellow, 4-7cm. Fr narrow-oblong in outline with a *short beak* on top, hairless. Woods, rocky and shady places, to 2000m. June-Aug. B,Cev,P.

1a Long-headed Poppy   **2a** *Papaver hybridum*

Cress Family Cruciferae

Annual or perennial herbs. Lvs usually alternate often tufted.Fls with 4 separate sepals and petals and 6 stamens. Fr a linear rounded or heart shaped, 2-parted capsule. **For more white- and yellow-flowered crucifers** see pp. 302-304. For the identification of crucifer fruits see p. 344.

1 LONDON ROCKET *Sisymbrium irio*. Short/med hairy or hairless *ann.* Lower lvs pinnately-lobed with a large end lobe, stalked; upper lvs smaller, scarcely lobed. Fls yellow, 4-6mm, in branched clusters. Fr *linear,* 25-65mm, hairless, over-topping fls when young. Waste places and fields, to 1400m. May-Sept. T. **1a False London Rocket** *S. loeselii* is hairier with larger fls; fr *not over-topping* fls.

2 AUSTRIAN ROCKET *Sisymbrium austriacum* (= *S. pyrenaicum*). Short/tall hairy or hairless bien/per. Lvs pinnately-lobed or unlobed, stalked. Fls yellow, 7-10mm. Fr linear, 15-50mm, *rather twisted*. Fields and waste places to 1600m. May-Sept. T, (B, S). **2a** *S.a. chrysanthum* has shorter contorted fr, 7-15mm. c, wP.

3 MURBECKIELLA *Murbeckiella pinnatifida* (= *Sisymbrium pinnatifidum*). Low/short per with *branched* hairs. Basal lvs toothed or untoothed, upper pinnately-lobed. Fls *white,* 5-6mm, in loose racemes, petals *notched*. Fr linear, 10-30mm, hairless. Rocky places to 3000m. June-Aug. c, eP, sw, wcA. **3a** *M. zanonii* has *all* lvs pinnately-lobed; fls larger, 7-10mm and fr longer. nAp.

4 TANSY-LEAVED ROCKET *Hugueninia tanacetifolia* (= *Sisymbrium tanacetifolium*). Med hairy or hairless, rather greyish, per. Lvs *2-pinnately-lobed,* the lower long-stalked. Fls yellow, 3-6mm, many, in loose clusters. Fr narrow-oblong, 6-15mm. Rocky and grassy places and streamsides, 1700-2500m. June-Aug. sw, wcA. **4a** *H.t. suffruticosa* is hairier with *less-lobed lvs.* P*.

5 ALPINE BRAYA *Braya alpina*. Low tufted, hairy per. Basal lvs lance-shaped, slightly toothed or untoothed, *stem lvs* one or few. Fls white or purplish, 4-6mm, in small clusters, petals *blunt-ended.* Fr broad-linear, 5-11mm, hairless. Dry rocky and stony places, usually on limestone, 2000-3000m. July-Aug. eA*. **5a**, p.335.

6 ALPINE WOAD *Isatis allionii*. Short hairless or slightly hairy per. Lvs greyish, *clasping stem* at base, untoothed, stalkless. Fls yellow, 5-6mm, in branched clusters. Fr elliptical *brown,* 15-25mm, often notched at top. Stony places, to 2350m. July-Aug. swA, n, cAp. **6a**, p.335.

TREACLE-MUSTARDS *Erysimum*. Anns or pers with branched hairs and narrow lvs. Fls yellow in small elongated clusters, sepals upright. Fr a long linear pod.

7 WOOD TREACLE-MUSTARD *Erysimum sylvestre*. Short/med tufted, *silvery-grey,* per. Lvs linear to narrow-lance-shaped, sometimes slightly toothed. Fls lemon-yellow, 12-16mm. Fr 40-90mm, grey-green with a short 1-2mm beak. Rocky, stony places and gravels to 1600m. May-Aug. c, eA, nAp. **7a** *E. helveticum* (= *E. pumilum*) is greener, fr with a longer beak, 2-3mm. To 2800m. P, A.

8 DECUMBENT TREACLE-MUSTARD *Erysimum decumbens* (= *E. dubium, E. ochroleucum*). Short/med green or greyish per. Lvs oblong or narrow-lance-shaped, untoothed. Fls pale yellow, 14-20mm. Fr 35-80mm, grey-green, with a *2-6mm beak.* Rocky and stony places, usually on limestone, to 2300m. May-June. P, swA.

9 HAWKWEED-LEAVED TREACLE-MUSTARD *Erysimum hieracifolium* (= *E. strictum, E. marschallianum*). Med/tall green or greyish bien/per. Lvs linear or oblong, margin *wavy, toothed.* Fls small yellow, 7-9mm, petals *hairy* on back. Fr 30-55mm, grey or greenish, beak short, 1-2mm. Woods, rocky and stony places, usually on limestone, to 1700m. May-Aug. A. **9a** *E. virgatum* has narrower *untoothed* lvs; *larger* fls, 10-12mm. wA.

10 HOARY TREACLE-MUSTARD *Erysimum incanum*. Short/med grey-green *annual.* Basal lvs pinnately-lobed, upper lance-shaped. Fls very small yellow, 3-4mm, petals hairy on back. Fr 30-55mm. grey. Rocky and stony places, to 1800m. P.

Cress Family *(contd.)*

1 ALPINE DAME'S VIOLET *Hesperis inodora.* Med/tall *downy-white* bien/per. Lvs lance-shaped, coarsely toothed, the upper ones *half-clasping* the stem. Fls white, 15-25mm, in loose branched racemes, *not* fragrant. Fr long, linear, to 100mm. Rocky and waste places, to 1500m. May-July. MA.

2 DAME'S VIOLET *Hesperis matronalis.* Med/tall *hairy* green bien/per. Lvs lance-shaped, lobed and toothed, the upper short-stalked or unstalked, *not* clasping. Fls violet-purple, 15-20mm, *very fragrant.* Pod long, linear, to 100mm, curving upwards. Rocky and waste places, roadsides, hedges, to 1400m. May-Aug. P,sA,Ap,(T). **2a** *H.m. candida* has *white* fls. P,A. **2b** *H. laciniata* has more deeply lobed lower lvs and *yellow fls* suffused with purple. To 1500m. sA,Ap.

3 SAD STOCK *Matthiola fruticulosa* (= *M. tristis*). Short/med slightly to densely *white hairy,* per. Lvs linear or oblong, untoothed, wavy-margined or lobed. Fls purple-red, occasionally yellow, 15-25mm, in loose spikes. Fr long, linear, 25-120mm. Rocky places, often on limestone, to 2000m. May-Aug. P,sA,Ap.

BITTERCRESSES *Cardamine.* Annuals or perennials with simple or pinnate lvs. Fls white or purple, rarely pale yellow; inner two sepals slightly pouched at base. Stigma slightly two-lobed. Fr a linear pod. Petal length includes the 'claw'.

4 SEVEN-LEAFLET BITTERCRESS *Cardamine* (= *Dentaria*) *heptaphylla.* Short/med per. Lvs pinnate with 3-5 pairs of narrow toothed lflets, slightly hairy. Fls white, pink or purplish, in a loose raceme, petals 14-20mm long. Woods, particularly of Beech, to 1800m May-July. P,A,Ap. **4a** *C. pentaphyllos* (= *Dentaria*) *pentaphyllos* □ has *palm-like lvs* with 3-5 lflets, and rather larger fls. Woods and rocky places, to 2200m. P,A.

5 KITAIBEL'S BITTERCRESS *Cardamine kitaibelii.* Short/med hairless per. Lvs *pinnate* with 2-6 pairs of lance-shaped, toothed lflets. Fls *pale yellow,* in loose racemes; petals 12-16mm. Woods, to 1900m. Apr.-June. A,Ap.

6 DROOPING BITTERCRESS *Cardamine* (= *Dentaria*) *enneaphyllos.* Short hairless per. Lvs *trifoliate,* in a *whorl* of two to four, lflets oval-lance-shaped, toothed. Fls pale yellow or white in a *drooping cluster;* petals 12-16mm. Woods, to 2150m. Apr-July. A–except France and Switzerland, Ap.

7 TRIFOLIATE BITTERCRESS *Cardamine trifolia.* Short, slightly hairy, per. Lvs *mostly basal, trifoliate,* lflets rounded or oblong, slightly lobed, purplish beneath. Fls white or pink, with *yellow* anthers; petals 9-11mm. Moist woods and shady places, often on limestone, to 1400m. Apr-June. A,n,eAp.

8 ASARUM-LEAVED BITTERCRESS *Cardamine asarifolia.* Short/med hairless per. Lvs *kidney-shaped,* slightly toothed, long stalked, the upper sometimes trifoliate. Fls white with *violet* anthers; petals 6-10mm. Damp pastures and streamsides, to 2000m. July-Aug. P,A,Ap.

9 LARGE BITTERCRESS *Cardamine amara.* Short/med slightly hairy per, basal lvs *not* in a rosette. Lvs pinnate with 2-5 pairs of oval, toothed, lflets. Fls white, rarely purplish, with violet anthers; petals 7-9mm. Damp pastures and woods, streamsides, to 2000m. Apr-June. B,S*,P*,A,Ap. **9a** *C. opizii* has *5-8 pairs* of lflets and smaller fls. eA.

10 RADISH-LEAVED BITTERCRESS *Cardamine raphanifolia.* Med/tall hairless per with a basal *rosette.* Lvs pinnate with 1-5 pairs of rounded lflets; upper lvs *with narrow* lflets. Fls reddish-violet, rarely white, with yellow anthers; petals 8-12mm. Damp pastures and streamsides, to 2500m. May-July. P,Ap.

4a *Cardamine pentaphyllos*

1

2

3

4

5

6

7

8

9

10

Cress Family (contd.)

1 CUCKOO FLOWER or LADY'S SMOCK *Cardamine pratensis.* Med hairless per, with basal lvs forming a rosette. Lvs pinnate with 1-7 pairs of lflets, stem lvs dissimilar from basal. Fls white tinged with violet or pink, anthers yellow; petals 8-13mm, slightly notched. Damp pastures and woods, streamsides, to 2600 m. Apr-July. T. **1a** *C. crassifolia* is smaller, *without* a distinct lf rosette; fls purplish, petals 6-8mm; lvs with 1-3 pairs of lflets. P. **1b Coral-root Bittercress** *C. bulbifera* has small brownish-purple *bulbils* at base of upper lvs. Rarely above 1500 m. T – except P.

2 IVY-LEAVED BITTERCRESS *Cardamine plumieri.* Low/short slightly hairy bien/per. Lower lvs *ivy-like* with 3-5 lobes, upper pinnate. Fls white, petals 6-8mm, often slightly notched. Damp rocky places, to 2200m. June-Aug. w,sA,n,cAp.

3 MIGNONETTE-LEAVED BITTERCRESS *Cardamine resedifolia.* Low/short hairless per. Lower lvs *spoon-shaped,* some above trifoliate, the upper with 3-7 lobes. Fls white, petals 5-6mm. Damp rocky places, to 3500m. June-Aug. A, Ap.

4 ALPINE BITTERCRESS *Cardamine bellidifolia.* Low hairless per. Lvs mostly in a basal rosette, *all spoon-shaped,* untoothed. Fls white in clusters of 2-5, petals 3·5-5mm. Fr purplish-brown. Damp meadows and gravels, usually on acid soils, to 2100m. July-Aug. S. **4a** *C.b. alpina* has 2-3 stem lvs and shorter brown fr. To 3000m. P,A.

5 TALL ROCKCRESS *Cardaminopsis arenosa.* Low/tall slightly hairy ann/per. Lvs *pinnately-lobed,* the upper ones narrower, toothed. Fls white, becoming pink or lilac, 8-10mm, in loose racemes. Fr linear 10-45mm long, strongly flattened. Sandy soils, usually calcareous, to 1400m. Apr-June. T. **5a** *C. petraea* is smaller with the upper lvs *untoothed.* B,S,C,eA. **5b** *C. halleri* like 5a but basal lvs *simple or pinnate* with rounded lflets. A,nAp.

6 BLUE ARABIS *Arabis caerulea.* Low/short tufted, hairy or hairless, per. Lvs oval, broadest above middle, toothed at apex, stalked. Fls pale blue 5-6mm, in small clusters. Damp rocks and moraines, on limestone, to 3500m. July-Aug. A.

7 PERENNIAL HONESTY *Lunaria rediviva.* Tall roughly-hairy per, branched. Lvs oval, pointed, sharply toothed, the upper stalked. Fls pale purple to violet, 15-20mm, *fragrant.* Fr *large* elliptical, 4-9cm, very *flattened,* pointed at both ends. Damp woods and shady places, often on calcareous soils, to 1400m. May-July. P,A. **7a Honesty** *L. annua* is bien with more rounded frs. Introduced in many areas, but seldom exceeding 1200m. (T).

8 ALYSSOIDES *Alyssoides utriculata.* Short/med starry-haired per, with a woody stock and *rosettes of lvs.* Lvs oblong-spoon-shaped, stalked except on stems, untoothed. Fls yellow, 8-14mm. Fr rounded, *inflated,* long beaked. Rocks and crevices, to 1500m. Apr-June. sw,sA,Ap.

9 WULFEN'S ALYSSUM *Alyssum wulfenianum.* Low/short prostrate or semierect, *grey or whitish* per. Lvs oval-oblong, untoothed, starry-haired. Fls yellow, 5-7mm, in rounded clusters. Fr *elliptical, inflated,* long beaked. Dry rocky places, to 2000m. June-Aug. seA. **9a** *A. ovirense* is greener with more rounded lvs and hairy-backed petals. To 2700m. seA.

10 MOUNTAIN ALYSSUM *Alyssum montanum.* Low/short prostrate or semierect, *often whitish,* per. Lvs oblong or spoon-shaped, starry-haired, the upper narrower. Fls yellow, 5mm, in elongated clusters; petals *notched.* Fr rounded, inflated, long beaked. Rocky places and gravels, to 2500m. May-July. P,A,Ap. (see also p.304).

11 PYRENEAN ALYSSUM *Ptilotrichum pyrenaicum* (= *Alyssum pyrenaicum*). Low/short *silvery,* cushion-forming, sub-shrub. Lvs oval-lance-shaped, narrowed at base, untoothed. Fls *white,* 5-7mm, in dense rounded clusters. Fr oval, flattened, hairy, long beaked. Limestone cliffs, to 1500m. June-July. eP– French*. **11a** *P. lapeyrousianum* has hairless, short beaked, frs. eP–Spanish and French.

Cress Family *(contd.)*

WHITLOW-GRASSES *Draba.* Low tufted perennials, sometimes annual. Lvs toothed or untoothed, margins sometimes with stiff bristles. Fls rather small in loose upright racemes or small clusters, yellow or white. Fr a round or elliptical-oblong pod with a short beak.

1 YELLOW WHITLOW-GRASS *Draba aizoides.* Low tufted hairless per, with lfless stems. Lvs linear-elliptical in rosettes, *margin bristly.* Fls yellow 4-5mm, clustered at end of 5-15cm long stems, petals oblong, *blunt.* Fr elliptical, with a *1·5-3mm beak.* Rocky and stony places, usually on limestone, to 3600m. Apr-July. T–exceptS. **1a** *D. hoppeana* is dwarfer the fr beak only 1mm long, 2200-3600m. July-Aug. w,cA. **ib** *D. aspera* □ like 1 but lvs *not more* than 1mm broad and fr beak *longer,* 3·5-7mm. P,AA,Ap. **1c** *D. alpina* like 1 but lvs broader, oval-lance-shaped, *slightly hairy,* hairs branched To 1650m. July. S.

2 WOOLLY-FRUITED WHITLOW-GRASS *Draba lasiocarpa.* More *robust* than 1 with lfless stems to 20cm. Lvs oblong, pointed, 3-4mm broad, margin bristly. Fls deep yellow, petals *rounded.* Fr. oblong with a short beak 1-1·5mm. Rocky and stony places, to 2000m. e,seA.

3 SAUTER'S WHITLOW-GRASS *Draba sauteri.* Low per forming loose cushions. Lvs linear or narrow lance-shaped, in rosettes, margin bristly. Fls yellow, clustered on *short* lfless stems 1-3cm stems. Fr short-oblong, *beak* 0·5mm. Limestone rocks and gravels, 1900-2900m. July-Aug. eA.

4 ENGADINE WHITLOW-GRASS *Draba ladina.* Low tufted per, slightly hairy. Lvs lance-shaped or elliptical, *covered in* star shaped hairs, margin slightly bristly. Fls pale yellow, in small clusters, on lfless stems 1-5cm long. Fr oblong, beak 1mm. Limestone rocks, 2600-3050m. July-Aug. cA– Engadine.

5 STARRY WHITLOW-GRASS *Draba stellata.* Low tufted, starry-haired, per; *stems lfy.* Lvs oval, broadest above middle; stem lvs one or two, *often* toothed. Fls white or cream, clustered. Fr elliptical, beak 1-2mm. Limestone rocks, to 2500m. June-Aug. eA.

6 CARINTHIAN WHITLOW-GRASS *Draba carinthiaca.* Low/short *starry-haired* per, forming loose cushions, stem lfy. Lvs lance-shaped, margin bristly *at base.* Fls white, clustered. Fr narrow-oblong, *unbeaked.* Rocks and screes, usually on limestone, 1500-3600m. June-Aug. P,A.

7 AUSTRIAN WHITLOW-GRASS *Draba dubia.* Low/short starry-haired per, stems usually lfy. Lvs narrow oval, *all untoothed.* Fls white, in small clusters. Fr oblong-elliptical, hairless, beak short, 0·5mm. Rocks and screes, 1300-3200m. May-July. P,A. **7a** *D. kotschyi* is laxer with broad oval lvs, the upper *toothed.* eA-Austrian. **7b** *D. tomentosa* like 7 but stems and *fr* starry-hairy. Limestone rocks and screes, to 3500m. June-Aug. P,A.

8 BALD WHITE WHITLOW-GRASS *Draba fladnizensis* (= *D. wahlenbergii*). Low tufted per, *stems lfless, hairless.* Lvs oblong-oval, margin *bristly.* Fls white, in small clusters. Fr elliptical, beak short, 0·5mm. Rocky and grassy slopes, prefering acid soils, 1600-3400m. June-Aug. S,P,A.

9 TWISTED WHITLOW-GRASS *Draba incana.* Short erect hairy bien, stems lfy. Lvs lance-shaped, blunt, sometimes toothed, *short-stalked.* Fls white, 3-4mm, in loose clusters, *petals* notched. Fr oblong-lance-shaped, *twisted.* Rocky and stony places, usually on limestone, to 2600m. May-July. B,S,P,A–except Italy and Yugoslavia. **9a** *D. nivalis* is sometimes per with pale bluish-green lvs; stems *usually lfless.* Fls creamy white. To 1920m. June. S.

10 WALL WHITLOW-GRASS *Draba muralis.* Short, slightly hairy, *ann;* stems erect, lfy. Lvs broad-oval, upper *partly-clasping* the stem. Fls tiny, white, 2mm, petals notched. Fr oblong-elliptical, hairless. Rocks and walls, to 1300m. Apr-June. T–exceptB. **10b** *D. nemorosa* has *pale yellow* fls and hairy fr. T–except B.

Cress Family (contd.)

1 SPRING WHITLOW-GRASS Erophila (= Draba) verna. Variable low/short, slightly hairy ann, stems lfless. Lvs lance-shaped or elliptic, sometimes toothed, in a basal rosette. Fls white or pinkish, 3-5mm, petals cleft to middle. Fr elliptical or rounded, on long stalks. Waste places, sandy or stony ground, to 1700m. Mar-May. T.

2 PYRENEAN WHITLOW-GRASS Petrocallis (= Draba) pyrenaica. Low densely tufted, greyish-hairy, per. Lvs wedge-shaped, lobed at tip, bristle margined. Stems lfless. Fls pale lilac or pink, rarely white, 6-7mm, petals rounded, in small clusters. Limestone rocks and screes, 1700-2900m. June-Aug. P,A.

3 COMMON SCURVY-GRASS Cochlearia officinalis. Variable low/med hairless bien or per. Lvs heart or kidney-shaped, fleshy, lower long stalked and in a rosette, upper clasping the stem. Fls white, 8-10mm, in loose clusters. Fr rounded or oval in outline. Dry banks, grassy and stony places, to 2200m. Apr-Aug. T−except Ap. **3a** C. pyrenaica has fr elliptical in outline. P,A.

4 KERNERA Kernera saxatilis. Variable low/short hairless per, usually branched. Basal lvs lance or spoon-shaped, toothed or not, stalked, in a rosette; upper lvs oval, often half-clasping the stem. Fls white, 3-5mm, many in racemes, petals rounded. Fr rounded. Rocky and grassy places on limestone, to 2000m. P,A.

5 ALPINE SCURVY-GRASS Rhizobotrya (= Kernera) alpina. Rather like a compact form of 3. Low cushion per with short stems. Lvs oblong-spoon-shaped, blunt, long-stalked. Fls tiny, white, 2mm, each with a small bract at base of fl stalk, in tight clusters. Fr oval. Dolomitic rocks, 1900-2800m. sA−Dolomites.

6 CHAMOIS CRESS Hutchinsia alpina. Low tufted per, stems lfless, hairy. Lvs pinnate, hairless, shiny. Fls white, 4-5mm, petals rounded, in small clusters. Fr 4-6mm, oval, pointed. Limestone rocks and screes, to 3400m. May-Aug. P,A,n,cAp. **6a** H.a. brevicaulis has hairless stems, smaller fls and blunt-ended 3·5mm fr. Basic or acid rocks. A,n,cAp.

7 HYMENOLOBUS Hymenolobus pauciflorus. Low, slender, slightly hairy ann or bien. Lvs spoon-shaped, 3-lobed or unlobed. Fls tiny, white, 2-3mm, in loose clusters. Fr 2-4mm, rounded or elliptical in outline. Rocky and stony places, waste ground, to 1600m. May-Aug. A.

8 AETHIONEMA Aethionema saxatile. Low/short hairless ann or per, branched. Lower lvs oval or oblong, blunt, untoothed, the upper narrower and pointed. Fls white, purplish or lilac, 3-5mm, in small clusters, petals rounded. Fr 5-9mm, rounded, winged. Rocks and screes, usually limestone, to 2300m. Apr-July. P,A,Ap.

2 Pyrenean Whitlow-grass

Cress Family (contd.)

PENNYCRESSES *Thlaspi*. Perennials or biennials, rarely annual. Stem lvs usually clasping. Fls in rounded clusters, elongating in fruit. Fr heart-shaped, flattened, usually winged.

1 SMALL-FLOWERED PENNYCRESS *Thlaspi brachypetalum*. Short/med hairless, bluish-green, bien. Lvs elliptical or oblong, toothed or untoothed; basal ones in a rosette, stalked. Fls white, 3-5mm, *petals* equal in length to sepals; anthers *whitish*. Woods and pastures, usually on acid soils, to 2000m. Apr-July. P, swA, Ap. **1a** *T. alpestre* has petals longer than sepals and reddish or dark violet *anthers*. T–often naturalised.

2 APENNEAN PENNYCRESS *Thlaspi stylosum*. Very low cushion-forming per. Lvs mostly in basal rosettes, elliptical-spoon-shaped, untoothed, fleshy. *Fls purplish*, 8mm; anthers violet. Rocks and screes, to 2000m. Apr-June. c,sAp.

3 EARLY PENNYCRESS *Thlaspi praecox*. Short, hairless, greyish-green, per; more or less cushion-forming. Lvs oblong or oval, toothed or not, often *violet beneath*. Fls white, 7-9mm; sepals usually violet-tipped. Grassy and stony places, shaded limestone rocks, to 2250m. Mar-May. seA, Ap. **3a** *T. goesingense* is a larger, greyer plant; fl stems *often branched*. seA.

4 MOUNTAIN PENNYCRESS *Thlaspi montanum*. Short *mat-forming*, hairless, per with erect fl stems. Lvs oval or rounded, not or slightly toothed, greyish, the basal ones in rosettes, stalked. Fls white, 7-9mm; anthers pale yellow. Fr wings more than 1mm wide, *notch deep*. Grassy and rocky places, screes and cliff ledges. June-Aug. A.

5 ALPINE PENNYCRESS *Thlaspi alpinum*. Low/short, mat or cushion-forming, hairless, per. Lvs oval to oblong, blunt, untoothed, the basal ones in rosettes, stalked. Fls white, 7-9mm; *anthers yellow*. Fr wings less than 1mm wide, *notch shallow*. Grassy and rocky places, screes, to 3000m. June-Aug. A. **5a** *T. kerneri* has *pointed* stem lvs and smaller fls. seA.

6 ROUND-LEAVED PENNYCRESS *Thlaspi rotundifolium*. Variable low tufted, hairless, per with *long rooting runners*. Lvs rounded or oblong, fleshy, rarely slightly toothed; basal ones in loose rosettes. Fls purple, 8-10mm, honey-scented. Fr keeled, *not winged*. Rocks and screes, to 3000m. June-July. A, Ap. **6a** *T.r. cepaefolium* is smaller, *without* basal lf rosettes, but stem lvs crowded. seA.

CANDYTUFTS *Iberis*. Like Pennycresses but fls with upper two petals small and lower two large. Fr heart-shaped, flattened, winged.

7 SPOON-LEAVED CANDYTUFT *Iberis spathulata* (= *I. nana*). Low hairy per, sometimes ann. Lvs fleshy, *spoon-shaped*, generally untoothed. Fls purplish to white, in flat-topped clusters. Fr narrowly winged. Rock crevices, screes and gravels, 1500-2800m. June-Aug. P. **7a** *I.s. nana* (= *I. nana*) is *hairless* with slightly toothed lvs. MA, Lig.

8 DAUPHINE CANDYTUFT *Iberis aurosica*. Low/short hairless per. Lvs rather fleshy, *narrow* oblong-spoon-shaped, sometimes with 1-2 teeth at tip. Fls purple or lilac, in flat-topped clusters. Rocks and gravels, usually on limestone, to 2600m. July-Aug. wA.

9 ANNUAL CANDYTUFT *Iberis stricta*. Med hairless ann, stems slender, branched. Lvs linear, pointed, toothed. Fls small pink to lilac, in flat topped clusters. Fr oval. Calcareous soils and rocks to 1300m. June-Aug. swA,–French. **9a** *I.s. leptophylla* has more or less untoothed lvs and rounded frs. swA, Lig.

10 ALPINE CABBAGE *Brassica repanda*. Variable short/med hairless per, forming small tufts. Lvs *all basal*, spoon-shaped or oval, toothed or pinnately lobed. Fls yellow, 8-15mm, in clusters of 2-12. Fr linear, narrowing into the beak, to 5cm long. Rocky and gravelly places, to 2500m. Apr-Aug. P–Spanish,swA. **10a** *B. gravinae* has lfy stems and smaller fls, 8-12mm. c,sAp.

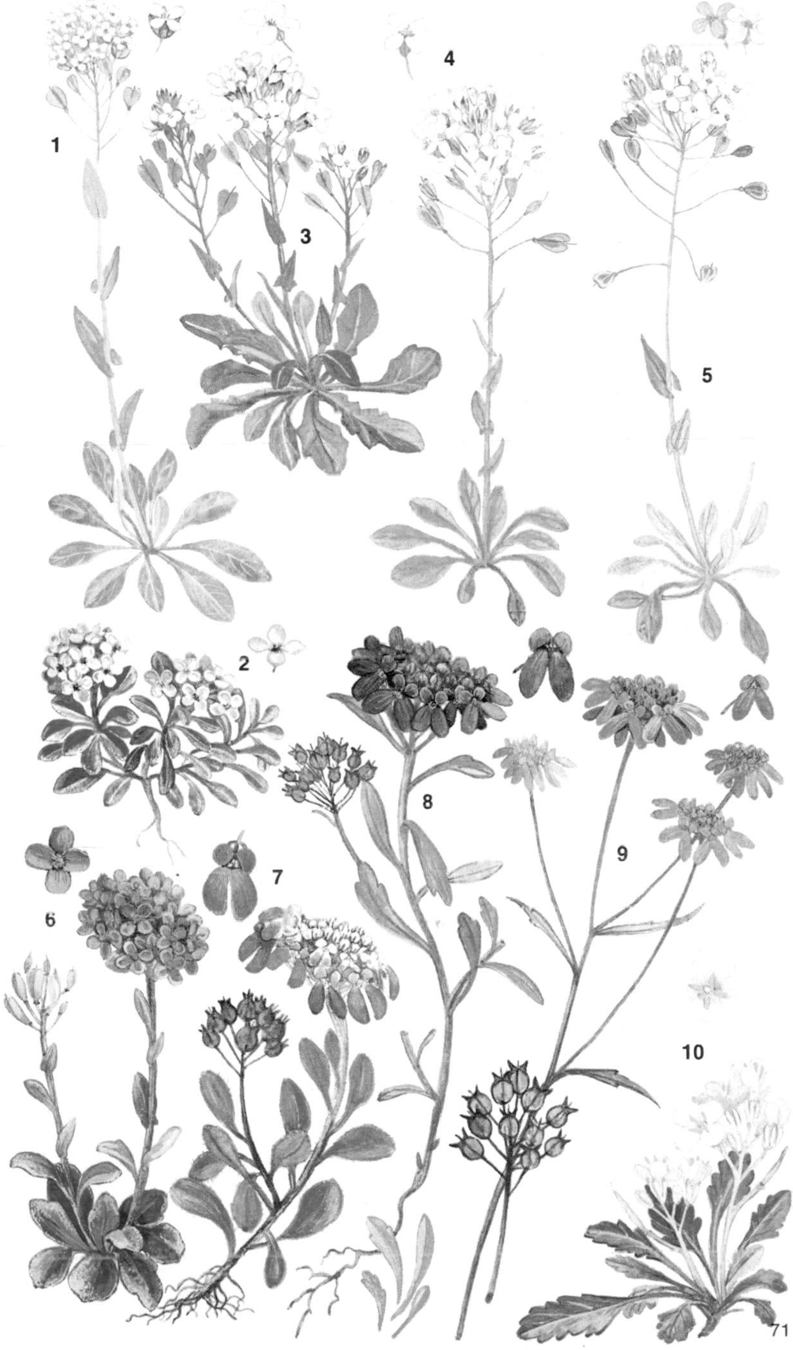
1
2
3
4
5
6
7
8
9
10

Cress Family *(contd.)*

1 BUCKLER MUSTARD *Biscutella laevigata*. Variable short/med *bristly-hairy*, sometimes hairless, per. Lvs mostly in a basal tuft, lance-shaped, *lobed* or toothed, stalked; stem lvs narrow-lance-shaped, untoothed. Fls yellow, 5-10mm, in branched clusters. Fr flattened, *with two almost rounded lobes* and a thin beak in between. Grassy and rocky places, open woods, to 2600m. May-Aug. P,A,Ap.

2 SCAPOSE BUCKLER MUSTARD *Biscutella scaposa*. Med hairy per, stems branched or not. Lvs in a basal tuft, lance or spoon-shaped, 5cm long or more, lobed and toothed; stem lvs *tiny or absent*. Fls yellow, 5mm. Fr as 1. Grassy and rocky places, to 1600m. May-July. eP. **2a** *P. flexuosa* has flexuous stems and *semi-clasping* stem lvs. c,eP.

3 ROSETTED BUCKLER MUSTARD *Biscutella brevifolia*. Short, slightly hairy, per. Lvs in a *dense basal rosette*, oblong-lance-shaped, 1-5cm long, pinnately-lobed usually, blunt-tipped. Fls yellow, 5-6mm. Fr as 1. Rocky places to 1600m. May-July. P–French. **3a** *B. intermedia* has *broader*, scarcely lobed, lvs. P.

4 CHICORY-LEAVED BUCKLER MUSTARD *Biscutella cichoriifolia*. Med hairy per. Lvs mostly in a basal tuft, oblong, lobed and toothed, long-stalked; stem lvs *clasping*. Fls yellow *larger* than 1-3, 12-20mm, in dense clusters, elongating in fr. Fr as 1. Rocky and dry places, to 1800m. May-July. P,A,Ap.

5 LEPIDIUM *Lepidium villarsii*. Short/med slightly hairy, grey-green, per. Lvs broadly elliptical, stalked, *untoothed*, the upper lvs more triangular, stalkless. Fls *white*, 4-6mm, anthers violet. Fr small, oblong, winged and notched at top. Rocky and grassy places, to 2500m. June-Aug. eP,swA–French*.

6 RHYNCHOSINAPIS *Rhynchosinapis richeri (Brassicella richeri)*. Med hairless, branched per. Lvs oblong, elliptical or spoon-shaped, toothed or not, long stalked. Fls yellow, 18-25mm, in dense *flat topped* clusters, elongating in fr. Fr linear, more than 40mm, with a tapered beak. Grassy and rocky places, to 2000m. June-Aug. swA. **6a Wallflower Cabbage** *R. cheiranthos* □ is *hairy* with pinnately-lobed lvs. Rocky, waste and dry places to 1500m. June-Aug. B,P,swA,Lig,n,cAp.

Mignonette Family   Resedaceae

Annuals or Perennials with alternate lvs. Fls tiny, in long spikes, with 4-8 sepals and petals; petals cut into narrow lobes; stamens numerous. Fr a capsule, open at top.

7 CORN MIGNONETTE *Reseda phyteuma*. Short/med downy ann or bien. Lvs spoon-shaped, often with 1-2 lobes on either side. Fls whitish, 5-7mm. Fr drooping. Cultivated and waste ground, banks, to 1600m. June-Sept. P,A,Ap,(B). **7a Wild Mignonette** *R. lutea* has 1-2-pinnate lvs and *yellow* fls; fr erect. To 2000m. B,P,A. **7b Weld** *R. luteola* is hairless with *unlobed lvs* and yellowish-green fls. Fr erect. (T).

8 PYRENEAN MIGNONETTE *Reseda glauca*. Short/med tufted, hairless, per. Lvs *linear*, generally untoothed, bluish-green. Fls whitish, 6-7mm, in long loose spikes. Fr erect. Grassy and rocky places, screes, to 2500m. July-Aug. c,eP.

9 SESAMOIDES *Sesamoides pygmaea* (= *Reseda sesamoides*). Low/short hairless per. Lvs lance-shaped, untoothed, the basal ones in *dense rosettes*. Fls like 7-8, whitish, 4-5mm, in long spikes. Fr *star-shaped*. Meadows, rocks and damp screes, to 2000m. May-Sept. P,Cev,Lig.

6a Wallflower Cabbage

Sundew Family Droseraceae

Insectivorous plants. Lvs in basal rosettes, covered in long sticky reddish hairs. Fls 5-6-parted, in long lfless spikes.

1 COMMON SUNDEW *Drosera rotundifolia.* Low per with solitary spreading rosettes. Lvs *rounded,* 5-8mm across, stalked. Fls white, 5mm, in 6-10-fld spikes. Peat or sphagnum bogs or moors, always acid, to 2000m. June-Aug. T.

2 LONG-LEAVED SUNDEW *Drosera anglica.* Low/short per with solitary, *erect,* rosettes. Lvs *narrow oblong,* 20-30mm long, tapered at base into stalk. Fls white, 6mm, in 3-6-fld, long-stalked spikes. Bogs, moors and wet heaths, sometimes slightly calcareous, to 1900m. July-Aug. T–except Ap. **2a** *D. intermedia* □ is similar, but rosettes grouped and lvs oblong, 7-8mm long, stalked; fl-spikes *short-stalked.* T–except Ap.

Stonecrop Family Crassulaceae

Succulent annuals or perennials with thick, generally untoothed, lvs. Fls usually starry, 5-parted or more; stamens alternating with petals or twice as many. Fr consisting of separate carpels, equal in number to the petals.

3 MUCIZONIA *Mucizonia sedoides.* Low hairless ann, forming dense tufts. Lvs alternate, oblong, 2-4mm, overlapping. Fls purplish pink, erect bells, 6-7mm, in crowded clusters; petals 5, *fused together* to the middle. Rocky places, gravels and screes on acid soils, 2000-3000m. June-Aug. P.

HOUSELEEKS *Sempervivum.* Perennials with dense succulent lfy rosettes, forming mats by production of runners, stem lvs alternate. Fls 8-16 parted, starry with narrow pointed petals; stamens twice as many.

4 WULFEN'S HOUSELEEK *Sempervivum wulfenii.* Short per with rosettes 40-50mm across. Lvs oblong-spoon-shaped, hairy margined, *otherwise hairless,* bluish-green. Fls lemon-yellow, 20-22mm, petals with a *purple spot* at base. Rocky places, acid rocks usually, 1700-2700m. July-Aug. c,eA.

5 LARGE-FLOWERED HOUSELEEK *Sempervivum grandiflorum* (= *S. gaudinii*). Short per with flat rosettes 20-50mm across, smelling of resin. Lvs oblong, narrowed at base, *hairy all over,* dark green with a *red-brown* tip. Fls yellow, 20-36mm; petals with a purple spot at base. Acid rocks, to 2500 m. July-Oct. csA. **5a**, p.335.

6 COBWEB HOUSELEEK *Sempervivum arachnoideum.* Low mat-forming per. Rosettes small, 5-15mm, covered in a *cobweb* of whitish hairs. Fls reddish-pink, 14-18 mm. Acid rocks, screes and alluvium, to 3100 m. July-Sept. A,Ap. **6a**, p.335.

7 MOUNTAIN HOUSELEEK *Sempervivum montanum.* Low mat or tuft-forming per. Rosettes small, 5-20mm, hairy, *resin-scented.* Lvs oval, broadest above the middle, dull green. Fls red-purple, 24-30mm. Acid rocks and screes, 1500-3200m. July-Aug. P,A,Ap. **7a, 7b**, p.335

8 DOLOMITIC HOUSELEEK *Sempervivum dolomiticum.* Low tufted per with globular rosettes, 20-40mm. Lvs oblong-lance-shaped, pointed, *slightly hairy,* bright green with a brownish tip. Fls deep pink, 18-20mm; petals with a central red-brown stripe. Dolomite and basal rocks, 1600-2500m. July-Sept. sA–Dolomites. **8a** *S. cantabricum* has larger *dark green* rosettes lvs, red tipped, and fls reddish-purple. cP–Picos de Europa.

9 COMMON HOUSELEEK *Sempervivum tectorum.* Variable short/med per with *large* flattened rosettes, 30-80mm across, blue-green tinged with red. Lvs oblong-lance-shaped, fine-pointed, *margin white-hairy,* otherwise hairless. Fls dull pink or purple, 18-20mm, in large clusters. Grassy and rocky places, screes, to 2800m. July-Oct. P,A. **9a**, p.335.

10 LIMESTONE HOUSELEEK *Sempervivum calcareum.* Like 9a but lfy rosettes more globular and the lvs broader, blue-green, *tipped with purple-brown.* Fls pale pink, 14-16mm. Limestone rocks, to 1800m. July-Sept. swA–French.

Stonecrop Family (contd.)

1 HEN-AND-CHICKENS HOUSELEEK Jovibarba sobolifera. Low/short succulent per with short runners. Lvs in globular rosettes, incurving, hairy margined, otherwise hairless, greyish-green, red-tipped; stem lvs more pointed, overlapping. Fls narrow yellow bells, 15-17mm; petals 6, margin toothed. Sandy and grassy places, usually on acid soils, to 1500m. July-Sept. c,eA–not Switzerland. **1a** J.allionii has pale yellowish-green, finely-hairy lvs and greenish-white fls. Rocks and screes, to 2000m. swA. **1b** J. arenaria like 1 but rosettes more open, bright green, the stem lvs narrower, long-pointed. To 1500m. eA. **1c** J. hirta like 1 but rosettes dark green and more open, lvs not red-tipped. To 1900m. e,seA.

STONECROPS Sedum. Fleshy perennials or annuals. Lvs usually alternate, rarely toothed, spaced along stem, not crowded into basal rosettes. Fls 5-8 parted, starry, in branched clusters. Fr usually 5-parted, thus distinguishing them from saxifrages (p. 80).

2 ORPINE Sedum telephium. Variable short/tall, often red tinged per. Lvs large, up to 7cm, flattened, rounded to narrow-oblong, toothed. Fls purplish-red, yellowish-green on whitish, 8-10mm, in dense flattish clusters. Woods, rocks and shady places, to 1800m. July-Sept. T.

3 REDDISH STONECROP Sedum anacampseros. Short, hairless, rather sprawling, per. Lvs flattish, elliptical-oval, bluish-green, untoothed. Fls deep red inside, bluish-lilac outside, 8-9mm, in dense rounded clusters. Acid rocks, 1400-2500m. July-Aug. P, swA, Ap.

4 CREAMISH STONECROP Sedum ochroleucum (= S. anopetalum). Short/med per. Lvs linear-cylindrical, pointed, 8-20mm. Fls cream to greenish-white, 12-14mm, erect in bud, petals sharp. Rocky places, walls and banks, to 2300m. June-July. P, A. **4a** S.o. montanum has bright yellow fls. eP, swA.

5 ROCK STONECROP Sedum reflexum. Like 4 but fl clusters drooping in bud. Fls yellow, usually 7-parted; sepals hairless. To 2000m. June-July. S–except north, P, A.

6 BITING STONECROP or WALLPEPPER Sedum acre. Low hairless, matforming, evergreen per. Lvs oval-cylindric, blunt, 3-6mm. Fls bright yellow, 10-12mm, in clusters of 2-4. Rocky and sandy places, banks, to 2300m. June-July. T. **6a** S. sexangulare □ has linear-cylindric lvs in 5-6 regular rows. T–except B, P. **6b** S. sartorianum hildebrandtii has dead lvs persisting, whitish with a black tip. eA. **6c**, p.336.

7 WHITE STONECROP Sedum album. Low/short, bright-green, often red-tinged, mat-forming per. Lvs cylindric, rather flattened above, 4-12mm. Fls white, 5-parted, 4-8mm, in loose, flat-topped clusters. Rocky places and walls, to 2500m. June-Aug. T.

8 ENGLISH STONECROP Sedum anglicum. Low mat-forming evergreen, greyish or reddish, hairless per. Lvs cylindrical-rounded, 3-5mm. Fls white or pink, 6-9mm, few to a cluster. Acid rocks and banks, to 1800m. June-Sept. B, P.

9 THICK-LEAVED STONECROP Sedum dasyphyllum. Like 8 but smaller. Lvs mostly opposite, downy, often slightly sticky. Fls white, streaked with pink, 5-6mm. Acid rocks and walls and banks, to 2500m. June-Aug. P, A, Ap, (B).

10 WHORLED-LEAVED STONECROP Sedum monregalense. Low/short downy-stemmed per. Lvs cylindric-oblong, 6mm, opposte or in whorls of 4, hairless. Fls white, in loose clusters; petals with hairy mid-vein. Shady rocks, to 1200m. June-Aug. swA, Ap.

11 CHICKWEED STONECROP Sedum alsinefolium. Low hairy per. Lvs alternate, rhombic or spoon-shaped, rather thin, 10-15mm. Fls white, 8-10mm, in loose lfy clusters. Shady rocks and caves, to 1450m. June-Sept. swA, MA, Ap. **11a**, p.336.

12 HAIRY STONECROP Sedum villosum. Low reddish, downy per, sometimes bien. Lvs alternate, narrow-oblong, flattened above, 4-7mm. Fls pink, 6-8mm, long-stalked, in loose clusters. Wet places, to 2450m. June-Aug. T, not Ap.

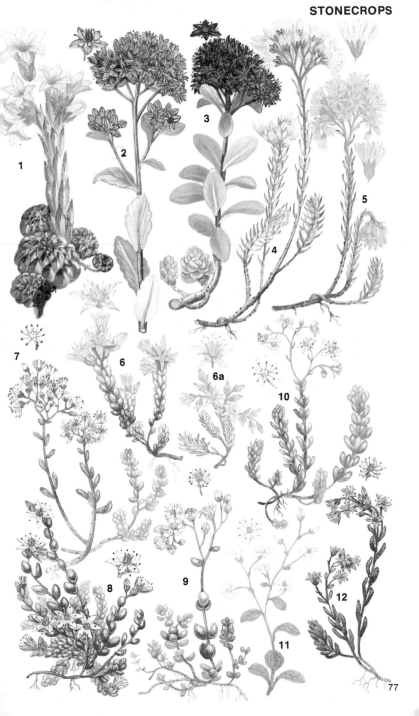

Stonecrop Family (contd.)

1 PINK STONECROP *Sedum cepaea.* Short hairy per/ann, stems thin. Lvs *opposite or whorled,* oblong to oblong-spoon-shaped, flat, lower stalked. Fls pale pink, 8-10 mm, five-parted, in loose, branched clusters. Shady places, to 1250 m. June-July. P,w,sA,Ap.

2 DARK STONECROP *Sedum atratum.* Low hairless ann, stems erect, often reddish. Lvs alternate, oblong or club-shaped, often reddish. Fls cream, red-lined, five-six-parted, in dense *flat-topped* clusters. Rocky and stony places, often on limestone, to 3200 m. June-Aug. P,A,Ap. **2a** *S.a. carinthiacum* is taller, greener, with *greenish-yellow* fls. eA.

3 ANNUAL STONECROP *Sedum annuum.* Low, branched, hairless, ann/bien, often red tinged. Lvs alternate, narrow-oblong. Fls yellow, five-parted, in loose branched clusters. Rocky and stony places, to 2900 m. June-Aug. T–except B. **3a** *S. rubens* □ has narrower, longer, lvs and *white or pink* fls. T. **3b** *S. hispanicum* like 3a but plant *often hairy;* fls six-nine parted. To 2250 m. A,Ap.

4 ROSEROOT *Rhodiola rosea* Low/short hairless, grey-green, per; stems often purple-tinged. Lvs alternate oval to oblong, *toothed,* thick. Fls dull-yellow, four-parted, in dense, *flat-topped,* clusters; male and female fls on different plants. Fr orange. Meadows, rocks and screes, on acid and limestone rocks, to 3000 m. May-Aug. T.

Grass of Parnassus Family Parnassiaceae

5 GRASS of PARNASSUS *Parnassia palustris.* Short hairless tufted per. Lvs heart-shaped, untoothed, long-stalked; stem lf *solitary,* stalkless, *clasping.* Fls solitary, white, 15-30 mm, five petals with transparent veins. Damp grassy places and marshy areas, to 2500 m. June-Sept. B,S,P*,A*.

Gooseberry Family Grossulariaceae

Deciduous shrubs with alternate, palmately-lobed, toothed lvs. Fls small, 5-petals and sepals. Fr an edible berry, developed below fls.

6 ROCK REDCURRANT *Ribes petraeum.* Bush to 3 m. Lvs rounded in outline, 3-5-lobed, hairy or hairless. Fls bell-shaped, pinkish, in *horizontal* or *drooping* racemes. Berry dark purple-red. Woods, stream banks and rocky places, to 2450 m. Apr-June. P,A,n,cAp.

7 WILD GOOSEBERRY *Ribes uva-crispa* (= *R. grossularia*). Spiny shrub to 1·5 m. Lvs 3-5-lobed, hairy of hairless. Fls with pinkish-green sepals and white petals, in *clusters* of one to three. Berry large green, yellowish or purple tinged, *stiff-hairy.* Woods, banks and rocky places, to 1800 m. T–but widely naturalised.

8 MOUNTAIN CURRANT *Ribes alpinum.* Bush to 2 m. Lvs 3-lobed, usually hairless. Fls greenish in *erect* racemes, male and female on different plants. Berry scarlet, hairless, rather tasteless. Open woods and rocky places, usually on limestone, to 1900 m. Apr-May. P,A.

9 REDCURRANT *Ribes rubrum* and **BLACKCURRANT** *R. nigrum* **(9a)**, are frequently naturalised in the area, especially in the Alps. The former is readily recognised by its hairless drooping-racemes of pale green fls, followed by strings of red berries, the latter by drooping racemes of reddish or brownish-green fls, followed by black, aromatic berries.

3a *Sedum rubens*

Saxifrage Family   Saxifragaceae

Perennials, sometimes annuals, often cushion-forming with simple or deeply lobed lvs, frequently in basal rosettes; stem lvs, when present, alternate except where stated; lvs sometimes lime-encrusted. Fls with 5 sepals and petals, 10 stamens and 2 stigmas. Fr a many seeded capsule. **(For lvs, see also p.350.)**

1 HAWKWEED SAXIFRAGE *Saxifraga hieracifolia*. Short/med hairy per. Lvs large, 3-7cm, all in basal rosettes, oval or oblong, slightly toothed, thick-stalked. Fls *greenish* tinged with purple-red, 4-7mm, in slender lfless spikes. Damp rocks, moraines and streamsides, to 2400m. July-Aug. S, Cev–Auvergne, eA.

2 STARRY SAXIFRAGE *Saxifraga stellaris*. Variable low/short, densely tufted, sparsely hairy per with lfless stems. Lvs in basal rosettes, oblong or rather spoon-shaped, toothed, short-stalked. Fls white, 10-15mm, each petal with *two yellow spots*, in lax clusters, anthers pink; sepals *downturned*. Damp places, streamsides and marshes, to 1900m. June-Aug. B, S.   **2a, 2b**, p.336.

3 FRENCH SAXIFRAGE *Saxifraga clusii*. Low/short hairy per. Lvs in a loose rosette, oval, irregularly toothed. Fls white, 10-16mm, in branched clusters; petals two short and three long, each with two yellow spots. Shady and damp places, streamsides, on acid rocks, to 2600m. June-Aug. P, Cev.   **3a, 3b**, p.336.

4 SPOON-LEAVED SAXIFRAGE *Saxifraga cuneifolia*. Short, loosely tufted per with lfless stems. Lvs fleshy, *spoon-shaped*, usually slightly toothed, hairless, stalks broad and hairy-edged. Fls white, 5-8mm, in small clusters; sepals downturned. Woods and shady rocks, to 2300m. June-Aug. P, Cev, A, nAp.

5 WOOD SAXIFRAGE *Saxifraga umbrosa*. Short/med tufted, somewhat hairy, per. Lvs rather leathery, oval-oblong, toothed, often hairless; stalks broad and flat, hairy margined. Fls white, each petal with a few *pale red spots and two yellow spots*, 8-10mm, in loosely branched clusters; sepals downturned. Woods, usually on limestone, to 1850m. June-July. w, cP.   **5a, 5b**, p.336.

6 ROUND-LEAVED SAXIFRAGE *Saxifraga rotundifolia*. Variable short/med loosely tufted, slightly hairy, per, with lfy stems. Lvs round to kidney-shaped, toothed, the lower long-stalked. Fls white, 12-20mm, each petal yellow spotted at base and red spotted near the tip, in laxly branched clusters; sepals erect. Damp or shady places, to 2500m. June-Oct. P, A, Ap.

7 ROUGH SAXIFRAGE *Saxifraga aspera*. Low/short, loosely matted per, almost hairless, with sprawling lfy stems. Lvs oblong-lance-shaped, pointed, with *bristly edges*, unstalked. Fls white or cream, the centre yellow, sometimes red spotted, 10-16mm, in loose clusters of 2-5. Rocky and stony places, often acid, to 2400m. July-Aug. c, eP, A, nAp.   **7a**, p.336.

8 MARSH SAXIFRAGE *Saxifraga hirculus*. Short loosely tufted, red-brown-hairy per, stems lfy in the lower half. Lvs lance-shaped, blunt, untoothed. Fls *bright yellow*, sometimes red spotted, 20-30mm, solitary or in clusters of 2-4. Wet places, bogs, to 1500m. June-Sept. B*, S, A–except s.

9 RUE-LEAVED SAXIFRAGE *Saxifraga tridactylites*. Low/short *stickily-hairy ann*, often reddish. Lower lvs spoon-shaped, stalked, untoothed, withered at flg time; upper lvs 3-5-lobed. Fls white, 4-6mm, in lax lfy clusters; petals *notched*. Walls and bare places, usually on limestone, to 1550m. June-Sept. T.

10 BIENNIAL SAXIFRAGE *Saxifraga adscendens* (= *S. controversa*). Low/short tufted, hairy bien. Basal lvs in a *dense rosette*, all lvs 2-5-lobed, *stalkless*. Fls white, rarely yellowish, 6-10mm, in small clusters; petals notched. Pastures, damp rocks and screes, 1800-3500m. June-Aug. S, P, A.   **10a, 10b**, p.336.

11 COBWEB SAXIFRAGE *Saxifraga arachnoidea*. Short tufted per covered with *sticky cobweb-hairs*. Lvs rhombic or oval, 3-7-lobed. Fls greenish-white or yellowish, 6mm, long-stalked, in branched clusters. Limestone caverns and underhangs, to 1700m. July-Aug. sA–near Lake Garda*.

12 FRAGILE SAXIFRAGE *Saxifraga paradoxa*. Low/short rather fragile, almost hairless, per. Lvs thin and shiny, kidney-shaped, 5-7-lobed usually, stalked. Fls *pale green* in lax clusters; petals *slightly shorter* than sepals. Shady crevices in acid rocks, rarely much above 1000m. seA–Karnten and Steiermark.

Saxifrage Family (contd.)

1 YELLOW MOUNTAIN SAXIFRAGE Saxifraga aizoides. Low/short lfy, loosely tufted per. Lvs *fleshy*, narrow-oblong, 10-25 mm, sometimes slightly toothed, stalkless. Fls bright yellow or orange, sometimes red spotted, in a lax head. Damp and stony places, stream-sides, to 3150 m. June-Sept. T−except sAp, B*

2 AWL-LEAVED SAXIFRAGE Saxifraga tenella. Low, almost hairless per, forming dense mats. Lvs *linear-pointed*, 8-10 mm, hairy on edges, stalkless. Fls creamy-white, 6 mm, in small clusters; flstems hairless, few lvd. Shady rocks and screes, usually limestone, to 2400 m. July-Aug. seA. **2a**, p. 336.

3 NEGLECTED SAXIFRAGE Saxifraga praetermissa Low/short mat-forming, slightly hairy per, with sprawling lfy shoots. Lvs oblong to almost rounded, to 10 mm, 3-5-lobed. Fls white, 9-10 mm, solitary, or 2-3 together, on slender stems. Damp or shady screes, snow patches, 1500-2500 m. July-Sept. P. **3a**, p.336.

4 WATER SAXIFRAGE Saxifraga aquatica. Med rather stout per, forming rounded tufts with *erect lfy stems*. Lvs shiny, 15-40 mm broad, divided into *numerous* pointed segments, stalked, slightly hairy. Fls white, 12-18 mm, in branched clusters. Stream margins and damp places, 1500-2200 m. July-Aug. c,eP. Hybrids (S. × capitata) occur between 3 and 4 where they grow together.

5 HAIRLESS MOSSY SAXIFRAGE Saxifraga pentadactylis. Low/short hairless per, in rounded cushions. Lvs slightly sticky, 3-5-lobed; stems sparsely lfy. Fls white, 8-9 mm. Acid rocks, screes, 1800-2900 m. July-Aug. eP. **5a, 5b**, p.336.

6 GERANIUM-LIKE SAXIFRAGE Saxifraga geraniodes. Low/short *stickily-hairy* per, forming rounded cushions of lvs. Lvs deeply divided into 9 or more narrow segments, hairy. Fls white, 22-24 mm, *slightly scented*, in close clusters, on sparsely lfy stems. Acid rocks and screes, to 2950 m. July-Aug. eP.

7 SCENTED-LEAVED SAXIFRAGE Saxifraga nervosa. Low hairy per forming loose, but rather hard, cushions. Lvs dark green with 3-5 narrow lobes, covered in short hairs, *scented*. Fls white, 8-10 mm, in small clusters. Exposed siliceous rocks, to 2700 m. May-Aug. cP.

8 PIEDMONT SAXIFRAGE Saxifraga pedemontana. Low/short, somewhat hairy per, forming fairly dense cushions. Lvs slightly fleshy, 5-11-lobed, broad stalked, with long hairs. Fls fairly large, white, 18-26 mm, in close clusters. Shaded siliceous rocks, 1500-2800 m. June-Aug. sw,cA. **8a**, p.336.

9 SCREE SAXIFRAGE Saxifraga androsacea. Low, slightly hairy cushion per. Lvs *all basal*, forming small rosettes, each with 3 short lobes or unlobed, hairy on edges. Fls white, 10-12 mm, 1-3 together; petals slightly notched. Damp screes and snow patches, 1500-3550 m. May-July. c, eP, Auvergne, A. **9a, 9b**, p.336.

10 EASTERN SAXIFRAGE Saxifraga sedoides. Low loose, mat-forming hairy per, stems lfy. Lvs lance or spoon-shaped 8-9 mm, *unlobed*. Fls small, dull yellow, 4-6 mm, solitary or 2-3 together; sepals *as long* as petals. Shady limestone screes and snow patches, to 3050 m. June-Sept. e, seA, Ap. **10a**, p.336.

11 BERGAMASQUE SAXIFRAGE Saxifraga presolanensis. Low/short dense cushion-forming, stickily-hairy, per. Lvs in dense columns, the lowest dead and whitish *but persisting*, upper pale green, narrow spoon-shaped, 12-15 mm. Fls greenish-yellow, 6-7 mm, in clusters of 2-4; petal narrow, notched. Shaded limestone cliffs and rocks, 1750-2000 m. Aug. sA−Bergamasque*. **11a, 11b**, p.336.

12 MUSKY SAXIFRAGE Saxifraga moschata. Variable slightly hairy per, in fairly dense cushions. Lvs 3-15 mm, *usually 3-lobed*, but sometimes 5-lobed or unlobed, lobes rounded, *not grooved*. Fls dull yellow or cream, rarely deep reddish, 5-8 mm, solitary or 2-7 clustered on almost lfless stems; petals oblong, not touching. Rocks, and stony places, 1200-4000 m. July-Aug. P,Cev,A,n,cAp. **12a**, p.336.

13 WHITE MUSKY SAXIFRAGE Saxifraga exarata. Variable low hairy per, forming fairly dense soft cushions. Lvs 3-5-lobed, broad-stalked, the lobes blunt and *grooved*. Fls white or pale yellow, rarely pink, 7-8 mm, in clusters of 3-8. Rocky and stony places, to 3600 m. June-Aug. A, Jura, Ap. Often confused with 12 but distinguished by the larger, usually white petals, and the grooved lvs. The two species hybridise readily. **13a**, p.336.

1

2

3

4

5

6

7

8

5b

9

13a

10

11

12

13

Saxifrage Family *(contd.)*

1 HAIRY SAXIFRAGE *Saxifraga pubescens.* Low loose cushion-forming per, with large lfy rosettes. Lvs 10-20mm long, *5-lobed*, sometimes 3-lobed, long-stalked. Fls white, 8-12mm, with rounded petals. Rocky and stony places, to 2800m. June-Aug. c, eP. **1a** *S.p. iratiana* □ is lower with denser cushions with *smaller lvs*, 4-10mm, overlapping along the stem. Fls often red veined. 2400-2800m. **1b** *S. cespitosa* is more compact, lvs usually 3-lobed, the lobes pointed; fls dull white or creamish. nB, S.

2 BULBOUS SAXIFRAGE *Saxifraga bulbifera.* Low short hairy bulbous per; stems unbranched. Lvs kidney-shaped, blunt-toothed, the lower long-stalked; each lf with a *small bulbil* at its base. Fls white, 14-20mm, in small clusters. Grassy, rocky and shady places, to 1400m. May-June. A.

3 MEADOW SAXIFRAGE *Saxifraga granulata.* Med, erect, hairy, bulbous per. Lvs most basal, *kidney-shaped*, blunt-toothed, long-stalked, rarely with stem lvs; *no stem bulbils*. Fls white, 10-18mm, in loose branched clusters; petals oblong. Meadows, avoiding lime, to 2200m. Mar-July. T.

4 DROOPING SAXIFRAGE *Saxifraga cernua.* Low, erect, thin-stemmed bulbous per. Lvs kidney-shaped with 5-7 pointed lobes, the lower long-stalked, the upper lvs and bracts *with small bulbils* at base. Fls white, 7-12mm, *usually solitary* or absent. Rocky, often shaded places, 1800-2500m. July. B*, A. **4a**, p.336.

5 PURPLE SAXIFRAGE *Saxifraga oppositifolia.* Low *mat-forming* per, with long trailing stems covered with small, 2-6mm, unstalked *opposite lvs*, bluish-green or green; each lf oval, broadest towards the top, with 1-5 lime pores, hairy in lower half. Fls solitary, pale pink to deep purple, 10-20mm, almost stalkless, anthers bluish. Rocky and stony places and screes, often on limestone, to 3800m. May-Aug. T. **5a** *S.o. rudolphiana* is very compact with *small*, 2mm long, closely overlapping lvs; fls 9-12mm, with *rather pointed* petals. eA. **5b** *S.o. blepharophylla* is compact with blunt *spoon-shaped lvs*, 3-4mm, with a margin of *long hairs*. eA–Austrian. **5c** *S.o. murithiana* has a single lime pore at the lf tip. P, A. **5d** *S.o. latina* □ like 5c but with *3 lime pores*. Ap. **5e** *S.o. speciosa* like 5 but lvs broad, with a *horny*, unhairy, apex; fls large, purple, 14-20mm. Ap– Abruzzi.

6 RETUSE-LEAVED SAXIFRAGE *Saxifraga retusa.* Rather like some compact forms of 5, but lvs *curved back* from middle and deep shiny green. Fls small, purplish-red, 8mm, in *clusters* of 2-3; petals narrow, oval, *anthers orange*. Rocky and stony places, on acid rocks, 2000-3000m. June-Aug. c, eP, A–except sw. **6a**, p.336.

7 TWO-FLOWERED SAXIFRAGE *Saxifraga biflora.* Low loose, matted, cushion per. Lvs broad-oval or rounded, 5-9mm, with a single lime pore, but not lime encrusted. Fls reddish-purple or whitish with a yellow centre, 16-20mm, in clusters of 2-8; *petals separate* from each other. Damp screes, moraines and river gravels, 2000-3200m. June-Aug. A.

8 COLUMNAR SAXIFRAGE *Saxifraga diapensioides.* Low cushion per, *stems columnar* with overlapping lvs. Lvs oblong, thick, 4-6mm, untoothed, bluish-green. Fls white, 14-16mm, in clusters of 2-6. Limestone rocks, 1500-2900m. July-Aug. swA. **8a** *S. tombeanensis* □ has shorter lance-shaped lvs, with an *incurved pointed tip*. Fls 18-22mm. To 2300m. May-June. sA–Italian. **8b, 8c**, p.336.

9 ONE-FLOWERED CUSHION SAXIFRAGE *Saxifraga burserana.* Low dense, bluish cushion per. Lvs lance-shaped, 5-12mm, pointed, mostly in basal rosettes. Fls white, 14-28mm, *solitary* on lfy, reddish, stems. Limestone rocks and screes, to 2200m. Mar-July. e, sA.

10 YELLOW SAXIFRAGE *Saxifraga aretioides.* Low dense, hard, cushion per; *stems columnar* with overlapping lvs, branched. Lvs narrow-oblong, 5-7mm, with a short incurved, pointed, tip. Fls *bright* yellow, 10-16mm, in clusters of 3-5. Rocky places to 2300m. June-Aug. P.

11 BLUE SAXIFRAGE *Saxifraga caesia.* Low cushion per with flattish grey- or bluish-green lf rosettes. Lvs oblong-spoon-shaped, curved backwards, 3-6mm, lime encrusted. Fls white, 8-11mm, in clusters of 2-5 on long *slender stems*. Limestone rocks and screes, to 3000m. July-Sept. P, A, n, cAp. **11a**, p.337.

85

Saxifrage Family *(contd.)*

1 THICK-LEAVED SAXIFRAGE *Saxifraga callosa* (= *S. lingulata*). Variable low clump-forming per. Lf rosettes *large*, untidy. Lvs long-linear or spoon-shaped, pointed, 25-90mm, greyish, lime encrusted. Fls white, 11-16mm, petals often *red-spotted* at base, in loose panicles. Limestone rocks and cliffs, to 2500m. June-Aug. MA, Lig A, Ap. **1a** *S. cochlearis* similar but smaller and more delicate, lvs narrow-spoon-shaped, *blunt,* 8-40mm. To 1900m. MA, Lig A. **1b** *S. valdensis* like 1a but smaller and densely cushioned, lvs linear to oblong-spoon-shaped, 4-12mm; fls *plain white*, in flattish clusters. To 2800m. July-Aug. swA – French and Italian.*

2 ENCRUSTED SAXIFRAGE *Saxifraga crustata.* Short tufted per; lf rosettes rather flat. Lvs *strap-shaped*, pointed, 15-60mm, greyish and lime encrusted. Fls white, 8-11mm, petals sometimes red-spotted at base, in narrow *short-branched* panicles. Limestone rocks, to 2200m. June-Aug. eA.

3 PYRAMIDAL SAXIFRAGE *Saxifraga cotyledon.* Short/med per, lf rosettes large, flattish. Lvs broad-oblong with *a point at flat tip*, 20-60mm, finely toothed, *not* lime encrusted. Fls white, 11-18mm, petals sometimes purple spotted, in *broad* panicles, branched from base of stem. Acid rock crevices, 1500-2600m. July-Aug. S, cP, sA.

4 HOST'S SAXIFRAGE *Saxifraga hostii.* Med tufted per; lf rosettes large, dark-green. Lvs oblong or oval, broadest above the middle, 20-100mm, tip often *down-curved, not* lime encrusted. Fls white, 9-15mm, petals red-spotted in panicles at top of stems. Limestone rocks, to 2500m. May-July. e, seA. **4a** *S.h. rhaetica* has narrower lvs, tapered to a *sharp point.* sA – Italian.

5 REDDISH SAXIFRAGE *Saxifraga media.* Low/short bluish-green per. Lvs in dense rosettes, 7-20mm, narrow-oblong, pointed, *pink at base.* Fls pinkish-purple, 4-6mm in slender spikes, petals short *surrounded by* a deep red hairy calyx. Rocks, usually limestone or shale, to 2510m. June-Aug. c, eP. **5a** *S. porophylla* □ has flatter rosettes, lvs 6-16mm. Ap.

5 Reddish Saxifrage   **5a** *Saxifraga porophylla*

1

2

3

4

Saxifrage Family *(contd.)*

1 PANICULATE or LIVELONG SAXIFRAGE *Saxifraga paniculata* (= *S. aizoon*). Variable short/med per, rather like a small version of Host's Saxifrage (p. 86) but lf rosettes more rounded and *lime encrusted*. Lvs oblong or oval, broadest above middle, *finely toothed*. Fls white or cream, rarely pale pink or red-spotted, 8-11mm, in loose panicles *at top* of stems. Rocky, stony places and screes, to 2700m. May-Aug. T−except B.

2 ORANGE SAXIFRAGE *Saxifraga mutata.* Med per, lf rosette loose, large. Lvs broad-strap-shaped, blunt, 10-70mm, *shiny dark-green*, not lime encrusted. Fls *starry orange*, 9-15mm; in large loose panicles. Damp stony places, on limestone, to 2200m. June-Aug. A.

3 PYRENEAN SAXIFRAGE *Saxifraga longifolia.* Med per, with *a single* large lfy rosette, long lived but dying after flg. Lvs linear-strap- shaped, pointed, 30-80mm, greyish, lime encrusted. Fls white, 9-11mm, in large narrow-panicles. Limestone cliffs and rocks, to 2400m. May-Aug. P.

4 THE ANCIENT KING *Saxifraga florulenta.* Med per with a single, many-leaved rosette, long lived but dying after flg. Lvs narrow-oblong, broadest above middle, very regularly arranged, green, *not lime encrusted*. Fls flesh-pink, 5-7mm, in long pointed, spike-like, racemes. Granite cliffs and crevices, 1900-3250m. July-Sept, but rarely seen in fl. cMA*.

3 Pyrenean Saxifrage

4 The Ancient King

Rose Family Rosaceae

A large family of trees, shrubs and herbs. Lvs alternate, often pinnate, usually with stipules at base of lf stalk. Fls small or large, clustered, spiked or solitary. Petals and sepals usually five; petals separate from each other; sepals fused to basal cup (receptacle). Stamens many. Fr variable; dry with separate fruitlets (achenes) or fleshy, hip or berrylike, often large and edible (apple, pear, strawberry etc.)

1 HAIRY SPIRAEA *Spiraea decumbens tomentosa.* Rather sprawling deciduous shrub to 0·5 m, thin stemmed. Lvs oblong, toothed, stalkless, *grey-hairy* below. Fls white, 5-7 mm, in rounded clusters. Limestone rocks and screes, to 1600 m. June-July. seA.

2 ELM-LEAVED SPIRAEA *Spiraea chamaedryfolia* (= *S. ulmifolia*). Densely branched deciduous shrub to 2 m, stems *brown, angular.* Lvs oval or oval-lance-shaped, pointed, toothed towards top, *hairless.* Fls white, 8-10 mm, in dense rounded clusters. Woodland and scrub, to 1200 m. May-June. seA.

3 GOATSBEARD SPIRAEA *Arancus dioicus* (=*A. vulgaris, Spiraea aruncus*). Tall hairless per with large pinnate lvs, *no stipules.* Fls white, 5 mm, in dense branched fingerlike spikes, forming a *pyramid.* Damp and shady places, usually on acid soils, to 1700 m. May-Aug. P, A, Ap.

4 DROPWORT *Filipendula vulgaris.* Short/med per rather like 3, but lvs with *stipules* at base. Lvs with 8 or more pairs of lflets. Fls pale cream, purplish beneath, 8-16 mm, in branched *flat-headed* clusters. Dry grassy meadows on limestone, to 1500 m. June-Sept. T. **4a Meadowsweet** *F. ulmaria* □ is a taller plant, the lvs with *not more* than 5 pairs of lflets. Fls creamy, 4-8 mm, *fragrant.* Wet meadows, marshes and woods, to 1500 m. June-Sept. T.

5 ROCK BRAMBLE *Rubus saxatilis.* Short/med hairy per, prostrate, prickly, with *annual* erect stems. Lvs trifoliate, lflets oval-elliptical, toothed. Fls white, 8-10 mm, with *narrow erect* petals, in loose clusters. Fr shiny red, with 2-6 large fleshy segments, edible. Woods, scrub and shady rocks, to 2400 m. May-June. T.

6 RASPBERRY *Rubus idaeus.* Tall erect or arching per, with biennial stems, armed with weak prickles. Lvs pinnate, with 5-7-toothed lflets, *white-hairy* beneath. Fls white, 10 mm, with oblong, *erect* petals, in loose branched clusters. Fr red or orange with many small fleshy segments, edible. Woods, heaths and banks, to 2300 m, but widely cultivated. June-Aug. T

7 BLACKBERRY or BRAMBLE *Rubus fruticosus* agg. A large and complicated group with over forty species in the area but mostly at rather low altitudes. Scrambling, often thicket forming shrub, much branched, usually with straight or curved *prickles.* Stems often very long and angled, rooting at the tip. Lvs prickly, pinnate with 3-5 toothed lflets. Fls white or pink, 15-30 mm, in stout branched clusters. Fr reddish at first but becoming *purple-black* when ripe, with many fleshy segments, edible. Woods, scrub, banks, rocky and waste places, seldom above 1700 m. May-Nov. T.

8 DEWBERRY *Rubus caesius.* Weak sprawling shrub; stems *round,* slightly prickly. Lvs trifoliate, lflets 2-3-lobed, toothed, hairy. Fls white, 20-25 mm, in small branched-clusters. Fr bluish-black with a waxy bloom, with a few segments, larger than 7, edible. Damp places, generally on limestone, to 1200 m. May-Sept. T.

9 ARCTIC BRAMBLE *Rubus arcticus.* Short creeping per, *not prickly.* Lvs trifoliate, lflts oval-elliptical, toothed. Fls *bright red*, sometimes pink, 15-25 mm, solitary. Fr dark red. Grassy places and thickets on moors, to 1150 m. July. S.

10 CLOUDBERRY *Rubus chamaemorus.* Low/short downy, creeping, per. Lvs *palmately-lobed.* Fls white, 15-20 mm, solitary; male and female on separate plants. Fr *orange* when ripe. Moors and bogs, to 1400 m. June-Aug. B.S.

Rose Family (contd.)

ROSES *Rosa.* Deciduous shrubs with thin, much branched, stems armed with thorns and/or bristles. Lvs pinnate, with toothed lflets and stipules. Fls large and showy, solitary or clustered, often sweetly scented; petals 5, notched; many stamens. Fr a berry-like hip.

1 FIELD ROSE *Rosa arvensis.* Scrambling shrub to 1m; stems green, thorns hooked. Lflts 5-7, slightly hairy, dull green above. Fls white, 25-45mm, styles *joined* into a column. Hip round or oval, red; sepals lobed, *falling*. Woods, scrub and hedges, to 2000m. June-Aug. T' except S.

2 BURNET ROSE *Rosa pimpinellifolia.* Suckering shrub to 1m; stems with *straight thorns* and stiff bristles. Lflets 5-11, hairless. Fls white, sometimes pink, 16-36mm, solitary. Hip rounded, *purple-black*; sepals unlobed, persisting. Dry open places and rocks, to 2000m. May-Aug. T.

3 CINNAMON ROSE *Rosa majalis* (= *R. cinnamomea*). Patch-forming shrub to 2m; stems reddish-brown, thorns slender, slightly curved. Lflets 5-7, *hairy, bluish-green* above. Fls purplish-pink, 32-50mm, solitary. Hip round or oval, red; sepals unlobed, persisting. Scrub, often along stream banks, to 2200m. May-July. A. **3a Blue-leaved Rose** *R. glauca* (= *R. rubrifolia*) ☐ has bluish-green or purplish, *hairless* lvs and fls usually *in clusters* of two to five. Stony places and woodland edges, to 2100m. June-Aug. P,A,n,cAp.

4 ALPINE ROSE *Rosa pendulina* (= *R. alpina*). Shrub to 2m; stems yellowish-green or purplish, *no thorns* usually. Lflets 7-11, slightly hairy, bright or yellowish-green. Fls deep purplish-pink, 26-46mm, solitary. Hips *narrow pear-shaped*, red, hairy; sepals unlobed, persisting. Woods and open places, to 2600m. May-Aug. P,A,Ap.

5 PROVENCE ROSE *Rosa gallica.* Patch forming shrub to 0·75m; stems with bristles and hooked prickles. Lflets 3-7, *hairy beneath*, dull bluish-green. Fls deep pink, 50-80mm, usually solitary. Hip round to elliptic, bright red, bristly; sepals lobed, *falling*. Dry open places and banks, to 2000m. May-Aug. A,Ap.

6 STYLED ROSE *Rosa stylosa.* Scrambling shrub to 3m; stems with *stout hooked* thorns. Lflets 5-7, hairy beneath, green. Fls white, sometimes pink, 26-56mm, solitary or clustered; styles *joined* into a column. Hip oval or round, red; sepals lobed, *falling*. Woods, scrub and hedgerows, to 2000m. June-July. B,P,A.

7 MOUNTAIN ROSE *Rosa montana.* Shrub to 3m; stems bluish-green or purplish with hooked or nearly straight thorns. Lflets 7-9, *hairless*, bluish-green. Fls pale pink, becoming whitish, 25-40mm, usually solitary. Hip oval to narrow-pear-shaped, red, *hairy*; sepals lobed, persisting. Rocks and screes to 2000m. June-July. A,Ap*. **7a** *R. jundzillii* ☐ has pale to deep pink fls and smaller hips; sepals *falling*. A.

1 Field Rose, fr.

Rose Family *(contd.)*

1 WHITISH-STEMMED BRIAR *Rosa vosagiaca* agg. Suckering shrub to 2m; stems with many hooked thorns, reddish, *whitish-bloomed* when young. Lflets 5-7, usually bluish-green, *hairy*. Fls bright pink, 25-50mm, solitary or clustered. Hip rounded or pear-shaped, deep red and smooth with *persisting erect sepals*. Scrub and lightly wooded areas, to 2200m. June-Aug. T.

2 DOG ROSE *Rosa canina.* Suckering shrub to 2m or more; stems green with many hooked thorns. Lflets 5-7, bluish-green or green, *hairless*. Fls pink or white, 30-50mm, solitary or clustered; styles hairy or hairless. Hips rounded or oval, red and smooth, with *downturned or no* sepals. Scrub, banks and grassy places, to 2200m. June-Aug. T.

3 BLUNT-LEAVED DOG ROSE *Rosa obtusifolia.* Suckering shrub to 2m; stems green with short hooked thorns. Lflts 5-7, bluish-green or green, *hairy*, gland-dotted beneath. Fls pink or white, 20-35mm, solitary or clustered. Hips rounded or oval, red and smooth, with *down-turned or no* sepals. Scrub, banks and grassy places, to 2200m. June-Aug. T.

4 DOWNY ROSE *Rosa tomentosa* agg. Variable shrub to 2m, stems pale green arching with *straight or slightly curved* thorns. Lflets 5-7, soft, densely downy, with a resinous smell. Fls pink, sometimes white, 30-45mm, solitary or clustered. Hips rounded or pear-shaped, *hairy*; sepals pinnately-lobed, downturned, not persisting. Scrub and hedgerows, to 2200m. May-July. T.

5 APPLE ROSE *Rosa villosa* agg. Like 4, but with slender straight thorns. Lflets 5-7, downy, bluish-green with a resinous smell. Fls pink, 30-45mm, solitary or clustered. Hip rounded to pear-shaped, hairy, with *erect persisting* pinnately-lobed sepals. Scrub, hedgerows and banks, to 2200m. June-July. P,A,Ap.

6 SWEET BRIAR *Rosa rubiginosa.* Shrub to 3m; stems with hooked thorns *mixed with* short bristles, especially below the fls. Lflets 5-7, yellowish-green often tinged with red or brown, sticky beneath with brown hairs, sweet smelling. Fls deep pink, 18-28mm, solitary or in clusters of 2-3; fl stalks hairy. Hips rounded or elliptical, bright red, smooth or hairy; sepals pinnately-lobed, *erect and persisting*. Scrub, dry banks and stony places, to 2100m. June-July. T. **6a** *R. elliptica* has no bristles and *smooth fl stalks;* fr smooth. June-Sept. B,P,A,Lig.

7 FIELD BRIAR *Rosa agrestis.* Shrub to 2m; stems arching, with hooked thorns. Lflets 5-7, dull green, stickily-hairy beneath. Fls white, 22-38mm, solitary or 2-3 clusters; fl stalks *hairless*. Hips rounded or elliptical, red, smooth; sepals down-turned but *not persisting*. Hedgerows, woodland margins and banks, to 2200m. May-July. T−except S. **7a** *R. micrantha* □ has *hairy fl-stalks* and hairy or smooth hips. T−except S.

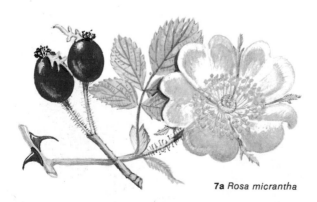

7a *Rosa micrantha*

Rose Family *(contd.)*

1 AGRIMONY *Agrimonia eupatoria.* Low/tall downy per with *pinnate lvs;* stipules present. Lflets deep green above, *whitish* beneath. Fls golden-yellow, 5-8 mm, in long spikes. Fr grooved, covered with small erect hooks. Dry and grassy places, usually on limestone, to 1800 m. June-Aug. T. **1a Fragrant Agrimony** *A. procera* is larger and aromatic, lvs green above and beneath; fls 8-10 mm; fr *hardly* grooved, with some hooks *bent back.* T–except the north.

2 BASTARD AGRIMONY *Aremonia* (= *Agrimonia*) *agrimonioides.* Low/short hairy per like 1 but stem lvs often *trifoliate.* Fls yellow, 7-10 mm, in short *lfy clusters.* Fr without hooks. Woods, shrub and waste places, to 1900 m. May-June. A, Ap, (B).

3 GREAT BURNET *Sanguisorba officinalis.* Short/tall hairless per. Lvs pinnate, with 3-7 pairs of lflets, bluish-green below. Fls tiny, dull crimson, in *dense oblong heads*, 10-30 mm long, no petals; stamens four, protruding. Damp grassy places and woods, to 2300 m. June-Sept. T. **3a Salad Burnet** *S. minor* ☐ has *greyish lvs* with 3-12 pairs of lflets; fls *greenish*, the upper with red styles, lower with yellowish anthers, in dense *rounded* heads. To 2200 m.

4 ITALIAN BURNET *Sanguisorba dodecandra.* Med/tall hairless per. Lvs pinnate, with 4-10 pairs of lflets, pale green below. Fls tiny, *greenish-yellow* to whitish in *dense oblong heads*, 40-70 mm long; stamens four to fifteen, protruding. Meadows and streambanks, to 2000 m. sA–Sondrio District.

5 MOUNTAIN AVENS *Dryas octopetala.* Low evergreen, carpeting, downy, sub-shrub. Lvs oblong-oval or heart-shaped, toothed, stalked, deep green above, whitish beneath. Fls white, 20-40 mm, solitary, with *seven or more* petals. Fr with long *feathery styles*, bunched. Meadows and rocks, usually on limestone, to 2500 m. May-Aug. T–except sAp.

AVENS or GEUMS *Geum.* Perennials with pinnate, trifoliate or tri-lobed lvs. Fls yellow or pinkish, with 5-7 petals. Fr with hairy, hooked or unhooked styles.

6 CREEPING AVENS *Geum reptans.* Low hairy per with long *reddish-runners.* Lvs pinnate, lflets toothed. Fls bright yellow, 25-40 mm, solitary; petals oval or rounded. Styles not hooked. Rocks, gravels and moraines, 1500-2800 m. July-Aug. A. **6a Alpine Avens** *G. montanum* ☐ has *no* runners and lvs with a very large *oblong* end lflet. Fls golden-yellow, petals often *slightly notched.* P, A, Ap.

7 WATER AVENS *Geum rivale.* Short downy per. Lvs pinnate, with 3-6 pairs of lflets; upper lvs trifoliate. Fls solitary *nodding-bells*, 8-15 mm, cream or pale pink with brownish-purple sepals. Styles hooked. Damp meadows, woods and stream-sides, to 2100 m. May-July. T.

8 PYRENEAN AVENS *Geum pyrenaicum.* Short hairy per, no runners. Basal lvs pinnate, with 4-6 pairs of lflets, the end lflet *very large, rounded.* Fls bright yellow, 18-24 mm, solitary; petals rounded. Styles *jointed* near top, not hooked. Stony meadows, to 1800 m. July-Aug. P.

9 HERB BENNET *Geum urbanum.* Short/med downy per. Lower lvs pinnate; upper trifoliate or trilobed with *lf-like stipules* up stems. Fls pale yellow, 10-15 mm, petals oval or oblong. Styles *hooked.* Woods and shady places, to 1850 m. May-Sept. T.

3a Salad Burnet

Rose Family *(contd.)*

1 WALDSTEINIA *Waldsteinia ternata.* Short creeping hairy per with *rooting runners*. Lvs trifoliate, lflets toothed, narrowed to the base. Fls white, 13-15mm, in loose clusters. Styles *falling* off in fr. Rocky and grassy places, seldom exceeding 1000m. seA.

CINQUEFOILS *Potentilla* are usually creeping perennials with digitate or pinnate lvs. Fls yellow, sometimes white or purple, usually 5-petalled, stamens numerous; calyx and epicalyx present. Fr a cluster of dry achenes, styles not persisting.

2 SHRUBBY CINQUEFOIL *Potentilla fruticosa.* Much branched, grey-hairy deciduous shrub, to 1m, but often dwarf at high altitudes. Lvs pinnate, lflets 5-7 oblong to elliptic, *untoothed*. Fls yellow, 16-24mm, solitary or in loose clusters. Rocky places, river banks, on calcareous soils, to 2600m. June-Aug. B*, S, P, MA.

3 MARSH CINQUEFOIL *Potentilla palustris* (= *Comarum palustre*). Short/med, slightly hairy, creeping per. Lvs pinnate, lflets 5-7 oblong, *toothed*. Fls purplish, starry, 20-30mm; sepals *much larger* than the linear, deep, purple, petals. Wet meadows, bogs and moors, preferring acid soils, to 2100m. May-July. T−except Ap.

4 SILVERWEED *Potentilla anserina.* Short creeping per with long rooting runners. Lvs pinnate *silvery* at least beneath, lflets toothed. Fls yellow, 14-20mm, solitary, with oval petals. Damp grassy places and banks, to 2400m. May-Aug. T.

5 ROCK CINQUEFOIL *Potentilla rupestris.* Short/med hairy per; no runners. Lvs *pinnate,* lflets oval to almost rounded, toothed. *Fls white*, 16-28mm, solitary or clustered. Rocky and wooded slopes, to 2200m. May-June. T−except Ap.

6 CUT-LEAVED POTENTILLA *Potentilla multifida.* Short/med downy per. Lvs pinnate, lflets 5-9 pinnately cut into *linear lobes*, green above, greyish-silky beneath. Fls yellow, 10-14mm, in clusters. Meadows and rocks, usually on acid soils, 2000-3000m. July-Aug. P, s, swA*.

7 PENNSYLVANIAN CINQUEFOIL *Potentilla pennsylvanica.* Short/tall *greyish-hairy* per. Lvs pinnate, lflets 7-19 oblong, coarsely toothed or lobed. Fls yellow, 16-24mm, in large clusters. Rocky and grassy places, to 1500m. May-July. swA.

8 SNOWY CINQUEFOIL *Potentilla nivea.* Low/short tufted per, stem often downy-white. Lvs *trifoliate*, lflets oval, toothed, green above, densely *white-downy* beneath. Fls yellow, 12-18mm, in small clusters. Rocky places and screes, usually limestone, 1600-2600m. June-Aug. S, A, nAp.

9 HOARY CINQUEFOIL *Potentilla argentea.* Variable short/med per; stems usually downy, erect or sprawling. Lvs *digitate*, lflets 5, pinnately-lobed, toothed, lower lflets with 2-7 teeth, dark green above, densely *white-hairy* beneath. Fls yellow, 8-12mm, in branched clusters. Rocky and stony places, open woods, to 1950m. June-Aug. **9a** *P. neglecta* has larger fls, 12-16mm, and lower lvs with 9-11 teeth to each lflet. T. **9b** *P. calabra* like 9 but lflets grey-green or whitish above. c, sAp.

4 Silverweed

1

2

3

4

5

6

7

8

9

Rose Family *(contd.)*

1 LARGE-FLOWERED CINQUEFOIL *Potentilla grandiflora.* Short/med hairy per. Lvs *trifoliate*; Iflets oval or rounded, toothed, green above, grey-hairy beneath. Fls yellow, 22-32mm, in branched clusters, petals slightly notched. Meadows and rocky places, on acid soils, to 3100m. June-Aug. Cev, ceP, A. **1a** *P. delphinensis* has lvs with *5-Iflets*, green above and below. On calcareous soils, 1500-2800m. July-Aug. SwA – Cottian and Dauphine.

2 PYRENEAN CINQUEFOIL *Potentilla pyrenaica.* Short/med per, hairs *pressed* against stems. Lvs digitate, Iflets 5, oblong, toothed, green above and beneath. Fls yellow, 22-32mm, in branched clusters. Meadows and rocky places to 2300m. July-Aug. P.

3 THURINGIAN POTENTILLA *Potentilla thuringiaca.* Short/med very hairy per. Lvs digitate, *Iflets 5-9* oblong to lance-shaped, toothed. Fls yellow, 15-22mm, in branched clusters. Meadows and rocky places to 2500m. Apr-Aug. S, s, swA, n, cAp.

4 DWARF CINQUEFOIL *Potentilla brauniana* (= *P. dubia*). Low tufted per, slightly hairy. Lvs small, trifoliate; Iflets oblong to oval, slightly toothed, hairless above. Fls yellow, 7-11mm, solitary or in clusters of 2-3, petals longer than sepals; epicalyx segments blunt. Grassy and rocky places, often by snow patches, on limestone, 1800-3150m. July-Aug. P, A, J. **4a** *P. frigida* is taller and much *hairier*, the petals shorter or as long as the sepals. Acid rocks, 2400-3700m. P, A.

5 GOLDEN CINQUEFOIL *Potentilla aurea.* Low *mat-forming* hairy per. Lvs digitate; Iflets 5, oblong, toothed, the end tooth *much smaller* than the adjacent ones, margin silkily-hairy, stipules oval to lance-shaped. Fls golden yellow, often orange in the centre, 14-24mm, in loose, long-stemmed, clusters; petals broad, notched. Grassy and rocky places, usually on acid soils, 1400-2600m. June-Sept. P, A, Ap. **5a Alpine Cinquefoil** *P. crantzii* ☐ is not mat-forming; Iflets with end tooth *equal* to the adjacent ones, margin not silkily-hairy. To 3000m. T. **5b** *P. tabernaemontani* like 5 but shoots *rooting down*, stipules of lower lvs *linear*. To 3000m. T–except Ap. **5c Creeping Cinquefoil** *P. reptans* has fl stems rooting at lf joints and *solitary fls.* To 1700m. T. **5d Tormentil** *P. erecta* ☐ has 4-petalled fls. To 2500m. T.

6 GREY CINQUEFOIL *Potentilla cinerea.* Low mat-forming, densely *grey-starry-hairy*, per; stems often rooting down. Lvs with 3-5 oblong or lance-shaped, toothed, Iflets, grey-green above, grey beneath. Fls yellow, 10-16mm, solitary or in clusters of 2-6. Dry meadows and rocky places, to 1600m. Apr-July. eP, A, Ap.

7 LAX POTENTILLA *Potentilla caulescens.* Low/short hairy per. Lvs digitate; Iflets 5-7, oblong, toothed. *Fls white*, 14-22mm, in branched clusters; petals narrow, slightly notched or not. Limestone rock crevices, to 2400mm. July-Aug. P, A, Ap.

8 EASTERN CINQUEFOIL *Potentilla clusiana.* Low tufted hairy per. Lvs digitate; Iflets 5, oblong, broadest above the middle, with *3-5 teeth* at the blunt apex. Fls white, 20-22mm, solitary or in clusters of 2-3; petals broad, notched. Limestone or dolomite rocks and screes, 1200-2400m. July-Aug. e, seA.

9 ALCHEMILLA-LEAVED CINQUEFOIL *Potentilla alchimilloides.* Short hairy per. Lvs digitate; Iflets 5-7, oblong-elliptical with 3 small teeth at the apex, hairless above but *silvery hairy beneath*. Fls white, 18-22mm, in branched clusters; petals notched, longer than sepals. Rocks and screes, to 2200m. July-Aug. P. **9a** *P. valderia* is taller and grey-hairy; Iflets toothed in upper half, petals *shorter* than sepals. MA*.

5d Tormentil

1

2

4

5a

3

5

7

9

6

8

Rose Family *(contd.)*

1 PINK CINQUEFOIL *Potentilla nitida.* Very low dense cushion-forming *silvery-grey* per. Lvs trifoliate; lflets oblong, broadest above the middle and toothed only at the apex. Fls *pink* or sometimes white, 22-26mm, solitary or 2-together. Limestone rocks and screes, 1200-3150m. June-Sept. sw, seA–including the Dolomites, nAp.

2 CREAMY CINQUEFOIL *Potentilla gammopetala.* Short hairy per. Lvs with 3-5 green lflets, oval, toothed in the upper half. Fls *cream or pale yellow*, 14-17mm, in branched clusters; petals as long as sepals and epicalyx, giving a *starry* effect. Acid rocks, usually gneiss, 1800-2100m. July-Aug. csA. **2a** *P. apennina* is small, all lvs trifoliate and *silvery-hairy*; lflets narrow, toothed only at the apex; fls white with a *cup-like* epicalyx. Rocks and screes, 1600-2200m. cAp.

3 TUFTED POTENTILLA *Potentilla saxifraga.* Low/short *densley tufted* hairy per. Lvs with 3-5 linear to lance-shaped lflets, *3-toothed* at apex, green and almost hairless above, silvery-hairy beneath. Fls white, 9-12mm, in small clusters; petals longer than sepals. Limestone crevices, to 1200m. May-June. MA.

4 CARNIC CINQUEFOIL *Potentilla carniolica.* Low hairy per. Lvs trifoliate; lflets oval, broadest above the middle, toothed, green above and grey-hairy beneath. Fls *small*, white, rarely pink, 6-8mm, solitary or 2-4 clustered. Calcareous rocks, to 1600m. Apr-June. seA–Yugoslavia.

5 SIBBALDIA *Sibbaldia procumbens.* Very low, stiffly-hairy, tufted per. Lvs trifoliate, lflets oval, broadest above the middle, 3-toothed at apex. Fls *small* yellow, 5mm, petals *shorter* than the sepals, sometimes absent; stamens five. Damp grassy and rocky places, often by snow patches, 2000-3200m. July-Aug. T.

6 WILD STRAWBERRY *Fragaria vesca.* Low/short hairy per with *long runners* rooting at intervals. Lvs trifoliate in basal tufts, bright green; lflets sharply toothed. Fls white, 15mm, clustered on a stalk not longer than the lvs; flstalks with *closely pressed* hairs. Fr the familiar small red wild strawberry, edible. Grassy places, woodland and scrub, to 2400m. May-June. T. **6a Hautbois Strawberry** *F. moschata* □ is larger with the main flstalks *much* longer than the lvs, seldom with runners; fls 20mm; fr without seeds at base. To 1800m T, but naturalised in the north. **6b** *F. viridis* □ like 6 but runners *short* and hairs of the flstalks *spreading*; fr without seeds at base. To 1900m. T–except B.

LADY'S MANTLES *Alchemilla.* A very confusing group with over fifty species in the area and here treated as aggregates for convenience. Tufted perennials. Lvs palmate with shallow or deep, usually toothed, lobes. Fls tiny green or yellowish, clustered, 4-parted with sepals and epicalyx, but no petals. Fr a single achene.

7 CUT-LEAVED LADY'S MANTLE *Alchemilla pentaphyllea.* Low slightly hairy per, stems often sprawling. Lvs small, not more than 3cm across, with 3-5 *deeply cut* wedge-shaped segments, separated almost to lf centre. Fls greenish-yellow, in 1-2 whorled clusters. Stony and gravelly north-facing slopes, often by snow patches, on acid soils, 1850-3000m. July-Aug. A.

8 ROCK LADY'S MANTLE *Alchemilla saxatilis.* Very low slightly hairy per. Lvs 5-lobed, the central lobe *separated* to the lf centre; lobes finely toothed at the apex, *hairless above*. Fls in whorled clusters on a long stalk. Acid rocks, to 2200m. P,s,swA,Ap. **8a** *A. transiens* often has *6-7-lobed* lvs which are more distinctly toothed.

9 ALPINE LADY'S MANTLE *Alchemilla alpina.* Variable low/short creeping hairy per. Lvs 5-7-lobed, the central one separated to the lf centre; lobes oblong, toothed at the apex, green above, *silvery-hairy beneath*. Fls pale green, in branched clusters on stems as long as lvs. Meadows, stony places and open woods, on acid soils, to 2600m. June-Aug. T. **9a** *A. basaltica* has fl stems *much longer* than the lvs. Acid rocks. P,w,swA. **9b** *A. subsericea* □ like 9 but lvs *greyish* and only slightly hairy beneath. P,A,Ap.

Rose Family *(contd.)*

1 HOPPE'S LADY'S MANTLE *Alchemilla hoppeana* agg. Low/short hairy per. Lvs 7-9-lobed, usually hairless above but slightly to silvery-hairy beneath; lobes elliptical, oblong to linear, *joined* near the lf centre, toothed in the upper half. Fls pale green, in whorled clusters, branched. Meadows, rocky and stony places, on limestone, to 2600m. June-Aug. A,Ap,(B). **1a** *A. plicatula* has the middle lobe *separated* to the lf centre. P,A,Ap. **1b** *A. conjuncta* has lvs dull blue-green and shiny above, hairy beneath swA,J,(B).

2 INTERMEDIATE LADY'S MANTLE *Alchemilla splendens* agg. Low/short hairy per. Lvs 7-11-lobed, hairy or hairless above; lobes usually cut halfway to lf centre, each with 10 or more teeth. Fls greenish, in branched, separated, clusters. Meadows, rocky places and woodland margins, to 2500m. June-Sept. P,A,J.

3 SMALL LADY'S MANTLE *Alchemilla glaucescens* agg. Low/short softly hairy per; stems with spreading hairs. Lvs *not more* than 6cm wide, 5-9-lobed, hairy *above and beneath*; lobes cut under halfway to lf centre, each with 8-10 teeth. Fls greenish, in dense clusters, branched. Meadows, rocky and stony places, to 2500m. June-Sept. T.

4 LADY'S MANTLE *Alchemilla vulgaris* agg. Very variable low/med, often rather densely hairy per; stems with spreading hairs. Lvs *often* more than 6cm wide, 7-11-lobed, green on both sides; lobes cut under halfway to lf centre, each with *12 or more* teeth. Fls pale green, 3-5mm, in loose branched clusters. Damp meadows and open woods, to 3100m. June-Sept. T.

5 DECEPTIVE LADY'S MANTLE *Alchemilla fallax* agg. Low/short per, stems and lfstalks *hairless* or hairs *pressed* to the surface. Lvs 7-9-lobed, hairless above, slightly hairy beneath; lobes cut under halfway to lf centre, each with 10 or more teeth. Fls greenish, 3-6mm, in branched clusters. Meadows and rocky places, to 3100m. June-Sept. P,A,Ap.

6 HAIRLESS LADY'S MANTLE *Alchemilla fissa*. Low/short, delicate, *hairless* per. Lvs small, 5-7-lobed; lobes cut halfway or more to lf centre, each with 8-12 *large* teeth. Fls pale green, 3·5-5mm, in small clusters. Wet rocks and by snow patches, to 2500m. P,A.

7 WILD COTONEASTER *Cotoneaster integerrimus*. Deciduous shrub, rarely more than 1m tall, often prostrate; young twigs downy, but soon becoming hairless. Lvs oval to almost rounded, *untoothed*, deep green above, *grey-downy* beneath. Fls pink in drooping clusters of 2-4; calyx hairless or slightly hairy along margin. Fr a small red berry. Dry rocky and stony places, open woods, to 2800m. Apr-June. T–B*. **7a** *C. nebrodensis* □ is taller with larger lvs and reddish fls *in larger groups* of 3-12; calyx hairy *all over*. Rocky and stony places, to 2400m. Apr-May. P,A,Ap.

7a Cotoneaster nebrodensis

Rose Family *(contd.)*

1 WILD PEAR *Pyrus pyraster*. Small/medium deciduous tree to 20 m; branches *usually spiny*, twigs grey to brown. Lvs rounded to oval, finely toothed, stalked, hairless when mature. Fls white, 25-35 mm, in clusters. Fr globular to pear-shaped, yellowish, brown or blackish, *hard*. Woods and scrub to 1700 m. Apr. T−except S, ?B. **1a Cultivated Pear** *P. communis* is usually non-spiny and with *reddish-brown* twigs; fr larger, *soft*, sweet tasting. Widely cultivated, but often naturalised in hedgerows, to 1600 m. T.

2 SOUTHERN PEAR *Pyrus nivalis* (= *P. communis nivalis*). Small/med deciduous tree to 20 m; branches usually non-spiny, twigs white-hairy when young, becoming blackish. Lvs oval, broadest above the middle, usually untoothed, *grey-hairy*. Fls white, 24-28 mm, in clusters. Fr globular, yellowish-green with purple dots. Sunny slopes and dry open woods, to 1600 m. Apr. sA, Ap. **2a** *P. austriaca* has *lance-shaped lvs*, finely toothed towards the apex, hairless above when mature. cA−Austria and Switzerland. **2b** *P. amygdaliformis* has narrower lvs than 2a, sometimes 3-lobed, and smaller 12-14 mm fls. To 1700 m. sA.

3 WILD CRAB *Malus sylvestris*. Small deciduous tree to 10 m; branches rather spiny. Lvs oval or elliptical, pointed, toothed, stalked, hairless when mature. Fls white or pink, 30-40 mm, in clusters. Fr globular, a yellowish-green apple, sometimes red flushed. Open woods and hedgerows, to 1600 m. May. T.

4 FALSE MEDLAR *Sorbus chamaemespilus*. Small deciduous shrub to 1·5 m. Lvs elliptical to oval, toothed, green above and beneath. Fls *pink* , 5-7 mm long, in small dense clusters; petals narrow, *erect*. Fr globular, scarlet, 10-13 mm. Open woods, stony places and cliffs, to 2500 m. May-July. P, A, Vosges, Ap.

5 MOUNTAIN ASH or ROWAN *Sorbus aucuparia*. Small slender deciduous tree to 15 m; branches smooth, silvery-grey. Lvs *pinnate*, lflets oblong, toothed, green above, greyish-hairy beneath at first. Fls creamy-white, 8-10 mm, in dense clusters. Fr a small orange or scarlet berry. Woods, scrub and rocky places, to 2400 m. May-June. T.

6 WHITEBEAM *Sorbus aria*. Small/med deciduous tree to 25 m; branches grey, smooth. Lvs oval to elliptical, widest below the middle, shallowly-lobed, toothed, stalked, green above, *white-hairy beneath*. Fls white, 10-15 mm, in dense flattish clusters. Fr 8-15 mm, oblong in outline, scarlet when ripe. Dry woods and rocky places, usually on lime, to 1700 m. May-June. T−except S. **6a** *S. torminalis* has more lobed lvs, *green beneath* at maturity; fr brown, finely dotted. T−except S. **6b** *S. mougeotii* like 6 but with more *prominently lobed* lvs, grey-hairy beneath and globular red frs. P, A−mainly in the w. **6c** *S. austriaca* has more rounded lvs than 6b and frs with many *large dots*. eA.

7 AMELANCHIER *Amelanchier ovalis*. Deciduous shrub to 3 m; bark blackish. Lvs oval, often broadest above the middle, toothed, white-downy beneath when young. Fls white, 16-20 mm, in *short-spiked clusters*; petals narrow, pointed. Fr a small bluish-black berry. Open woods and rocky places, usually on limestone, to 2400 m. Apr-June. P, A, Ap.

8 HAWTHORN or MAY *Crataegus monogyna* (= *C. oxyacantha*). Shrub or small tree to 10 m; branches spiny. Lvs oval to rhombic, deeply *3-5-lobed* more than halfway to the midrib; stipules *untoothed*. Fls white, 8-15 mm, in broad clusters; styles one. Fr a dark to bright red berry (haw), 6-10 mm. Hedges and thickets, to 1700 m. May-June. T. **8a** *C. macrocarpa* is a spreading shrub with finely toothed lf lobes and toothed stipules; fr 12-15 mm, styles 2-3. s, c, eA. **8b** *C. laevigata* like 8 but lvs *less deeply lobed*, the lobes finely toothed. Fr 8-12 mm; styles 2-3. T−except S and Ap.

Rose Family *(contd.)*

1 MARMOT PLUM *Prunus brigantina* (= *P. brigantiaca*). Small spreading decid-uous tree or shrub, to 6m; young twigs *glossy, hairless*. Lvs oval to elliptical, pointed, toothed, glossy above, hairy on veins beneath. Fls white, 15-20mm, in clusters of 2-5, appearing with the young lvs. Fr a glossy yellow plum, 25-30mm. Dry stony slopes, scrub, 1200-1800m. Apr.-May. w,swA.

2 BLACKTHORN or SLOE *Prunus spinosa.* Dense deciduous spiny shrub to 4m; bark black, young twigs usually hairy. Lvs oval, broadest above the middle, finely toothed, dull green. Fls white, 10-15mm, *usually solitary*, before the lvs. Fr a small bluish-black plum, 10-15mm. Scrub and hedges, to 1600m. Mar-May. T

3 WILD CHERRY or GEAN *Prunus avium.* Med spreading tree to 20m; bark reddish-brown, twigs hairless. Lvs oblong pointed, toothed, often reddish, sparsely hairy beneath. Fls white, 14-22mm, in clusters of 2-6. Fr a bright red, *usually bitter*, cherry, 9-12mm. Woods and hedges, to 1700m. Apr-May. T. Various varieties are frequently cultivated.

4 ST. LUCIE'S CHERRY *Prunus mahaleb.* Deciduous shrub or small tree, to 10m; *young twigs* glandular-hairy. Lvs oval, toothed, usually hairless beneath, *with glands* along the edges. Fls white, 8-12mm, in clusters of 3-10. Fr a black bitter cherry, 8-10mm. Open woods, thickets and dry slopes, to 1700m. Apr-May. P,A,Ap. Sometimes cultivated.

5 BIRD CHERRY *Prunus padus.* Deciduous tree to 17m; young shoots hairless; bark brown, peeling, foetid. Lvs elliptical-oblong, almost hairless, finely toothed, dull green. Fls white, 10-16mm, in *long arching or drooping spikes*, heavy scented. Fr a shiny black cherry, 6-8mm. Woods, hedges and moors, to 2200m. May. T. **5a** *P.p. borealis* is a shrub to 3m with *hairy young shoots;* fls scarcely scented. A,Vosges.

Pea Family   Leguminosae

A large and highly distinctive family. Lvs usually trifoliate or pinnate, sometimes simple, alternate, occasionally with spines or tendrils. Fls 5-petalled, the upper the 'standard', often broad and erect, overlapping the two side petals or 'wings' which lie on either side of the lower two united-petals or 'keel', which conceals the 10 stamens and style; sepal tube with 5 short or long teeth. Fr a pod, splitting when ripe in most instances.

6 LABURNUM *Laburnum anagyroides.* Deciduous shrub or small tree to 7m; bark smooth, twigs greyish-green with *hairs pressed* to the surface. Lvs trifoliate, lflets untoothed, grey-green beneath. Fls golden-yellow, in drooping stalked spikes. Pod 40-60mm, flattened, pressed-hairy. Woods and scrub, often on limestone, to 2000m. Apr-May. A,Ap. Frequently cultivated. **6a** *L. alpinum* has *hairless* green twigs and hairless pods; lvs pale green beneath. sA,Ap.

7 LUGANO BROOM *Cytisus emeriflorus* (= *Genista glabrescens*). Small shrub to 60cm; branches rigid, angular, often rather gnarled, young twigs hairy. Lvs trifoliate, lflets lance-shaped, *silvery-hairy beneath*. Fls yellow 1-4 in lfy clusters. Pods 25-35mm, hairless. Thickets and stony places, to 1850m. June. sA–Lugano and Como areas.

8 ARDOIN BROOM *Cytisus ardoini.* Small shrub to 60cm; young twigs downy with 8-10 *winged ridges*. Lvs trifoliate, lflets oblong, hairy above and beneath. Fls yellow, in clusters of 1-3 on short *side shoots*. Pod 20-25mm, densely hairy. Calcareous rocks, to 1500m. Apr-May. MA–French. **8a** *C. sauzeanus* has *5-angled* twigs, the pod usually hairy only along the edges. wA–French. **8b** *C. decumbens* like 8a but with *simple* oblong or lance-shaped lvs and hairy pods. s,swA,Ap.

9 PYRENEAN BROOM *Cytisus purgans* (=*Sarothamnus purgans, Genista purgans*). Bluish-green shrub to 1m, usually less; twigs ridged, *scarcely lfy*, hairy when young. Lvs sparse, trifoliate, unstalked, simple on fl stems. Fls deep yellow, *vanilla-scented*, solitary or paired at stem tips. Pod black when ripe, 15-30mm, hairy. Dry stony slopes, on acid rocks, to 1900m. May-July. P. **9a**, p.337.

Pea Family *(contd)*

1 PURPLE BROOM *Chamaecytisus purpureus* (=*Cytisus purpureus*). Small almost hairless, subshrub, to 30 cm. Lvs trifoliate, lflets oblong, pointed. Fls *lilac-pink or purplish*, 15-25 mm, in clusters of 2-3, forming lfy spikes. Pod 15-25 mm, hairless. Scrub and rocky places, on lime, to 1400 m. May, s,seA.

2 HAIRY BROOM *Chamaecytisus hirsutus* (*Cytisus hirsutus, Cytisus pumilus*). Spineless hairy subshrub, to 1 m, usually less; stems more or less erect. Lvs *trifoliate*, lflets oval to elliptical. Fls yellow or pinkish-yellow, 20-25 mm, the standard sometimes brown spotted, in clusters of 1-4. Pod linear, 25-40 mm, hairy. Grassy places and scrub, often on acid soils, to 1900 m. Apr-June. A. **2a** *C. polytrichus* forms low mats, not more than 30 cm tall; pod densely hairy. s,seA,MA,Lig,Ap.

3 DYER'S GREENWEED *Genista tinctoria*. Variable small to medium spineless deciduous shrub to 1 m, but often less, slightly hairy. Lvs elliptical to lance-shaped, *not trifoliate*. Fls yellow, 8-15 mm, in lfy stalked spikes. Pod 25-30 mm, *hairless*. Meadows, scrub and open woods, to 1800 m. May-Aug. T. **3a Black Broom** *Lembotropis nigricans* has *trifoliate lvs* and smaller fls in lfless spikes. A−except w.

4 SILVERY BROOM *Genista sericea*. Much branched hairy shrub, to 40 cm. Lvs narrow-elliptical, *not trifoliate*, green and hairless above, but *silvery hairy* beneath. Fls yellow, 10-14 mm, in clusters of 2-5. Pods hairy. Rocky slopes, to 1300 m. May-July. seA. **4a** *G. cinerea* is erect with *long spikes* of fls; bracts in groups up the spike. To 1900 m. Apr-July. P, swA. **4b** *G. pilosa* like 4a but bracts *alternate* up the spike. Woods, heaths and open places. Apr-Oct. T−except S and sAp.

5 GERMAN GREENWEED *Genista germanica*. Small *spiny shrub*, to 60 cm, hairy. Lvs lance-shaped, hairy beneath. Fls yellow, 10-12 mm, in short spikes, the standard shorter than the keel petal. Pod *short oval*, 5-8 mm, hairy. Grassy places, scrub and heath, to 2300 m. May-Sept. eP,A,n,cAp.

6 SOUTHERN GREENWEED *Genista radiata* (=*Cytisanthus radiatus*). Small bushy shrub to 50 cm, *spineless*. Lvs *opposite trifoliate*, lflets linear-lance- shaped, silvery-hairy beneath. Fls yellow, 8-14 mm, in small clusters. Pod short oval, 5-6 mm. Rocky places, woods and scrub on lime, to 2200 m. May-July. sw,s,seA,n,cAp.

7 SPANISH BROOM *Genista hispanica*. Rather like 5 but more densely spiny. Lvs lance-shaped, often broadest above the middle, hairy beneath. Fls yellow, 6-8 mm, in dense, almost *rounded, clusters*, the standard as long as the keel petal. Pod short oval, 5-6 mm, almost hairless. Rocky and stony places, to 1500 m. Apr-Sept. eP. **7a** *G.h. occidentalis* has larger fls, 8-11 mm. wP*.

8 WINGED GREENWEED *Chamaespartium sagittale* (=*Genista sagittalis, Genistella sagittalis*). Low spineless subshrub, to 30 cm, with prostrate branches giving rise to erect *green-winged* fl branches, slightly hairy. Lvs small, elliptical, sparse. Fls yellow, 10-12 mm, in dense terminal clusters. Pod oblong, 14-20 mm, hairy. Open woods, grassy and rocky places, to 1950 m. May-July. P,A,Ap.

9 ECHINOSPARTUM *Echinospartum horridum*. (=*Genista horrida*). Low densely spiny shrub, to 40 cm, *branches opposite*. Lvs trifoliate, lflets oblong, broadest above the middle, silvery-hairy beneath. Fls yellow, 12-16 mm, *solitary or two together* at stem tips; calyx slightly inflated. Pod oblong, 9-14 mm, silkily-hairy. Limestone rocks, stony meadows, to 1800 m. June-Sept. P.

Pea Family *(contd.)*

1 HEDGEHOG BROOM *Erinacea anthyllis* (=*E. pungens*). Short *spiny* cushion shrub, thickly branched. Lvs and branches opposite; lvs small solitary or trifoliate, *quickly falling*. Fls blue-violet, 16-18mm, in groups of two to three. Pod linear-oblong, hairy. Dry rocky slopes, usually on limestone, to 2000m. May-June. eP*.

2 GORSE *Ulex europaeus.* Dense evergreen shrub to 2·5m, hairy. Lvs stiff *furrowed spines*. Fls golden yellow, 14-18mm, in spiny clusters, almond-scented; calyx 2-lipped, greenish-yellow, hairy. Pod oval-oblong, hairy. Heaths, banks and grassy places, to 1200m. Fls most of the year. T—except S. **2a** *U. minor* is small, often prostrate, with shorter, weaker, spines and smaller, paler, fls. July-Nov. P, Cev.

3 BLADDER SENNA *Colutea arborescens.* Deciduous shrub to 4m, hairy, much branched. Lvs *pinnate*. Fls deep yellow, often red-marked, 20-24mm, in loose racemes. Pod large, *inflated*, papery brown when ripe. Dry slopes and open woods, often on limestone, to 1600m. June-Aug. P, A, Ap, (B).

MILK-VETCHES *Astragalus.* Perennials, often tufted, with pinnate lvs, usually with a terminal lflet, sometimes spiny. Fls in loose racemes or dense clusters at base of lvs, keel-petal blunt-tipped; calyx with short teeth.

4 WILD LENTIL *Astragalus cicer.* Short/med straggling, hairy, per. Lflets 10-15 pairs, lance-shaped or oval. Fls *pale yellow*, in dense rounded, long-stalked clusters. Pods rounded, inflated, pointed, *black and white* hairy. Meadows and scrub, to 1800m. June-July. P, A.

5 PURPLE MILK-VETCH *Astragalus danicus.* Short slender hairy per. Lflets 6-13 pairs, oval or oblong, *blunt*. Fls purple or bluish-violet, in dense rounded, long-stalked clusters. Pods oval, inflated, *white* hairy. Meadows, usually on limestone, 1800-2400m. May-July. B, S, sw, eA.

6 PURPLE VETCH *Astragalus purpureus.* Short slender hairy per. Lflets 7-15 pairs, elliptical-oblong, *notched*. Fls purplish, sometimes whitish, in dense rounded, long-stalked clusters. Pods oval, inflated, *white* hairy. Stony places and scrub, usually on limestone, to 1800m. May-July. P, sA*, Ap.

7 PALLID MILK-VETCH *Astragalus frigidus* (=*Phaca frigida*). Short *almost hairless* per, unbranched. Lflets 3-8 pairs, broad-elliptical. Fls *yellowish-white* in loose, long-stalked clusters. Pod brownish, elliptical, black or white-hairy at first. Meadows and stony places, usually on limestone, 1700-2800m. July-Aug. S, A.

8 MOUNTAIN LENTIL *Astragalus penduliflorus.* Short/med hairy per, branched. Lflets 7-15 pairs, elliptic or oblong-lance-shaped. Fls *yellow*, in loose, long-stalked clusters. Pod oval, *inflated*, black-hairy at first. Meadows, woods and stony places, to 2850m. July-Aug. S, c, eP, A.

9 ALPINE MILK-VETCH *Astragalus alpinus* (=*Phaca alpina, P. astragalina*). Rather like a slender version of 5. Lflets 7-12 pairs, elliptic, blunt or pointed. Fls whitish or pale-violet with a *bluish-violet* keel, in loose, long-stalked clusters. Pod oblong, *blackish-hairy*. Meadows rocky and stony places, 1900-3100m. July-Aug. B*, S, P, A.

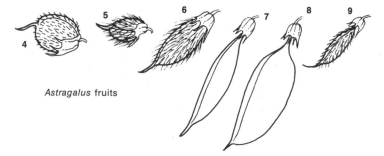

Astragalus fruits

1

2

3

4

5

6

7

8

9

Pea Family (contd.)

1 SPRAWLING MILK-VETCH *Astragalus depressus*. Variable low *tufted*, hairy, per. Lflets 6-14 pairs, oblong or heart-shaped, broadest above the middle. Fls whitish or bluish-purple, in *short-stalked* clusters. Pod linear-lance-shaped, hairless. Sunny, dry, limestone rocks, to 2700 m. May-July. P, A, Ap.

2 NORWEGIAN MILK-VETCH *Astragalus norvegicus* (=*A. oroboides*). Short erect per, hairless or almost so. *Lflets* 5-8 pairs, oblong-oval, notched. Fls pale violet, in dense, long-stalked clusters. Pod *egg-shaped*, blackish-hairy. Meadows, 1900-2500 m. July-Aug. S, eA.

3 SOUTHERN MILK-VETCH *Astragalus australis* (=*Phaca australis*). Low /short hairy per, more or less erect. Lflets 4-8 pairs, narrow-elliptic to oval-lance-shaped. Fls yellowish-white, often with violet tips, in loose, long-stalked racemes. Pod oblong-oval, *inflated*, hairless. Meadows and stony places, 1800-3100 m. May-July. P, A, n, cAp.

4 WILD LIQUORICE *Astragalus glycyphyllos*. Med/tall straggling per, slightly hairy. Lflets 4-6 pairs, oval or broad-elliptical, blunt. Fls pale cream, in dense, *short-stalked* racemes. Sepal tube *hairless*. Pod linear-oblong, slightly *curved*, hairless. Grassland, open woods and scrub, to 2000 m. June-Aug. T.

5 STEMLESS MILK-VETCH *Astragalus exscapus*. Low tufted, *very hairy* per. Lflets 12-19 pairs, elliptic oval. Fls bright yellow, in clusters *amongst* lvs. Sepal tube very hairy. Pod oblong, hairy. Meadows and open woods, usually on limestone, to 2200 m. May-July. c, sA.

6 CENTRAL-ALPS MILK-VETCH *Astragalus centralpinus*. Med/tall erect per, stems *very hairy*. Lflets 20-30 pairs, elliptic to oval-lance-shaped. Fls yellow, in oblong *unstalked* clusters. Sepal tube very hairy. Pod oval, hairy. Grassy places, to 1500 m. July-Aug. swA.

7 MOUNTAIN TRAGACANTH *Astragalus sempervirens*. Low/short *spiny*, tufted, per; greyish-hairy. Lflets 4-10 pairs, linear-oblong, broadest above the middle. Fls white to pale purple, in short-stalked clusters *amongst* lvs. Pod egg-shaped, very hairy. Rocky places and gravels, usually on limestone, to 2750 m. May-Aug. eP, c, swA, Ap.

8 AUSTRIAN MILK-VETCH *Astragalus austriacus*. Short/med, more or less erect, slightly hairy per. Lflets 5-10 pairs, *linear*. Fls blue and violet, in loose, long-stalked racemes. Pod linear-oblong, pointed, hairy. Dry grassy and rocky places, scrub, usually on limestone, to 1700 m. June-Aug. eP, sw, c, eA.

9 TYROLEAN MILK-VETCH *Astragalus leontinus*. Low/short *pale-green*, hairy per. Lflets 5-10 pairs, oval to narrowly-elliptic. Fls violet or pale-purplish, in rounded, long-stalked clusters. Pod oblong-egg-shaped, hairy. Meadows and rocky places, usually on limestone, to 2650 m. July-Aug. A.

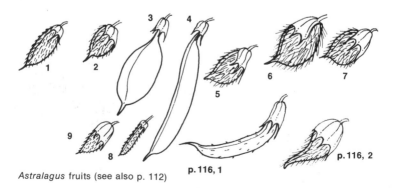

Astralagus fruits (see also p. 112)

Pea Family (contd.)

1 FALSE VETCH *Astragalus monspessulanus.* Low/short tufted, slightly hairy, per, stemless. Lflets 10-20 pairs, rounded to oblong, blunt. Fls purplish-violet, sometimes whitish, in loose, long-stalked clusters. Pod linear, curved and pointed, *almost hairless* (see p.114). Meadows and stony ground, usually on limestone, to 2600m. Apr-Aug. P,w,c,sA,Ap.

2 INFLATED MILK-VETCH *Astragalus vesicarius.* Low hairy per. Lflets 5-10 pairs, linear-lance-shaped to oblong. Fls whitish with a purple or violet *standard*, in loose, long-stalked, clusters. Pod oblong, pointed, *very hairy* (see p.114). Meadows and stony places to 2000m. P,sw,c,eA,n,cAp. **2a** *A.v. pastellianus* is taller with *yellowish* fls. eA–Italian.

MILK-VETCHES *Oxytropis* is similar and often confused with *Astragalus* but keel of fls ending in a small point, *not* blunt. Fls in dense-rounded, long-stalked, clusters.

3 NORTHERN MILK-VETCH *Oxytropis lapponica.* Low tufted, hairy, per. Lflets 8-14 pairs, lance-shaped or oblong; stipules joined together. Fls violet-blue. Pod narrow-oblong, with *short hairs.* Meadows, stony places and screes, 1800-3050m. July-Aug. S,P,A.

4 MOUNTAIN MILK-VETCH *Oxytropis jacquinii* (= *Astragalus montanus* in part). Low/short, slightly hairy, tufted per. Lflets 14-20 pairs, lance-shaped or narrow-oval; stipules *hardly* joined. Fls purplish-violet. Pod oval, pointed, slightly inflated, *short-stalked,* slightly hairy. Meadows and stony places, usually on limestone, 1500-2900m. July-Aug. A, Jura.

5 GAUDIN'S MILK-VETCH *Oxytropis gaudinii* (= *Astragalus triflorus* var. *gaudinii*). Low sprawling, stemless, *silvery-hairy* per. Lflets 10-12 pairs, lance-shaped. Fls lilac-blue. Pod narrow-oblong, hairy. Meadows and stony places, 1800-3100m. July-Aug. sw,wcA*. **5a** *O. amethystea* has *13-20 pairs* of lflets, pale purplish fls and *oval*, densely-hairy, pods. eP,swA. Forms natural hybrids with 4 where the two overlap in distribution.

6 SAMNITIC MILK-VETCH *Oxytropis pyrenaica* (= *O. montana samnitica*). Low tufted, downy per, stemless. Lflets 12-20 pairs, oblong-elliptic to lance-shaped. Fls purplish or bluish-violet. Pod narrow-oval, pointed, slightly hairy. Limestone rocks to 3000m. July-Aug. A,Ap. **6a** *O. triflora* (=*Astragalus triflorus*) is more slender with *less than* 12 pairs of lflets and only 3-5 fls to a cluster. eA.

7 MEADOW or YELLOW MILK-VETCH *Oxytropis campestris* (= *Astragalus campestris*). Low, tufted, downy per, stemless. Lflets 10-15 pairs, elliptical or lance-shaped. Fls *pale yellow.* Pod oval, pointed, hairy. Meadows and rocky places, to 3000m. July-Sept. T. **7a** *O.c. tiroliensis* has *pale violet* or *whitish* fls. B*,S,c,eA.

8 FOUCAUD'S MILK-VETCH *Oxytropis foucaudii.* Low hairy per, stemless. Lflets 12-16 pairs, oval-lance-shaped. Fls lilac. Pod narrow-elliptical, *very hairy.* Meadows and rocky places, to 2600m. July-Aug. P.

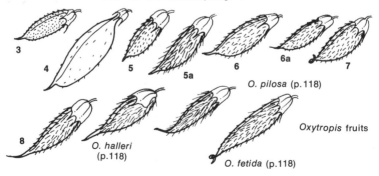

O. pilosa (p.118)

O. halleri (p.118)

O. fetida (p.118)

Oxytropis fruits

Pea Family *(contd.)*

1 SILKY MILK-VETCH *Oxytropis halleri* (= *O. sericea, Astragalus sericeus*). Low *silky-hairy* per, stemless. Lflets 10-14 pairs, oval-lance- shaped. Fls bluish-purple. Pod oval to narrow-elliptical, densely short-hairy (see p.116). Dry meadows and stony places, usually on acid soils, 1500-2950 m. July-Aug. B*,S,P,A. **1a** *O.h. velutina* has *much hairier* stalks and pale purplish fls. cA*.

2 WOOLLY MILK-VETCH *Oxytropis pilosa* (= *Astragalus pilosus*). Short/med tufted per, *densely long-hairy*, with a stem. Lflets 9-13 pairs, oblong. Fls *pale yellow*. Pod oval or oblong, pointed, very hairy (see p.116). Grassy, stony places and gravels, to 2600 m. June-Aug. A,n,cAp.

3 STINKING MILK-VETCH *Oxytropis fetida* (= *Astragalus fetidus*). Low/short stemless, tufted per, *stickily-hairy*, unpleasant smelling. Lflets 15-25 pairs, lance-shaped or oblong. Fls yellowish. Pod oblong, slightly curved, hairy (see p.116). Grassy slopes and screes, 1800-3000 m. July-Aug. sw,wcA*.

VETCHES *Vicia*. Perennials, sometimes annual, climbing or scrambling; stems *not* winged. Lvs pinnate, ending in a clasping tendril or lflet. Fls in loose, stalked or unstalked, one-sided clusters, rarely solitary. Pods narrow-oblong, splitting and brown when ripe.

4 SILVERY VETCH *Vicia argentea*. Short, *silvery-hairy* per. Lvs ending in a lflet; lflets 7-9 pairs, linear, blunt. Fls white with violet veins, 18-25 mm long, in stalked-clusters. Pod oblong, brown-hairy. Meadows, stony places and screes, 1600-2300 m. July-Aug. P*.

5 TUFTED VETCH *Vicia cracca*. Clambering per to 1 m, slightly hairy. Lvs *with tendrils*; lflets 6-15 pairs, linear to oblong. Fls bluish-violet, 8-12 mm, in stalked clusters of 10-30. Pod oblong, brown, *hairless*. Meadows, scrub, banks and hedges, to 2200 m. June-Aug. T. **5a** *V. incana* has *densely hairy* stems and clusters of 20-40 fls. P,A,Ap. **5b** *V. tenuifolia* has *larger* purple, pale lilac or bluish-lilac fls, 12-18 mm. T. **5c** *V. onobrychioides* like 5 but fls much larger, 17-24 mm, violet with a paler keel. P,swA,MA. **5d Wood Vetch** *V. sylvatica* ☐ is *hairless* with white, purple veined fls, 12-20 mm. Woods. T.

6 PALE VETCH *Vicia oroboides*. Short/med, slightly hairy, per. Lvs *not ending* in a tendril or lflet; lflets 1-4 pairs, oval, pointed. Fls pale yellow, 14-19 mm, in short-stalked clusters. Pod oblong, hairless, *black*. Meadows and woods, to 1600 m. May-July. e,seA.

7 PYRENEAN VETCH *Vicia pyrenaica*. Low/short almost hairless, per. Lvs ending in a *tendril*; lflets 3-6 pairs, rounded or oblong. Fls bright violet-purple, 16-25 mm, solitary. Pod oblong, hairless, black. Pastures and screes, 1400-2500 m. June-Aug. P,swA–French.

8 BUSH VETCH *Vicia sepium*. Variable med/tall downy per. Lvs ending in a *branched tendril*; lflets 3-9 pairs, oval to oblong. Fls dull bluish-purple, 12-15 mm, in *short-stalked* clusters. Pod oblong, hairless, black. Fields, scrub and hedgerows, to 2150 m. Apr-Oct. P,A,Ap.

9 HAIRY TARE *Vicia hirsuta*. Short, slender, hairy *ann*. Lvs ending in a branched tendril. Lflets 4-10 pairs, linear to oblong. Fls *small*, dirty-white, tinged purple, 2-5 mm, in short-stalked clusters. Pod oblong, black, *downy*. Grassy and waste places, to 1800 m. May-Aug. T. **9a** *V. tetrasperma* has *pale purple* fls, 4-8 mm and brown, *hairless* pods. T.

The **Common Vetch**, *Vicia sativa*, and **Broadbean**, *V. faba*, are often cultivated to 2200 m.

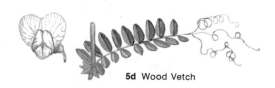

5d Wood Vetch

Pea Family (contd.)

PEAS *Lathyrus*. Similar to vetches but stems *winged* or *angled;* lvs often with fewer lflets and fls larger, usually in one sided, stalked clusters.

1 SPRING VETCHLING *Lathyrus vernus*. Short/med, tufted, usually hairless, per, *stems angled*. Lvs *without* tendrils; lflets 2-4 pairs, oval or lance-shaped, pointed. Fls reddish-purple, becoming blue, 13-20mm. Pod oblong, hairless, brown. Woods and scrub, usually on limestone, to 1900m. Apr-June. T−except B. **1a** *L. venetus* has *more rounded* lflets and pods covered in *brown glands*. c,s,eA,Ap.

2 YELLOW PEA *Lathyrus laevigatus*. Short/med hairless per. Lvs *without* tendrils; lflets 2-6 pairs, elliptical, or oval. Fls *yellow,* 15-25mm. Pod hairless, brown. Meadows and woods, usually on limestone, to 2300m. June-July. **2a** *L.l. occidentalis* is slightly hairy with narrower lflets. P,sw,cA. **2b Yellow Vetchling** *L. aphaca* ☐ has large *lfy stipules* and a tendril, *no* lflets; with *solitary* fls. T−except S.

3 SLENDER VETCH *Lathyrus filiformis* (= *L. canescens*). Short/med hairless per; stems *ridged*. Lvs *without* tendrils; lflets 2-4 pairs, linear-lance-shaped, pointed. Fls bright reddish-purple, 14-22mm. Pod hairless, brown. Limestone rocks, to 1450m. May-July. sw,sA. **3a** *L. bauhinii* has *larger fls*, 20-27mm long. P,A,Jura.

4 FELTED VETCH *Lathyrus pannonicus* agg. Variable short/med hairy per; stems narrow-winged or ridged. Lvs without tendrils; lflets 1-4 pairs, linear to oblong-lance-shaped. Fls pale cream, tinged reddish-purple, 12-18mm. Pod hairless, pale brown. Grassland and scrub, to 1200m. May-June. P,A,Ap.

5 BITTER VETCH *Lathyrus montanus* (= *L. macrorrhizus*). Short/med hairless per; stems winged. Lvs without tendrils; lflets 2-4 pairs, linear to elliptic. Fls crimson, *turning* greenish-blue, 10-16mm. Pod hairless, red brown. Pastures, woods and scrub, usually on acid soils, to 2200m. Apr-June. T.

6 MEADOW VETCHLING *Lathyrus pratensis*. Short/tall scrambling, hairless or downy per; stems *angled*. Lvs with tendrils; lflets *one pair,* linear or lance-shaped. Fls *yellow,* 10-16mm. Pod sometimes hairy, black. Grassland and scrub, to 2150m. May-Aug. T.

7 CIRRHOSE VETCH *Lathyrus cirrhosus*. Tall scrambling, hairless, per; stems winged. Lvs with tendrils; lflets 2-3 pairs, elliptic or oblong-lance-shaped. Fls pink, 12-17mm. Pod hairless, pale brown. Dry grassy places, scrub and banks, to 1600m. May-July. P,Cev.

8 BROAD-LEAVED EVERLASTING PEA *Lathyrus latifolius*. Variable clambering per to 3m, hairy or hairless; stems *broad winged*. Lvs with tendrils; lflets *one pair,* oval to elliptic. Stipules *broad*. Fls *large* purple-pink, 20-30mm. Pod hairless, brown. Scrub and hedgerows, to 1500m. July-Sept. P,A,Ap,(B). **8a** *L. heterophyllus* has 2-3 pairs of lflets on upper lvs and smaller fls, 12-22mm. To 1900m. T−except B. **8b** *L. sylvestris* similar to 8 but with *narrower* lvs and stipules and smaller fls, 13-20mm. T−except B. **8c** *L. tuberosus* ☐ like 8 but stems angled, *not winged* and fls small, bright crimson, 12-20mm. P,A,Ap,(B).

9 BROWN VETCH *Lathyrus setifolius*. Short/med hairless ann. Lflets linear. Fls *solitary,* orange-red or brownish, 8-11mm, on long slender stalks. Grassy places and banks, to 1300m. May-July. P,sA,Cev,sA,Ap.

The **Chickling Pea**, *Lathyrus sativus*, is commonly cultivated up to 1800m.

2b Yellow Vetchling   **8c** *Lathyrus tuberosus*

Pea Family (contd.)

RESTHARROWS Ononis. Perennials or small shrubs, often glandular-hairy and spiny. Lvs trifoliate or single on same plant. Fls pink, sometimes yellow. Pod oval or oblong, hairy, splitting.

1 ROUND-LEAVED RESTHARROW Ononis rotundifolia. Med hairy, much-branched, subshrub. Lflets elliptic to rounded, *blunt,* toothed. Fls pink or whitish, 16-20mm long, in long-stalked clusters of one to three. Woods and rocky places, usually on limestone, to 1750m. May-Aug. P,A (except se),n,cAp.

2 SHRUBBY RESTHARROW Ononis fruticosa. Med/tall *erect,* hairy, subshrub. Lflets oblong-lance-shaped, *pointed,* saw-toothed, *hairless.* Fls pink, 10-20mm, in stalked clusters of one to three. Dry rocky places, to 1600m. May-Aug. cP,swA–French.

3 MT. CENIS RESTHARROW Ononis cristata (= O. cenisia) Low/short, *often mat-forming,* hairy, subshrub. Lflets oblong, toothed. Fls pink, 10-14mm, *solitary,* on slender stalks. Meadows, rocky places and screes, usually on limestone, to 1800m. June-Sept. eP,swA,cAp.

4 YELLOW RESTHARROW Ononis striata. Low/short downy per. Lflets rounded to oblong, *broadest* above the middle, toothed. Fls *yellow,* 10-13mm, solitary or in small, short-stalked clusters. Limestone rocks, to 1800m. June-Aug. P,swA,MA. **4a** O. aragonensis has *larger lflets and fls,* 12-18mm, in loose terminal clusters. Meadows and rocks, to 2500m. June-July. P*. **4b Large Yellow Restharrow** O. natrix □ is taller than 4a, much branched and *stickily hairy.* Fls deep yellow, often veined red or purple, 14-16mm. Dry rocky places on limestone, to 2100m. May-July. sA.

5 SPINY RESTHARROW Ononis spinosa. Variable short/med erect or sprawling, hairy per; stems *stiffly-spiny.* Lflets narrow oval, blunt or pointed, toothed. Fls pink or purplish, 6-20mm, usually solitary. Meadows, dry slopes and waste places, often on limestone, to 1800m. Apr-Sept. T. **5a Restharrow** O. repens is lower and more spreading with *spineless or softly spiny* stems; lflets *notched* at tip. T.

6 TALL MELILOT Melilotus altissima. Med/tall hairless bien/per. Lfs trifoliate; lflets oblong-oval, blunt, toothed. Fls yellow, 5-7mm, in slender, stalked, spikes. Pod oblong, pointed, *downy,* black when ripe. Damp waste places and fields, to 2000m. June-Sept. T. **6a Common Melilot** M. officinalis □ has brown *hairless* pods; fls with keel shorter than wings. To 2000m. T.

7 WHITE MELILOT Melilotus alba. Med/tall hairless bien. Lvs trifoliate; lflets oval to oblong, toothed. Fls white, 4-5mm, in stalked spikes. Pod *hairless,* greyish-brown when ripe. Fields and waste places, to 1850m. July-Oct. T.

8 PYRENEAN MEDICK Medicago hybrida. Short hairless per. Lvs trifoliate; lflets broad-oval to rounded, pointed, finely toothed. Fls *small,* yellow, 2-3mm, in stalked clusters of two to five. Pod oblong, *curved,* pointed, black when ripe. Woods and rocky places, on limestone, to 1800m. June-Aug. nP,Corbières. **8a Black Medick** M. lupulina is *hairy* with ten to fifteen fld racemes; pod *kidney-shaped,* black. To 2300m. T.

9 SPRAWLING MEDICK Medicago suffruticosa. Low/short slightly downy per, rather sprawling. Lvs trifoliate; lflets oval or heart-shaped, broadest above middle, toothed. Fls yellow, 3-6mm, in stalked clusters of three to eight. Pod *twisted* into 2-4 spirals with a hole in the middle. Rocky places, to 2300m. July-Aug. P,Corbières. **9a Lucerne** M. sativa □ is very variable with larger fls, 6-11mm, blue, purplish or yellow; pod curved *or* spiralled. To 1800m. T.

4b Large Yellow Restharrow   **9a** Lucerne

1

2

3

4

5

6

7

8

9

123

Pea Family (contd.)

CLOVERS *Trifolium.* Perennials, sometimes annuals with trifoliate lvs, usually stalked; lflet edges slightly toothed. Fls in dense rounded or oblong clusters; wings longer then keel petal. Pod small, often more or less covered by the dead fl remains.

1 ALPINE CLOVER *Trifolium alpinum.* Low tufted hairless per. Lflets lance-shaped or linear. Fls *large,* pink or purplish, rarely cream, 18-25mm, in 3-12-fld clusters. Meadows and pastures, on acid soils, 1700-2500m. June-Aug. P,A,n,cAp.

2 MOUNTAIN CLOVER *Trifolium montanum.* Short/med unbranched per, stems woolly. Lflets oval, lance-shaped or elliptical, hairless above. Fls *white or yellowish,* 7-9mm, in dense rounded clusters. Dry grassy places, open woods, often on limestone, to 2600m. May-July. S,P,A,Ap.

3 WHITE or DUTCH CLOVER *Trifolium repens.* Low/short creeping, more or less hairy per; stems *rooting* at lf joints; lflets bright green, often with a *white crescent zone,* oval or elliptical. Fls white or pale pink, rarely purple, 7-10mm, in dense rounded clusters, *sweetly scented.* Meadows, grassy and waste places, to 2750m. May-Oct. T, frequently cultivated.

4 PALE CLOVER *Trifolium pallescens.* Low tufted *hairless* per. Lflets bright green, elliptical or oval, broadest above the middle. Fls yellowish-white or pink, *becoming dark brown,* 6-10mm, in dense rounded clusters, sweetly scented; angle between sepal teeth sharp. Damp pastures and screes, on acid rocks, 1800-2700m. July-Sept. P,A,Ap. **4a** *T. thalii* has *dull green* lvs and white or red fls in oval clusters. Meadows and pastures on limey soils, to 3100m. P,A,Ap. **4b Alsike Clover** *T. hybridum* like 4 but taller with white and pink fls; angles between sepal teeth blunt. To 2150m. T,(B,S).

5 BROWN CLOVER *Trifolium badium.* Short tufted hairy or hairless per. *Uppermost lvs opposite;* lflets elliptical or diamond-shaped. Fls golden yellow, *becoming* bright chestnut-brown, 7-9mm, in dense rounded clusters. Meadows and damp stony places, usually on calcareous soils, 1400-2800m. July-Aug. P,A,Ap.

6 LARGE BROWN CLOVER *Trifolium spadiceum.* Rather like 5. Lflets oblong. Fls golden yellow, becoming very dark brown, 6mm, in *dense oblong clusters.* Grassy places, on acid soils, to 2200m. June-Aug. T−except B.

7 CRIMSON CLOVER *Trifolium incarnatum.* Short/erect hairy *ann.* Lflets rounded to oblong, broadest above the middle. Fls *blood-red,* pink or white, 10-12mm, in dense oblong clusters. Fields and waste places to 1500m. May-July. T,(S) but widely cultivated. **7a** *T. saxatile* is greyish-hairy, with small whitish or pinkish fls, 3-4mm, few in *stalkless clusters.* Gravels and moraines, to 3100m. A.

8 RED CLOVER *Trifolium pratense.* Variable low/tall erect hairy tufted per. Lflets green, often with a *white crescent zone,* oblong-lance-shaped or rounded; stipules triangular, bristle-tipped. Fls reddish-purple, sometimes pink or white, 12-15mm, in dense rounded clusters, sometimes paired. Meadows, fields and waste places, to 2700m. May-Oct. T, widely cultivated.

9 CREAM CLOVER *Trifolium noricum.* Low/short tufted hairy per. Lvs mostly basal; lflets elliptical or oblong; stipules whitish, pointed. Fls green, 15mm, in dense rounded heads. Meadows and pastures, on calcareous soils, 1600-2600m. July-Aug. eA,Ap.

10 ZIGZAG CLOVER *Trifolium medium.* Short/med slightly hairy per; stems usually *rather zigzagged.* Lflets oval, oblong or elliptic, scarcely toothed; stipules triangular, hairless. Fls pale purple-red, 12-20mm, in dense rounded or oval clusters. Poor grassland, woods and scrub, usually on calcareous or clayish soils, to 2100m. May-July. T. **10a** *T. alpestre* has small *unstalked* flheads and purple fls; stipules linear to narrow-lance-shaped, *hairy at top.* To 2300m. P,A,Ap.

Pea Family (contd.)

1 RED TREFOIL *Trifolium rubens.* Short *hairless,* erect, per. Lflets oblong lance-shaped or elliptical, sharply toothed; stipules oval or lance-shaped. Fls *purple,* sometimes white, 15mm, in dense oblong heads. Dry open woods, scrub and stony places, to 2050m. June-Aug. P,A,n,cAp.

2 HUNGARIAN TREFOIL *Trifolium pannonicum.* Short/med erect, hairy, per. Lflets oblong-lance-shaped or elliptical; stipules *linear.* Fls yellowish-white, 20-25mm, in dense rounded or oblong, *long-stalked,* clusters. Meadows and open scrub, to 1200m. May-Aug. sw,sA,MA,Ap. **2a Sulphur Clover** *T. ochroleucon* has smaller fls, 15-20mm, in *short-stalked* clusters. Shady and damp places, to 1800m. T−exceptS.

3 ALPINE BIRDSFOOT TREFOIL *Lotus alpinus* (= *L. corniculatus* var. *alpinus*). Low, rather tufted, hairless per. Lvs *pinnate,* lflets 5, lance-shaped or rounded. Fls yellow marked red, 12-18mm, in stalked-clusters of 1-3. Meadows, grassy and stony places, 2000-3100m. May-Aug. P,A. **3a Birdsfoot Trefoil** *Lotus corniculatus* □ agg. is a much *larger* sprawling plant, with 3-6-fld clusters. To 1600m. May-Sept. T.

4 MOUNTAIN KIDNEY-VETCH *Anthyllis montana.* Short densely-tufted *sub-shrub,* hairy. Lvs pinnate, lflets narrow-elliptical to narrow-oblong, untoothed. Fls purple, in dense rounded, stalked, clusters; bracts shorter than fls. Meadows, rocky and stony places, usually on limestone, to 2400m. June-July. A(except se),Ap. **4a** *A.m. jacquinii* has *pink* fls and bracts *longer* than the fls. e,seA.

5 COMMON KIDNEY-VETCH *Anthyllis vulneraria.* Variable low/med, silkily-hairy, bien/per. Lvs pinnate, the lowest with 5-7 lflets, the end lflet *larger* than the others. Fls yellow or red, in rounded clusters; sepal tube red-tipped. Dry meadows, rocky and stony places, usually on limestone, to 3000m. May-Aug. B,S. The following subsp. are recognised – **5a** *A.v. vulnerarioides* like 5 but lowest lvs with 1-5 lflets; fls yellow; calyx red-tipped, 7-9mm. P,swA,cAp. **5b** *A.v. forondae* is like 5a but *more robust,* lowest lvs with 9-13 lflets; calyx 11-13mm. P,swA. **5c** *A.v. carpatica* is less hairy than 5a and all lvs with 1-7 lflets. Fls pale or deep yellow or reddish; calyx *white,* 9-12mm. w,c,sA. **5d** *A.v. alpestris* like 5c but calyx *longer,* 13-15mm, with greyish-black hairs. A. **5e** *A.v. pyrenaica* like 5c but calyx *red-tipped;* fls pink or red. P.

6 FALSE SENNA *Coronilla emerus.* Small loosely-branched deciduous shrub, 1-2m; twigs green. Lvs pinnate, *greyish-green;* lflets 2-4 pairs, oval, broadest above the middle. Fls pale yellow, often red-tipped, 14-20mm, in stalked-clusters of 2-4. Pod linear, pointed, 5-11cm long. Rocky places, woodland edges, to 1800m. Apr-May. P,A,Ap.

7 SMALL SCORPION VETCH *Coronilla vaginalis.* Small deciduous shrub to 0·5m. Lvs pinnate, green; lflets 2-6 pairs, oblong to rounded, *white margined.* Fls small, yellow, 6-10mm, in 4-10fld clusters. Pod linear-oblong, 1·5-3·5cm, *constricted* between seeds. Dry grassland, scrub and open woods, on limestone, to 2250m. June-Aug. A,Ap.

8 HORSESHOE VETCH *Hippocrepis comosa.* Short hairy per. Lvs pinnate; lflets 3-8 pairs, linear to oblong. Fls yellow, 6-10mm, in long-stalked clusters. Pod twisting, with *horseshoe-shaped segments.* Meadows and stony places, to 2800m. Apr-June. T−except S.

3a Birdsfoot Trefoil

Pea Family (contd.)

1 ALPINE SAINFOIN *Hedysarum hedysaroides.* Low/short per, hairless or almost so. Lvs pinnate; Iflets 4-6 pairs, oval or elliptical. Fls reddish-violet, 13-25 mm, in long-stalked spikes. Pod oblong, *constricted between* the seeds into 2-5 segments. Pastures, stony places and screes, to 2800 m. July-Aug. P, A. **1a** *H.h. exaltatum* is taller with 6-10 pairs of Iflets. sA.

2 WHITE SAINFOIN *Hedysarum boutignyanum.* Med erect hairless per. Lvs pinnate; Iflets 4-8 pairs, elliptical to oval, broadest above the middle. Fls *cream or white,* sometimes bluish-veined, 15-25 mm, in long-stalked spikes. Pod oblong, constricted between the seeds into 2-5 segments. Pastures and stony places, to 2800 m. July-Aug. swA.

SAINFOINS *Onobrychis* have short fruits, not splitting or jointed, with *netted sides* and often *toothed* along the edges.

3 SILVERY SAINFOIN *Onobrychis argentea hispanica.* Short silkily-hairy per. Lvs pinnate; Iflets 5-8 pairs, elliptical to narrowly-oblong. Fls pink with darker veins, 10-14 mm, in long-stalked spikes. Pod rounded, 4-8 mm, *hairy.* Meadows, rocky and stony places, to 2000 m. June-Aug. P. **3a** *O. pyrenaica* is *smaller* and less hairy; fls 8-10 mm long. P. **3b Rock Sainfoin** *O. saxatilis* is more tufted and *white downy* with 6-15 pairs of Iflets; fls pale yellow veined with pink. Pod *not tooth-edged.* To 1800 m. eP, swA.

4 MOUNTAIN SAINFOIN *Onobrychis montana.* Short/med *slightly hairy* per. Lvs pinnate; Iflets 5-8 pairs, elliptical to oval or oblong. Fls pink with purple veins, 10-14 mm, in long-stalked spikes. Pod rounded, 7-12 mm, tooth-edged. Pastures and stony places, 1400-2500 m. July. A, n, cAp.

5 SMALL SAINFOIN *Onobrychis arenaria.* Short/tall erect hairy or hairless per. Lvs pinnate; Iflets 3-12 pairs, linear-oblong to elliptical. Fls pink with purple veins, *smaller* than 4, 8-10 mm, in long-stalked spikes. Pod *small,* rounded, 4-6 mm, tooth-edged. Pastures and stony places, to 2500 m. June-Sept. A, n, cAp. **5a** *O.a. taurerica* has larger fls, 10-12 mm, the standard petal *very* pale pink on back. seA.

Wood-sorrel Family   Oxalidaceae

Perennials with trifoliate lvs. Fls 5-parted, wide cup-shaped; stamens 10; opening in sunshine. Fr a capsule.

6 WOOD-SORREL *Oxalis acetosella.* Low creeping slightly hairy per. Lvs bright green; Iflets heart-shaped. Fls white, veined with lilac, sometimes tinged purple, *solitary.* Woods and shady places, to 2100 m. Apr-July. T.

7 YELLOW OXALIS *Oxalis corniculata.* Low/short creeping per; stems rooting at lf junctions. Lvs green, often purplish tinted, Iflets heart-shaped. Fls yellow, 4-7 mm, in *long-stalked umbels.* Dry open and bare places and cultivated ground. Occasional weed above 1000 m. May-Oct. sS, P, A, Ap.

4 Mountain Sainfoin

Geranium Family Geraniaceae

Hairy perennials, sometimes annuals, with opposite or alternate palmately (*Geranium*) or pinnately (*Erodium*) cut or lobed lvs. Fls cup-shaped, long-stalked with 5 petals and sepals and 10 stamens. Fr of 5-basal seeds with a long 'cranesbill' beak, which splits apart when ripe.

1 ROCK CRANESBILL *Geranium macrorrhizum.* Short/med per, *strongly aromatic* when crushed. Lvs deeply cut into 5-7 lobes. Fls pink to purplish-red, 20-25 mm, in pairs or small clusters; petals rounded. Shaded limestone rocks and crevices, to 2500 m. July-Aug. MA, s, seA, Ap.

2 ASHY CRANESBILL *Geranium cinereum.* Low tufted per. Lvs all basal, cut almost to middle into 5-7 lobes. Fls in pairs, pale lilac with darker veins, 25-30 mm, petals slightly notched. Grassy or rocky places, 1500-2400 m. July-Aug. w, cP. **2a** *G.c. subcaulescens* has larger lvs and reddish-purple fls. c, sAp. **2b** *G. argenteum* similar to 2 but lvs *silvery-grey* and the fls pale rose-pink. Limestone rocks and screes 1700-2200 m. July-Aug. AA, s, swA, nAp.

3 BLOODY CRANESBILL *Geranium sanguineum.* Low/short tufted per. Lvs mostly on the stems, cut almost to middle into 5-7 lobes. Fls bright reddish-purple, rarely pink or white, solitary or paired; petals slightly notched. Dry rocky or sandy places and open woods, to 1900 m. June-Sept. T.

4 MEADOW CRANESBILL *Geranium pratense.* Med/tall per. Lvs cut to the base into 5-7 lobes. Fls in pairs, bright violet-blue, 25-30 mm; petals *rounded.* Fl stalks *bent down* after flg, but erect when fr ripens. Meadows and grassy places, usually on limestone, to 1900 m. June-Sept. T−except sAp.

5 WOOD CRANESBILL *Geranium sylvaticum.* Low/med per. Lvs like 4. Fls in pairs, blue, reddish-mauve or pink, 12-25 mm; petals rounded. Fl stalks *always upright.* Meadows, woods and stony places, usually on acid soils, to 2400 m. June-Aug. T. **5a** *G.s. rivulare* □ has *white fls* veined with red. w, sA.

6 WESTERN CRANESBILL *Geranium endressii.* Med/tall per. Lvs cut almost to middle into 5-lobes. Fls solitary pink, 24-28 mm, long-stalked; petals not or slightly notched. Wet meadows and streambanks, to 1200 m. June-July. wP, (B).

7 KNOTTED CRANESBILL *Geranium nodosum.* Low/med per. Lvs cut *half*-way into 3-5 lobes. Fls bright pink or violet with *darker veins,* 20-30 mm, in small clusters; petals notched. Open woods, to 1600 m. May-Sept. P, A, Ap, (B).

8 DUSKY CRANESBILL *Geranium phaeum.* Short/tall per. Lvs cut half-way into 5-7 lobes. Fls in pairs, *blackish-purple to brownish-purple,* 16-20 mm; petals slightly notched, *often pointed.* Damp meadows and open woods, to 2400 m. May-Aug. P, A, n, cAp, (B).

9 MARSH CRANESBILL *Geranium palustre.* Short/med per. Lvs cut almost to middle into 5-7, wedge-shaped, lobes. Fls in pairs, *pale purple* or reddish-purple, 20-30 mm; petals not or slightly notched. Fr stalk bent down. Damp meadows, to 1500 m. June-Aug. S, P, A, nAp.

10 SPREADING CRANESBILL *Geranium divaricatum.* Short/med branched ann. Lvs cut *almost to middle* into 3-5 lobes. Fls pink, 8-12 mm, in small clusters; petals notched. Woods, hedgerows and stony places, to 2000 m. May-Aug. P, w, sA, nAp.

5a *Geranium sylvaticum rivulare*

Geranium Family *(contd.)*

1 PYRENEAN CRANESBILL *Geranium pyrenaicum.* Short/med branched ann. Lvs cut *halfway* into 5-7 lobes. Fls in pairs, purplish-pink or lilac, 14-18mm; petals *deeply notched.* Meadows, open woods and waste places, to 1900m. May-Oct. T,(B).

2 DOVESFOOT CRANESBILL *Geranium molle.* Low/short, rather sprawling, *grey-green* ann/bien; stems *long-hairy.* Lower lvs cut less than half-way into 5-7 lobes; upper lvs alternate, more deeply cut. Fls in pairs, pinkish-purple, 5-12mm; petals deeply notched. Dry grassy and waste places, to 2000m. Apr-Sept. T. **2a** *G. pusillum* □ has stems with short hairs and pale lilac fls, 4-7mm. T. **2b** *G. rotundifolium* □ like 2 but lvs rounded, *scarcely cut;* fls pink, 9-13mm, petal not or only slightly notched. To 1600m. T−except S.

3 HERB ROBERT *Geranium robertianum.* Short/med ann/bien, rather strong smelling, often reddish. Lvs cut to middle into 3-5 *stalked* lobes. Fls in pairs, bright pink, 15-22mm; petals blunt-ended, pollen orange. Shady places, rocks and banks, to 2000m. May-Sept. T. **3a** *G. purpureum* has *smaller* purplish-pink fls, 8-16mm, and yellow pollen. B,P,w,sA,Ap.

4 COMMON STORKSBILL *Erodium cicutarium.* Low/med ann/bien, often stickily-hairy. Lvs 2-pinnate. Fls purplish-pink, lilac or white, 7-18mm, in long-stalked umbel-like heads; petals oval-elliptical, often unequal. Fields and waste places, to 2100m. June-Sept. T.

5 ROCK STORKSBILL *Erodium petraeum.* Low/short tufted, *stemless,* per; strong-smelling. Lvs 2-pinnate, ferny, greyish or silvery. Fls pink, 15-25mm, in umbel-like heads of two to five, rarely solitary. Rocky and stony places on limestone, to 1700m. May-June. cP. **5a** *E.p. lucidum* □ has more or less hairless, *shiny-green,* lvs and white fls veined with pink. Generally on acid rocks. e,cP. **5b** *E.p. glandulosum* has very hairy, sticky lvs and violet or purple fls, one or two upper petals larger with a *black blotch* at base. P. **5c** *E.p. crispum* □ like 5b but fls white, pale pink or lilac with red or purple veins, upper petals blotched; lvs *not* sticky. eP.

6 LARGE PURPLE STORKSBILL *Erodium manescavi.* Low/short tufted, *stemless* per. Lvs large, 2-pinnate, hairy. Fls *large,* purple or carmine-red, 25-35mm, in umbel-like heads of five to twenty. Bracts *rounded.* Meadows and pastures, to 2300m. July-Aug. c,wP.

7 ALPINE STORKSBILL *Erodium alpinum.* Low, slightly hairy, per, with *erect stems.* Lvs 2-pinnate, green or greyish. Fls violet, 14-24mm, in umbel-like heads of two to nine; petals rounded or wedge-shaped. Bracts narrow-lance-shaped or triangular. Stony places to 2000m. July-Aug. n,cAp.

2a *Geranium pusillum*   **2b** *Geranium rotundifolium*

Spurge Family Euphorbiaceae

SPURGES *Euphorbia*. Stems with milky juice. Lvs alternate and usually stalkless and untoothed. Fls small in broad, branched, umbel-like clusters, each surrounded by two conspicuous usually green or yellowish-green, floral bracts; no sepals or petals, but a solitary 3-styled female fl surrounded by several 1-stamened male fls partly enclosed in a calyx-like cup with 4-5 glands along its perimeter. Fr a 3-parted capsule. **Poisonous.**

1 IRISH SPURGE *Euphorbia hyberna*. Med little-branched per. Lvs oblong, downy beneath, turning pinkish-red. Fls and floral bracts *yellowish-green;* glands kidney-shaped. Fr stalked, covered in long and short warts. Damp and shady places, to 2000 m. Apr-July. B, P, s, swA. **1a** *E.h. canuti* has fr capsule *hardly stalked.* MA.

2 CARNIAN SPURGE *Euphorbia carniolica*. Short/med hairy or hairless per, forming tufts; stems *scaly* at base. Lvs oblong, large 40-70 mm, narrowed to the base. Fl cups long-stalked; glands *brownish-yellow.* Fr covered with short warts. Woods and shrubby slopes, to 1900 m. Apr-June. e, seA.

3 PYRENEAN SPURGE *Euphorbia chamaebuxus*. Low hairless creeping per; stems erect, scaly at base. Lvs small, 10-30 mm, elliptical to oblong, broadest above the middle. Umbels few fld, fls *often solitary,* glands 4, *reddish.* Fr covered with flap-like warts. Rocks and screes, to 2000 m. wP.

4 GLAUCOUS or BLUE SPURGE *Euphorbia myrsinites*. Low/short *bluish or greyish-green,* hairless, per; stems rather sprawling, thick, fleshy. Lvs crowded, oval to *almost rounded* with a pointed tip. Fls yellowish-green, glands with long *club-shaped horns,* often reddish-brown. Fr smooth. Rocky and grassy places, to 2260 m. Mar-June. Ap.

5 ROCK SPURGE *Euphorbia saxatilis*. Low/short rather bluish-green, hairless, creeping per. Lvs small, 2-25 mm, in *basal rosettes* and along stems; rosette lvs narrow-oblong, the others oblong to almost rounded. Fls greenish-yellow; glands with two horns. Fr smooth. Limestone rocks and screes, to 1200 m. June-July. e, seA. **5a** *E. kerneri* is taller with broader lvs *not notched* at the end. seA.

6 VALLINO'S SPURGE *Euphorbia valliniana*. Rather like 5 but a more bluish-green per *without* basal lf rosettes. Lvs small, 3-19 mm, rounded to oblong. Fls greenish-yellow, glands kidney-shaped, *not horned.* Fr almost smooth. Rocky calcareous slopes and screes, to 2000 m. swA – France and Italy.

7 CYPRESS SPURGE *Euphorbia cyparissias*. Short/med tufted, bright greenish-yellow, hairless per. Lvs *linear,* crowded, becoming red eventually. Flheads *golden-yellow;* glands with short horns, brownish. Fr almost smooth. Grassy, rocky and waste places, to 2650 m. Apr-June. T, (B, S).

8 ANNUAL MERCURY *Mercurialis annua*. Short/med slightly hairy or hairless ann; stems branched. Lvs *opposite,* oval to elliptical, *toothed,* short-stalked. Fls tiny, green, in stalked or unstalked tassels, male and female often on separate plants. Waste places, banks and cultivated ground, to 1800 m. Apr-Nov. T, (S).

9 DOG'S MERCURY *Mercurialis perennis*. Low/short unbranched, downy, foetid per, often forming drifts. Lvs opposite, *crowded* towards stem tops, lance-shaped, toothed. Fls tiny greenish, male in long, stalked, tassels, female in small almost stalkless clusters, on separate plants. Woods and shady places, to 1800 m. Feb-May. T. **9a** *M. ovata* has broader more rounded lvs, *not crowded* at stem tops. A.

4 Glaucous Spurge

Flax Family Linaceae

Slender perennials or annuals with untoothed lvs; no stipules. Fls with 5-petals overlapping spirally in bud and 5-sepals; opening in sunshine. Fr a capsule.

1 PERENNIAL FLAX *Linum perenne* agg. Short/med hairless, tufted, per. Lvs alternate, linear to narrow lance-shaped, 1-veined. Fls in branched clusters, pale to bright blue, 18-25 mm, flstalks *erect;* inner sepal often blunt. Dry grassy places, to 1700 m. May-Aug. B, P, A, Ap. **1a, 1b**, p. 337.

2 PYRENEAN FLAX *Linum suffruticosum salsoloides.* Low/short rather sprawling, hairless, per, stem branched, *some none-flowering.* Lvs alternate *linear,* greyish-green, margins rolled *under.* Fls white with a pink or violet centre, 20-30 mm. Grassy and rocky places, to 1750 m. May-July. P, Cev, MA.

3 YELLOW FLAX *Linum flavum.* Variable med hairless, tufted per. Lvs alternate, spoon-shaped to lance-shaped, 3-veined. Fls in branched clusters, *yellow,* 25-30 mm. Dry grassy places, to 1600 m. May-Aug. P, A.

4 STICKY FLAX *Linum viscosum.* Med *hairy* per, forming tufts, upper lvs and bracts *sticky.* Lvs velvety, oval to lance-shaped, 3-5-veined. Fls in branched clusters, pink, 25-35 mm. Meadows and grassy places, to 1900 m. June-Aug. P, sA, Ap.

5 PURGING FLAX *Linum catharticum.* Low/short, slender, hairless ann. Lvs *opposite,* oblong to lance-shaped, 1-veined. Fls small, in loose branched clusters, white with a yellow centre, 4-6 mm. Grassy places, especially on lime, to 2350 m. May-Sept. T.

Milkwort Family Polygalaceae

Hairless perennials, with usually alternate, stalkless, untoothed lvs; no stipules. Fls with 5-sepals, the inner 2 large and petal-like (called wings) on either side of the 3 true petals (the keel), which are joined at the base; stamens 8 are often fused into a tube. Fr flat, heart-shaped, often winged.

6 SHRUBBY MILKWORT *Polygala chamaebuxus.* Low more or less mat-forming, *evergreen subshrub.* Lvs oval to lance-shaped, shiny green, leathery. Fls with a yellow keel and white, pink or purple wings, 12-18 mm, in lfy spikes. Woods, pastures and rocky slopes, to 2500 m. Apr-Sept. A, Cev, Ap.

7 PYRENEAN MILKWORT *P. vayredae.* Like 6 but lvs *linear* to *narrow* lance-shaped and fls pinkish-purple. Seldom above 800 m. Apr-May. eP–Spanish.

8 NICE MILKWORT *Polygala nicaeensis* agg. Variable low/short per, somewhat sprawling but with erect flstems. Lvs lance-shaped to linear, the lower broadest above the middle. Fls blue or pink, 8-11 mm, in slender spikes; wings *3-5-veined.* Dry grassy and stony places, to 1700 m. Apr-July. P, Cev, A, Ap.

9 TUFTED MILKWORT *Polygala comosa.* Rather like 8 in habit and lf, but lower lvs *falling* before fls open. Fls lilac-pink, 4-6 mm, in long spikes; wings 1-3 veined. Dry grassy places and open woods, to 2200 m. May-July. P, A.

10 MOUNTAIN MILKWORT *Polygala alpestris.* Low/short slender tufted per. Lvs lance-shaped, *increasing in size upwards,* the uppermost crowded. Fls blue or white, 4-6 mm; wings 1-3-veined. Meadows, to 2200 m. May-Aug. P, A, Ap.

11 THYME-LEAVED MILKWORT *Polygala serpyllifolia.* Low/short slender per. Lower lvs *opposite,* oblong to elliptical, the upper lvs narrow lance-shaped, usually alternate. Fls usually blue, 4·5-5·5 mm, the upper petals *longer* than the wings. Acid meadows, to 1800 m. May-Aug. B, P, Cev, A. **11a**, p. 337.

12 BITTER MILKWORT *Polygala amara.* Low/short per with numerous stem arising from a basal lfy rosette. Lvs *bitter,* elliptical to oblong; stem lvs pointed, alternate. Fls blue, violet, pink or white, 4·5-8 mm, in long spikes. Damp meadows on lime, to 2600 m. Apr-July. e, seA. **12a**, p. 337.

13 ALPINE MILKWORT *Polygala alpina.* Low prostrate per with *lfy rosettes.* Lvs *not* bitter, oblong to linear, stem lvs much smaller, alternate. Fls bright blue, 4-4·5 mm, in short spikes. Pastures, usually on limestone, 1500-3000 m. July-Aug. P, w, swA*.

Mallow Family Malvaceae

Rather leafy, downy, or softly hairy, annuals or perennials. Fls in clusters at base of lvs, or solitary; petals five, notched; sepals in two rings; stamens numerous, bunched together on the end of a short stalk. Fr a flat disc, surrounded by sepals.

1 CUT-LEAVED MALLOW *Malva alcea*. Med/tall erect per; stems with *branched hairs.* Lvs rounded-heart-shaped, with five, toothed lobes; upper lvs more deeply cut and further lobed. Fls large, *solitary,* bright-pink 3·5-6cm; outer sepals *broad.* Grassy and scrubby places, woods, to 2000m. June-Sept. T−except S. **1a** *M. moschata* ☐ has more deeply cut lvs with 5-7 narrow lobes; stems with *unbranched* hairs; outer sepals *narrow.* To 1500m. T.

2 COMMON MALLOW *Malva sylvestris.* Med/tall bien or per; stems with branched or unbranched hairs. Lvs rounded or kidney-shaped, with 3-7, toothed, lobes. Fls pink or purple with darker veins, 2-5cm, *two or more* together. Meadows and waste places, to 1800m. May-Sept. T.

3 DWARF MALLOW *Malva neglecta.* Rather sprawling or *prostrate* ann. Lvs rounded or kidney-shaped, with 5-7, toothed, lobes. Fls pale lilac to whitish, 1·5-2cm, in clusters of three to six. Fields and waste places, to 1900m. May-Sept. T.

Balsam Family Balsaminaceae

Succulent hairless annuals with alternate lvs and spurred fls. Fr exploding when ripe.

4 TOUCH-ME-NOT *Impatiens noli-tangere.* Short/tall ann. Lvs oval-elliptical to lance-shaped, stalked. Fls large yellow with brownish spots, 3-6 in *short-stalked* clusters; spur *downcurved,* 10-20 mm. Damp shady places, to 1500 m. July-Sept. T. **4a Small Balsam** *I. parviflora* has small pale-yellow fls with *short* 3-5 mm spurs. (T), native of Central Asia.

Oleaster Family Elaeagnaceae

5 SEA BUCKTHORN *Hippophae rhamnoides.* Deciduous shrub to 4m, branches brown, thorny. Lvs narrow-lance-shaped, untoothed, silvery-grey when young. Fls before lvs, tiny, greenish-yellow, no petals; male and female on different plants. Fr a bright orange berry, edible. River gravels and alluvium, to 2000m. Apr-May. T.

Daphne Family Thymelaeaceae

Small shrubs, sometimes herbs, with alternate, untoothed lvs. Fls tubular with four spreading sepal-lobes, no petals. Fr a fleshy berry or a small nut.

6 ANNUAL THYMELAEA *Thymelaea passerina.* Low/med erect, usually hairless, ann. Lvs narrow-lance-shaped, pointed. Fls tiny, greenish, solitary or two to three together. Fr a hairy nut. Dry rocky and waste places, to 1200m. July-Sept. P, Cev, A, Ap.

7 TWISTED THYMELAEA *Thymelaea tinctoria* (= *Passerina tinctoria*). Dwarf shrub to 50cm, stems rough, *twisted; young shoots* hairy. Lvs narrow-oblong. Fls solitary, yellow 5-6mm long. Fr hairless. Rocky woods and scrub on limestone, to 1200m. Apr-June. P. **7a** *T. dioica* has narrower lvs, *hairless* shoots and *fls* 6-9mm long; fr hairy. Limestone rocks, to 2000m. May-June. P, swA*−French, MA. **7b** *T. calycina* like 7 but lf margin rolled under towards tip. Stony places, to 2500m. June-Sept. w, cP. **7c Hairy Thymelaea** *Thymelaea pubescens.* Short, hairy per, base *woody.* Lvs narrow to broadly-elliptical. Fls yellow, 6mm, in *clusters* of two or three. Fr hairless. Dry rocky places, to 1700m. June-July. eP.

4a Small Balsam

1

1a

2

3

4

5

6

7

Daphne Family *(contd.)*

1 MEZEREON *Daphne mezereum. Deciduous* shrub to 1 m. Lvs oblong-lance-shaped, pale green. Fls pinkish-purple, 7-9 mm long, in clusters of two to four, fragrant. Fr a bright-red shiny berry. Woods and pastures, usually on limestone, to 2600 m. Feb-July. T.

2 SPURGE LAUREL *Daphne laureola. Evergreen* upright shrub to 1 m. Lvs leathery oblong-oval, broadest above middle, shiny dark green. Fls *yellowish-green*, 6-10 mm, in drooping clusters, slightly fragrant. Fr a *black* berry. Woods and clearings, to 1600 m. Jan-Apr. T–except S. **2a** *D.l. philippi* is a lower, more spreading, shrub with smaller fls, 5-6 mm long. P.

3 ALPINE MEZEREON *Daphne alpina.* Dwarf deciduous shrub to 50 cm, branches twisted. Lvs narrow-oblong, greyish-green, *clustered* at shoot tips. Fls *white*, 4-6 mm, in small clusters. Limestone rocks, to 2000 m. Apr-June. eA, nAp.

4 GARLAND FLOWER *Daphne cneorum.* Evergreen *prostrate* bush, branches straight, young shoots *hairy.* Lvs oblong or narrow spoon-shaped, blunt. Fls pink with a 6-10 mm tube, petal-lobes rounded, in clusters of six to ten, fragrant. Fr a brownish-yellow berry. Stony places and pastures, usually on limestone, to 2150 m. Apr-Aug. P, A, nAp. **4a** *D. striata* □ is *completely* hairless, the lvs crowded at shoot tips; fls reddish-purple. 1500-2900 m. May-Aug. A.

5 ROCK MEZEREON *Daphne petraea.* Dwarf mat-forming evergreen shrub, branches *twisted.* Lvs lance or wedge-shaped, keeled below. Fls glistening, bright pink, with a *hairy,* 9-15 mm long tube, in clusters of three to five, fragrant. Limestone rock crevices, to 2000 m. June-July. sA–Lake Garda area.

Rockrose Family   Cistaceae

Small thinly branched subshrubs with opposite, untoothed lvs. Fls 5-petalled, open cups in short one-sided racemes; sepal unequal; stamens numerous. Fr egg-shaped.

6 COMMON ROCKROSE *Helianthemum nummularium.* Very variable, more or less prostrate, subshrub. Lvs oblong, lance-shaped or oval, *downy-white* beneath, one-veined. Fls yellow, orange or white, 14-22 mm. Dry meadows and rocky places, usually on limestone, to 2800 m. June-Sept. T. There are a number of subspecies that can be recognised. **6a** *H.n. glabrum* □ Lvs green above and below, hairless; fls yellow. P, A, Cev. **6b** *H.n. semiglabrum* □ like 6a but fls pink. MA, nAp. **6c** *H.n. grandiflorum* □ like 6a but lvs hairy. P, A, Cev. **6d** *H.n. pyrenaicum* □. Lvs whitish-hairy beneath like 6 but *fls pink.* P. **6e** *H.n. berterianum* □ like 6d but lvs greyish-hairy beneath. MA, Ap. **6f** *H.n. ovatum* □ Like 6c but fls small, S, A.

7 APENNINE or WHITE ROCKROSE *Helianthemum apenninum.* Loose spreading subshrub, lvs linear to narrow-oblong, grey-white hairy beneath, margin *rolled under.* Fls white, with a *yellow base* to each petal, 18-20 mm. Grassy and stony places, on limestone, to 1800 m. May-July. B*, S, P, w, sA, Ap.

8 ALPINE ROCKROSE *Helianthemum oelandicum* subsp. *alpestre* (= *H. alpestris*). Dwarf shrub to 20 cm. Lvs oblong to narrow-lance-shaped, green, *usually hairless.* Fls yellow, 14-20 mm. Dry meadows and stony places, often on limestone, 1600-2900 m. May-Sept. T–except B.

9 HOARY ROCKROSE *Helianthemum canum.* Erect or sprawling subshrub to 20 cm. Lvs elliptical to narrow-lance-shaped, *clustered* at shoot tips, green or grey-hairy above, grey-white hairy below. Fls small, yellow, 8-15 mm. Dry meadows and stony places on limestone, to 1650 m. May-July. T. **9a** *H.c. piloselloides* has *broader* lvs spaced *all along* the non flg shoots. P.

10 SHRUBBY ROCKROSE *Helianthemum lunulatum.* Dwarf subshrub to 20 cm, stems *twisted,* prickly when old. Lvs elliptical-lance-shaped, green, slightly hairy. Fls yellow, petals with an *orange-base,* 14-16 mm, solitary or two to three. Rocky places, to 1600 m. June-Aug. MA–Italian.

St. John's Wort Family Hypericaceae

Hairless, rarely hairy, perennials with opposite or whorled, untoothed, lvs which have transparent veins, often gland dotted, usually stalkless. Fls yellow, in branched clusters, sometimes solitary; petals and sepals 5, stamens numerous. Fr a capsule.

1 YELLOW CORIS *Hypericum coris.* Low/med subshrubby per, stems erect. Lvs *in whorls* of four, linear, the margins rolled under. Fls bright yellow, 20mm, in elongated clusters, rarely solitary. Sunny limestone rocks, to 2000m. May-July. sw,wA,n,cAp.

2 WESTERN ST. JOHN'S WORT *Hypericum nummularium.* Short creeping per, stems erect. Lvs opposite, oval or rounded, bluish-green beneath, *with two* black dots near top. Fls yellow, sometimes red veined, 20-30mm, in small clusters or solitary. Limestone rocks and crevices, to 2500m. June-Sept. P,swA – French. **2a** *H. hirsutum* □ is *downy* and with narrower lvs. Woods and riverbanks, to 1600m. T.

3 MOUNTAIN ST. JOHN'S WORT *Hypericum montanum.* Short/tall erect per. Lvs opposite, oval to oblong-elliptical, margin beneath with *a row* of black dots. Fls pale yellow, 10-15mm, in flat-topped clusters, fragrant. Fields, woods and thickets, often on limestone, to 1900m. June-Sept. T.

4 ALPINE ST. JOHN'S WORT *Hypericum richeri.* Low/med erect per, patchforming. Lvs opposite, oval to elliptical or almost triangular. Fls bright yellow, 20-25mm, in flat-topped clusters; petals *covered* by tiny black dots. Meadows, screes and woods, usually on limestone, to 2500m. June-Sept. sw,cA,J,Ap. **4a** *H.r. burseri* (= *H. burseri*) has lvs *clasping* stem at their base and *larger fls,* 30-45mm. P.

5 IMPERFORATE ST. JOHN'S WORT *Hypericum maculatum.* Short/med per; stems four-sided, erect. Lvs opposite oval or oblong, *without* translucent dots usually. Fls golden yellow, 20mm in branched clusters; petals with tiny black dots and streaks in centre. Damp meadows, wood margins and stream banks, to 2650m. June-Sept. T.

6 PERFORATE ST. JOHN'S WORT *Hypericum perforatum* agg. Short/tall per; stems erect with *two raised lines* running down them. Lvs oval to linear with numerous *translucent dots.* Fls golden-yellow, 20mm; petals with tiny dots in the centre and along margin. Dry fields, woods and scrub, to 2000m. May-Sept. T.

7 TRAILING ST. JOHN'S WORT *Hypericum humifusum.* Slender, *usually prostrate,* per. Lvs opposite oblong or lance-shaped, *with* translucent dots. Fls small, yellow, 8-10mm. Open woods and scrub on acid rocks, to 1800m. July-Oct. T probably.

Tamarix Family Tamaricaceae

8 MYRICARIA *Myricaria germanica.* Hairless evergreen shrub, to 2m. Lvs *tiny, scale-like,* overlapping along the slender stems, bluish-green. Fls pink, 5-6mm long, in dense, *catkin-like,* spikes. Seeds with fluffy tufts of hairs. Rocky places and gravels along rivers and streams, to 2400m. May-Aug. S,P,A,n,cAp.

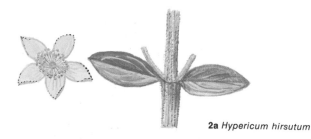

2a *Hypericum hirsutum*

143

Violet Family Violaceae

VIOLETS and PANSIES are tuft forming perennials, sometimes annuals, with or without stolons. Lvs alternate, often in basal clusters, toothed, with *stipules* at base. Fls solitary with 5 petals, the lower-most largest, lip-like and *spurred behind.* Fr a small 3-valved capsule, splitting. Hybrids frequently occur.

1 SWEET VIOLET *Viola odorata.* Low creeping hairy per, with long *rooting runners.* Lvs round-heart-shaped, long-stalked; stipules oval. Fls dark violet or white, 15mm, *fragrant;* sepals blunt. Woods scrub and hedgerows, to 1400m. Mar-May and sometimes Aug-Sept. T. **1a** *V. suavis* has short thick runners; fls larger, violet with a *white throat.* T.

2 WHITE VIOLET *Viola alba scotophylla.* Variable low creeping, slightly hairy per, with long *non-rooting* runners. Lvs dark green, oval or triangular-heart-shaped, long-stalked; stipules narrow lance-shaped, *hairy edged.* Fls white, sometimes violet, 15-20mm, fragrant. Woods and hedgerows, to 1200m. Mar-June. c,s,eA.

3 HAIRY VIOLET *Viola hirta.* Low downy per, *no runners.* Lvs pale green, heart-shaped; stipules broad lance-shaped, hairy edged. Fls violet, 15mm, *not* fragrant; spur dark violet. Pastures and open woods, usually on calcareous soils, to 2000m. Mar-June. T−except S. **3a Hill Violet** *V. collina* has narrow stipules and pale blue fragrant fls, spur *whitish.* Mar-Apr. S,P,A.

4 AUSTRIAN VIOLET *Viola ambigua.* Low slight hairy per, no runners. Lvs *oblong-oval,* long-stalked; stipules *dark green,* broad lance-shaped, short toothed. Fls dark violet, 10-15mm, fragrant. Meadows, heaths and rocky places on acid soils, to 2300m. Apr-Aug. e,neA. **4a** *V. thomasiana* has lvs slightly heart-shaped at base and narrow lance-shaped stipules; fls lilac or almost white, to 2300m. c,sA.

5 PYRENEAN VIOLET *Viola pyrenaica.* Low slightly hairy per. Lvs *shiny-green,* heart-shaped; stipules lance-shaped, short toothed. Fls pale violet with a *white throat,* 15mm, fragrant. Meadows, open woods and rocky places, to 2250m. Mar-July. P,A,J, cAp*.

6 TEESDALE VIOLET *Viola rupestris* (= *V. arenaria*) Low tufted, hairy, per. Lvs heart-shaped, long-stalked; stipules oval-lance-shaped, toothed or not. Fls reddish-violet, pale blue or white, 10-15mm, spur short, pale violet. Dry meadows, gravels and heaths on limestone, to 3100m. Mar-July. T−B*. **6a** *V. mirabilis* is larger with pale violet *fragrant* fls, 20mm; spur *long* whitish. Woods, to 1800m. T−except B.

7 COMMON DOG VIOLET *Viola riviniana.* Variable low/short, almost hairless, tufted per. Lvs heart-shaped, long-stalked; stipules lance-shaped, toothed. Fls bluish-violet, 14-25mm, *unscented,* petals broad; spur short, whitish or pale purple. Dry grassy places and woods, to 1800m. Apr-June. T. **7a Early Dog Violet** *V. reichenbachiana* □ has narrow lvs and stipules, fls violet, darker in centre, 12-18mm, petals *narrow,* spur *deep violet.* T.

8 HEATH DOG VIOLET *Viola canina* agg. Variable low/short hairless or slightly downy per. Lvs *not in a tuft,* oval to lance-shaped, heart-shaped at base; stipules usually toothed. Fls blue or white, 15-25mm, spur white or greenish-yellow; sepals *pointed.* Open woods and heaths on acid soils, to 2500m. Apr-July. T.

9 BOG VIOLET *Violet palustris.* Low, usually hairless, slightly tufted per. Lvs *kidney-shaped;* stipules oval-lance-shaped, toothed or not. Fls pale lilac, 10-15mm, unscented. Bogs and marshes, 1200-2600m. Apr-July. T−except c,sAp.

10 YELLOW WOOD VIOLET *Viola biflora.* Low fragile, *creeping,* slightly hairy per. Lvs tufted, pale green, kidney or heart-shaped; stipules small oval, slightly toothed. Fls *bright yellow,* 15mm. Damp or shady places, to 3000m. May-Aug. T−except B.

Violet Family (contd.)

1 FINGER-LEAVED VIOLET Viola pinnata. Low slightly hairy tufted per. Lvs fan-shaped with finger-like lobes; stipules whitish, lance-shaped. Fls pale violet, 10-20mm, fragrant. Meadows, rocky places and screes, to 2500m. June-Aug. A*.

2 DIVERSE-LEAVED VIOLET Viola diversifolia. Low stiff-hairy tufted per. Lower lvs rounded to broadly-oval, untoothed, stalked, the upper narrow-oblong or oval; stipules oblong, toothed. Fls violet, 15-20mm, fragrant. Meadows and rocky places to 2600m. June-July. c,eP.

3 MT. CENIS PANSY Viola cenisia. Low hairy or hairless per. Lvs small, oval or oblong, untoothed, long-stalked; stipules like lvs, but smaller. Fls bright violet, 20-25mm; spur slender, 5-8mm long. Limestone rocks and screes, to 2900m. June-Sept. sw,cA. **3a** V. comollia has bright violet fls with an orange or deep yellow spot; spur short, 3-5mm. sA–Alpi Orobie.

4 MARITIME ALPS PANSY Viola valderia. Rather like 3 but lvs larger and stipules deeply toothed at base. Fls bright violet, 20mm; spur slender, 7-10mm. Acid rocks and screes, to 2900m. June-Sept. MA. **4a** V. magellensis has stipules slightly toothed at base and lvs with short stout stalks; fls violet or pink, the upper two petals often dark reddish-violet. Limestone pastures and screes, to 2900m. cAp.

5 ALPINE PANSY Viola alpina. Low tufted, stemless hairless per. Lvs shiny-green, oval-heart-shaped, blunt-toothed, long-stalked. Fls violet, 20-30mm; spur short, 3-4mm. Meadows and screes, on limestone, 1600-2200m. June-July. neA. **5a** V. nummulariifolia has short stems and smaller bright blue fls, 10-12mm across. Meadows and acid rocks. MA.

6 LONG-SPURRED PANSY Viola calcarata. Low tufted hairy or hairless per. Lvs rounded, oval or lance-shaped, blunt-toothed, stalked; stipules oblong, slightly toothed. Fls 1-2 together, violet, 20-30mm; spur long, 8-15mm. Meadows and screes, 1300-2400m. Apr-Oct. A,J,Ap. **6a** V.c. zoysii ☐ has yellow fls. Limestone rocks. seA–Karawanken. **6b** V.c. villarsiana has deeply toothed stipules and fls in groups of 1-4, yellow, blue or white. swA. **6c Horned Pansy** V. cornuta ☐ rather like 6 but stipules oval-triangular, deeply tooth-cut; fls fragrant, violet or lilac, 20-35mm, narrow-petalled; spur long and pointed, 10-15mm. P.

7 BERTOLONI'S PANSY Viola bertolonii (= V. heterophylla). Low/short usually hairless per. Lvs variable, rounded, oval or lance-shaped, blunt-toothed, long-stalked; stipules pinnately-lobed. Fls violet or yellow, 20-30mm; spur long, 9-12mm. Meadows, to 2150m. July-Aug. MA,nAp*.

8 DUBY'S PANSY Viola dubyana. Low hairy or hairless per. Lvs rounded, blunt-toothed, stalked, the upper linear to lance-shaped; stipules pinnately-lobed. Fls violet with a yellow central spot, 20-25mm; spur slender short, 5-6mm. Dry meadows and rocky places, on limestone screes, to 2100m. May-July. sA–Italian.

9 HEARTSEASE Viola tricolor. Variable low/short, hairless or downy ann/bien. Lvs heart-shaped to oval or lance-shaped, blunt toothed; stipules pinnately-lobed, with a long end-lobe. Fls violet or yellow or bicoloured, 10-25mm, petals narrow; spur short, 3-6·5mm. Grassy and waste places, to 2700m. Apr-Oct. T. **9a** V.t. subalpina is usually per with yellow fls, the upper petals sometimes violet. P,A,Ap. **9b** V. eugeniae has larger violet or yellow fls with broad petals; lvs rounded, long-stalked. Ap.

10 MOUNTAIN PANSY Viola lutea. Low/short hairy or hairless per, with slender creeping stems. Lvs oval, oblong or lance-shaped, toothed, long-stalked; stipules pinnately-lobed, the end lobe not larger than the others. Fls yellow, violet or bicoloured, 15-30mm; spur short, 3-6mm. Grassy and rocky places, on acid rocks, to 2000m. May-July. B,P,A. **10a** V. bubanii has violet fls with a long, 10mm, spur. P.

9 Heartsease, colour forms

147

Willowherb Family Onagraceae

Generally hairy perennials with opposite or alternate lvs. Fls in racemes with 4 sepals and 4 notched-petals (2 only in *Circaea*) and 8 stamens. The fr of Willowherbs are long narrow pods, splitting when ripe to release the seeds which are covered in silky plumes. Hybrids are frequent.

1 ALPINE ENCHANTER'S NIGHTSHADE *Circaea alpina*. Short/med slightly hairy per. Lvs opposite, *heart-shaped,* toothed, stalked. Fls tiny, white, with 2 deeply notched petals. Fr oblong, *covered in* short hooked bristles. Damp woods, stony places and streambanks, usually on acid soils, to 2200m. June-Aug. T.

2 ROSEBAY WILLOWHERB *Epilobium* (= *Chamaenerion*) *angustifolium*. Robust tall, almost hairless, patch-forming per. Lvs *alternate,* lance-shaped, slightly toothed. Fls large, bright pink-purple, 20-30mm, petals *scarcely notched.* Fr pinkish-purple. Open woods, banks and waste places, to 2500m. June-Sept. T.

3 ALPINE WILLOWHERB *Epilobium* (= *Chamaenerion*) *fleischeri*. Short/med, patch-forming, hairless per. Lvs alternate, narrow lance-shaped, slightly toothed. Fls bright pink, 20-25mm, petals oblong-elliptic, *not notched.* Gravels, moraines and riverbanks, usually on acid soils, to 2700m. July-Sept. A. **3a** *E. dodonaei* is taller with narrower, slightly hairy lvs. To 1500m. A, Ap.

4 WESTERN WILLOWHERB *Epilobium duriaei*. Short/med slightly hairy per. Lvs opposite, oval, slightly toothed, upper lvs *alternate.* Fls pink, petals notched 6-10mm long. Stigma *4-lobed.* Acid rocks and banks, to 2500m. June-Sept. P,sw,wA,J. **4a Greater Willowherb** *E. hirsutum* □ is much taller and very hairy with lvs *clasping* stem at base, fls larger, purple pink; petals 10-16mm long. Damp and waste places. T. **4b** *E. parviflorum* like 4a but lvs narrower, *not clasping;* petals 4-9mm long. T.

5 MOUNTAIN WILLOWHERB *Epilobium montanum*. Short/med slightly hairy per; stems *round.* Lvs opposite oval, toothed, upper lvs alternate. Fls purplish-pink, petals notched 6-10mm long. Stigma *4-lobed.* Shady and waste places, to 2600m. May-Aug. T. **5a** *E. collinum* is *smaller* with pale purplish-pink fls, petals 3-6mm long. T. **5b** *E. lanceolatum* is like 5 but stems slightly *4-angled* and lvs mostly alternate, longer stalked. *Fls white,* becoming pink. T.

6 WHORLED-LEAVED WILLOWHERB *Epilobium alpestre* (= *E. trigonum*). Short/med hairy per. Lvs usually *in whorls* of 3-4. Fls pink-violet, petals 6-12mm long. Stigma *club-shaped.* Rocky and waste places, open woods, to 2400m. June-Aug. P,A,Ap. **6a** *E. tetragonum* has *opposite stalkless* lvs and purplish-pink fls, petals 3-7mm long. T.

7 NODDING WILLOWHERB *Epilobium nutans*. Low/short per. Lvs *all opposite,* oval or elliptic, scarcely toothed, *stalkless.* Fls pale violet, petals 3-6mm long. Stigma club-shaped. Moors and waysides, to 2500m. June-Aug. P, A, Ap. **7a** *E. palustre* has *untoothed* lvs and pale pink or white fls. Wet places, to 2300m. T. **7b** *E. roseum* has distinctly *stalked* lvs and white fls, becoming pink streaked. T.

8 PIMPERNEL-LEAVED WILLOWHERB *Epilobium anagallidifolium* (= *E. alpinum*). Low creeping, almost hairless per. Lvs opposite, oval or elliptic, scarcely toothed, upper lvs alternate. Fls *small,* pale purplish, petals 3-4·5mm long. Stigma club-shaped. Wet places on acid soils, to 3000m. June-Sept. T.

9 CHICKWEED WILLOWHERB *Epilobium alsinifolium*. Low/short creeping, slightly hairy per, rather like 8 but larger. Lvs toothed. Fls purplish-pink, *petals* 7-11mm long. Wet places, to 2900m. June-Aug. T.

4a Greater Willowherb

1

2

3

4

5

6

7

8

9

Dogwood Family Cornaceae

Deciduous perennials, shrubs or small trees with opposite, untoothed, stalked lvs. Fls in clusters, 4-petalled. Fr a berry.

1 COMMON DOGWOOD *Cornus sanguinea* (= *Thelycrania sanguinea*). Well branched shrub to 3-4m with straight *reddish twigs.* Lvs broadly-elliptical or oval, pointed, pale green, but reddening in the autumn. Fls white 7-12mm, in dense clusters. Fr a black berry. Banks and hedgerows on calcareous soils, to 1550m. May-July. T.

2 CORNELIAN CHERRY *Cornus mas.* Much branched shrub or small tree to 7-8m with *greenish-yellow twigs.* Lvs oval or elliptical. Fls yellow, 3-4mm, in small clusters, appearing *before* the lvs. Fr a large shiny red berry, edible. Woods, shrub and banks, to 1500m. Feb-Mar. A, Ap, but widely planted elsewhere.

3 DWARF CORNEL *Chamaepericlymenum suecicum* (= *Cornus suecica*). Short creeping per, often carpeting the ground. Lvs rounded to elliptical, unstalked. Fls tiny, purplish-black in a cluster surrounded by 4 conspicuous *white petal-like bracts.* Fr a small red berry. Heather and bilberry moors and heaths, to 1200m. June-Aug. B, S.

Ivy Family Araliaceae

4 IVY *Hedera helix.* Variable evergreen creeper or climber. Lvs shiny deep green, usually triangular in outline with 3-5 shallow lobes, but those of fl shoots unlobed. Fls usually 5-petalled, pale green with yellow anthers, in *dense umbels.* Fr a small berry, black when ripe. To 1800m. Sept-Nov. T.

Carrot Family Umbelliferae

A large and readily recognised family with fls usually in flat-topped umbrella-like clusters (umbels) with all the fl stalks (rays) converging on one point, the primary rays often carrying smaller secondary umbels. Lvs alternate, without stipules. Fls small 5-petalled, the petals often of differing sizes; stamens 5. Bracts often present at base of primary rays. Fr dry, often ridged in various ways and flattened, splitting into 2 seeds – very important in identification. **(See p.348.)**

5 SANICLE *Sanicula europaea.* Short/med hairless per; stems often reddish. Basal lvs rounded in outline, *palmately* 3-5-lobed, shiny, long-stalked. Fls whitish or pale pink in clusters of small tight umbels. Fr rounded with hooked spines. Woods, often on calcareous soils, to 1600m. May-July. T.

6 HACQUETIA *Hacquetia epipactis.* Low/short hairless creeping per. Basal lvs rounded in outline, palmately *3-lobed,* the lobes wedge-shaped, toothed, shiny green. Fls yellow, in small umbels surrounded by a *ruff* of large shiny-green, toothed, bracts. Woods and shrub, to 1500m. Apr-May. eA.

MASTERWORTS *Astrantia.* Readily recognised by their small pin-cushion-like umbels surrounded by many narrow pointed, papery, petal-like bracts. Lvs 3-9-palmately lobed, toothed, the lower long-stalked.

7 GREAT MASTERWORT or MOUNTAIN SANICLE *Astrantia major.* Med/tall per, usually unbranched. Basal lvs 3-5-lobed, lobes broadly oval, coarsely toothed. Umbels 20-30mm with greenish-white or pinkish fls; bracts green and pink or purplish above, whitish beneath equalling the umbel. Meadows and woods, generally on calcareous soils, to 2000m. June-Sept. P, A, Ap, (B). **7a**, p.337.

8 LESSER MASTERWORT *Astrantia minor.* A shorter, more delicate plant than. Basal lvs 5-9-lobed, the lobes lance-shaped or elliptical, *separated* to lf centre. Umbels 10-15mm with pinkish fls; bracts thin, greenish flushed pink above, whitish beneath, exceeding the umbel. Meadows, woods and stony places, often on acid soils, to 2700m. July-Aug. P, swA, nAp.

9 BAVARIAN MASTERWORT *Astrantia bavarica.* Short/med slender per. Basal lvs 5-lobed, *only* the middle lobe separated to lf centre. Umbels 10-20mm, with white fls; bracts whitish, exceeding the umbel, *without* cross veins. Meadows, shrub and woods on calcareous soils, to 2300m. June-Aug. eA. **9a**, p.337.

2

3

4

7

5

8

9

6

Carrot Family (contd.)

1 ALPINE ERYNGO or QUEEN OF THE ALPS *Eryngium alpinum.* Med/tall erect bluish-green per, hairless. Basal lvs oval to heart-shaped, *spine-toothed,* long-stalked. Fls small steely-blue in dense egg-shaped clusters surrounded by a ruff of narrow, spiny-toothed, *violet-blue bracts.* Meadows and grassy places, usually on limestone, to 2500 m. July-Sept. A,J.

2 SILVER ERYNGO *Eryngium spinalba.* Short stout hairless per. Basal lvs rounded to heart-shaped, irregularly lobed and with large spiny-teeth, long-stalked. Fls bluish-white in egg-shaped heads surrounded by a ruff of narrow, spiny, *greenish-white bracts.* Dry rocky places and screes, usually on limestone, 1400-1700 m. June-July. swA*.

3 PYRENEAN ERYNGO *Eryngium bourgatii.* Short/med, erect, hairless per, stems usually flushed with steely-blue. Basal lvs deeply cut and lobed, spine-toothed, green with *white-patterning* along veins. Fls bluish in dense globular heads, surrounded by a ruff of steely-blue, scarcely toothed, spiny bracts. Dry rocky and stony places, to 2000 m. July-Aug. P.

4 HAIRY CHERVIL *Chaerophyllum hirsutum.* Med/tall deep-green, softly-hairy, per. Lvs 2-3-pinnate, segments wedge-shaped, pointed. Fls white to pinkish, in loose umbels; petals *edged* with tiny hairs. Fr narrow-oblong, 8-12 mm, tapering to the top. Damp meadows, woods and shady places, to 2500 m. July-Aug. Cev,A,Ap. **4a** *C. elegans* has *narrower,* long-pointed, lf segments. A. **4b** *C. villarsii* is like 4a but with fewer, stiff, hairs and larger frs, 8-20 mm. A.

5 SWEET CICELY *Myrrhis odorata.* Tall, stout, very hairy per, *aromatic.* Lvs 2-3-pinnate, segments oblong-lance-shaped, toothed. Fls white with *unequal petals* in umbels, 1-5 cm across; bracts absent. Fr narrow-oblong, beaked, 15-25 mm, dark shiny-brown when ripe. Grassy and stony places, woods, to 2000 m. June-Aug. P,A,Ap,(B,S).

6 MOLOPOSPERMUM *Molopospermum peloponnesiacum.* Tall, robust, hairless per; *unpleasant smelling.* Lower lvs large, 2-4-pinnate, segments lance-shaped, sharply-toothed. Fls white, in loose umbels; *bracts present.* Fr oval, ridged, with narrow wings, 12 mm. Stony places and scrub, usually on calcareous soils, to 2000 m. May-Aug. P,s,sw,seA.

7 BUNIUM *Bunium alpinum petraeum.* Med tuberous rooted hairless per. Basal lvs 2-3-pinnate, segments narrow-elliptical, blunt. Fls white, petals oval, *notched;* umbels 3-5-rayed; bracts linear. Fr oblong, 3·5-5·5 mm, ridged. Rocky and grassy places to 2500 m. July-Aug. cAp.

8 CONOPODIUM *Conopodium pyrenaicum.* Short/med erect, tuberous-rooted per, hairless. Basal lvs 2-3-pinnate, with *sheathing lfstalks,* segments oval, blunt. Fls white, petals oval, notched, *brown-veined* on back; umbels 6-16-rayed. Fr oblong, 3-4 mm, narrowly ridged. Grassy and rocky places, to 1500 m. June-July. wP.

1 Alpine Eryngo
basal lf

3 Pyrenean Eryngo
basal lf

2

3

4

5

6

7

8

Carrot Family *(contd.)*

1 GREATER BURNET SAXIFRAGE *Pimpinella major.* Med/tall hairless per; stems deeply grooved, hollow, branched in the upper half. Lower lvs very large, usually 1-pinnate with 3-9 segments; stem lvs smaller with *sheath-like* bases. Fls white to deep pink, in dense umbels; no bracts; rays 10-15. Fr egg-shaped, slightly flattened, 2-3·5mm, whitish with prominent ridges. Grassy, stony places and banks, to 2300m. May-Sept. T. **1a** *P. siifolia* is shorter with narrower lf segments and *narrowly-winged* frs, 5-6mm. Mountain pastures. wP. **1b Burnet Saxifrage** *P. saxifraga* ☐ is like 1 but stems solid and only slightly grooved, usually softly hairy. T.

2 DETHAWIA *Dethawia tenuifolia.* Short/med, slender stemmed, hairless per. Lvs mostly basal, 3-pinnate with crowded lobes. Fls white with elliptical petals; rays 4-10; bracts 1-3 unequal. Fr egg-shaped, 4-6mm, with prominent ridges. Rocks and screes, to 2000m. July-Aug. P.

3 PYRENEAN SESELI *Sesili nanum.* Low *cushion-forming,* hairless per. Lvs 2-pinnate with elliptical-oblong lobes. Fls white or pink in dense rounded umbels; rays 5-8; bracts absent usually. Fr oblong, flattened, *bristly.* Rock crevices and screes, to 2400m. July-Sept. P.

4 ATHAMANTA *Athamanta cretensis.* Med hairy per; stems erect. Lvs 3-5-pinnate with linear to narrow-oblong lobes. Fls white, the *petals notched;* rays 5-15; bracts present or absent, sometimes pinnately-cut. Fr oblong, 6-8mm, finely hairy. Rocky places and screes, to 2700m. June-Aug. eP, c, sA, Ap. **4a** *A. cortiana* is low and *much branched;* umbels dense with 15-20 rays. AA*.

5 GRAFIA *Grafia golaka* (= *Hladnikia golaka*). Med/tall hairless, *bluish-green,* per. Lvs 3-4-pinnate, the lobes oval or diamond-shaped. Fls white with notched petals; rays 12-22; bracts numerous, oval, pointed. Fr oblong, 8-13mm, with prominent ridges. Limestone rocks, to 1800m. July-Aug. seA, cAp.

6 XATARDIA *Xatardia scabra.* Short thick hairless, usually unbranched per. Lvs 2-3-pinnate, the lobes narrow-triangular, toothed; lf stalk *broadly* sheathing the stem at base. Fls *greenish-yellow;* the petals lance-shaped; rays very unequal in length, bristly. Fr egg-shaped, 6-7 mm, prominently ridged. Screes of limestone and schist, 1600-2300m. July-Sept. eP*.

7 TROCHISCANTHES *Trochiscanthes nodiflora.* Tall hairless per; stems with *opposite or whorled* branches. Lvs 3-4-pinnate, the lobes oval-toothed. Fls greenish-white in small but numerous loose umbels; rays 4-8; bracts 1 or absent. Fr egg-shaped, 6mm, with prominent, slender, ridges. Mountain woods, to 1600m. May-July. sw, sA.

8 BALDMONEY or SPIGNEL *Meum athamanticum.* Short/med, *strongly aromatic,* hairless per. Lvs 3-4-pinnate with numerous *thread-like* lobes, mostly basal. Fls white or purplish. the petals oval; rays 3-15; bracts 0-2. Fr oblong-egg-shaped, 4-10mm, with prominent stout ridges. Rocks and screes, usually on limestone, to 2700m. June-Aug. B, P, A, Ap.

9 PLEUROSPERMUM *Pleurospermum austriacum.* Tall slightly hairy bien/per; stems *ridged,* hollow. Lvs 2-3-pinnate, *triangular,* the lobes oval, pinnately-lobed. Fls white with rounded petals; rays 12-20; bracts numerous, downturned. Fr egg-shaped, ridged. Damp grassy and stony places, open woods, to 2000m. June-Sept. A.

1b Burnet Saxifrage

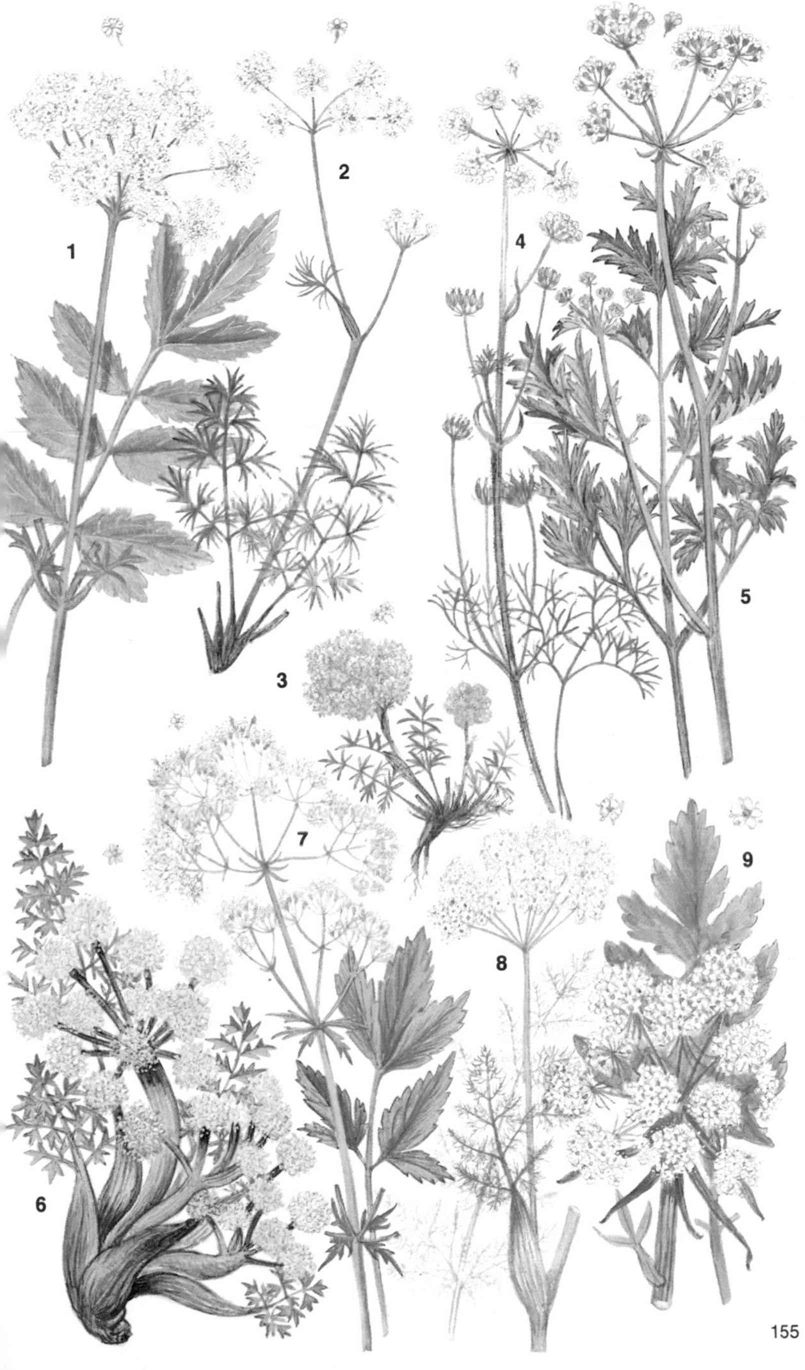
1
2
3
4
5
6
7
8
9

Carrot Family *(contd.)*

HARE'S-EARS *Bupleurum.* Hairless annuals or perennials with undivided lvs. Fls small, yellow or purplish, in small umbels; secondary umbels surrounded by a ruff of oval or rounded 'leaf-like' bracts (except in 5). Fr egg-shaped.

1 LONG-LEAVED HARE'S-EAR *Bupleurum longifolium.* Med/tall yellowish or purple-tinged per. Lower lvs *elliptical-spoon-shaped* or lance-shaped, long-stalked, the upper lvs oval or rounded, clasping the stem. Fls yellowish, the umbels with 5-12 rays. Meadows, open woods and stony places, to 2000 m. July-Aug. A*.

2 PYRENEAN HARE'S-EAR *Bupleurum angulosum.* Short/med per. Basal lvs linear to lance-shaped, stalked, the upper lvs 3-5, heart-shaped, clasping the stem. Fls yellowish, the umbels with 3-6 rays. Limestone rocks, 1500-2300 m. July-Aug. P. **2a** *B. stellatum* □ has a *single,* narrower, stem lf. Meadows, scrub and rocky places, on acid soils, to 2650 m. A*.

3 ROCK HARE'S-EAR *Bupleurum petraeum.* Short/med per. Basal lvs *linear,* grassy, the stem lvs usually *absent.* Fls yellowish, the umbels with 5-15 rays. Limestone rocks and screes, 1300-2300 m. July-Aug. s,e,seA*.

4 THREE-VEINED HARE'S-EAR *Bupleurum ranunculoides.* Short/med per. Lower lvs linear to narrow lance or spoon-shaped, *3-veined,* flat-stalked; upper lvs oval, pointed, partly clasping the stem. Fls yellowish, the umbels with 3-10 rays. Meadows and rocky places, often on limestone, to 2600 m. July-Aug. P, A, Ap. **4a** *B.r. gramineum* has lower lvs with *inrolled edges,* upper lvs linear-lance-shaped. P, A, Ap.

5 SICKLE HARE'S-EAR *Bupleurum falcatum.* Med/tall per. Basal lvs oblong-elliptical, *often curved,* stalked, the stem lvs lance-shaped to linear, partly clasping the stem. Fls yellowish, the umbels with 3-15 thread-like rays; bracts of secondary umbels *linear-lance-shaped,* unequal in size. Grassy and waste places, to 1600 m. July-Oct. B*, P, A, Ap. **5a** *B.f. cernuum* has *linear* stalkless lvs. P, sA, Ap.

6 ENDRESSIA *Endressia pyrenaica.* Low/short, usually unbranched, hairless per with *lfless* stems. Lvs all basal 2-pinnate, lobes narrow oblong-lance-shaped. Fls white, the umbels with 9-25 rays, *no bracts.* Fr oval. Meadows and fields, to 2200 m. Aug-Sept. eP.

7 CARAWAY *Carum carvi.* Med/tall hairless per, with *lfy branched* stems. Lvs 2-3-pinnate, lobes linear-lance-shaped. Fls white, the umbels with 5-16 rays, usually without bracts. Fr oblong, ridged, *aromatic* of Caraway when crushed. Meadows and waste places, to 2200 m. May-July. T—except B. **7a** *C. rigidulum* has broad, rather than narrow, lower leaf segments; fls often yellowish-white. AA.

8 PYRENEAN ANGELICA *Selinum pyrenaeum* (= *Angelica pyrenaea*). Short/med, often slightly branched, hairless per; stem with 1-2 lvs or lfless. Lvs 2-3-pinnate, lobes linear to lance-shaped. Fls *yellowish-white,* petals heart-shaped, the umbels with 3-9 rays; bracts and sepals *absent,* secondary bracts linear. Meadows and pastures, to 2300 m. June-Aug. P, Cev, Vosges.

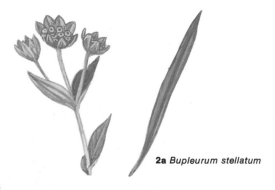

2a *Bupleurum stellatum*

157

Carrot Family *(contd.)*

1 UNBRANCHED LOVAGE *Ligusticum mutellinoides.* Low/short almost hairless per; stems *unbranched.* Lvs all basal, 2-3-pinnate, oval in outline, with narrow oblong pointed lobes. Fls white or pink, the petals notched; rays 8-20; lower bracts *numerous,* linear. Fr elliptical, 3-5mm, ridged, usually smooth. Grassy and stony places, often exposed, 1900-3350m. July-Aug. A. **1a** *L. ferulaceum* has *branched* stems, narrower lflets and *pinnately cut bracts.* swA,J.

2 ALPINE LOVAGE *Ligusticum mutellina.* Short/med almost hairless per; stems with *1-2 alternate branches.* Lvs mostly basal, 2-3-pinnate triangular in outline, with narrow lance-shaped lobes. Fls red or purple usually, with notched petals; rays 7-10; lower *bracts 1-2,* lance-shaped or absent. Fr oblong, 4-6mm, ridged. Damp meadows and open woods, to 3000m. July-Aug. A. **2a** *L. lucidum* is much taller, the stems with *opposite or whorled branches;* umbels large with 20-50 rays.

3 WILD ANGELICA *Angelica sylvestris.* Tall robust, almost hairless per; stems often purplish. Lvs 2-3-pinnate, with broad toothed lobes; upper lvs much smaller with *inflated sheathing* stalks. Fls white or pinkish in large umbels; petals lance-shaped; lower bracts often absent, upper threadlike. Fr oval flattened, 4-5mm, with *broad membranous wings.* Damp and shady places, to 1800m. July-Sept. T. **3a** *Angelica* A. *archangelica* has *green stems* usually and cream or greenish-white fls; fr with *corky wings.* Widely grown as a confection and naturalised especially in the north. **3b** *A. razulii* is like 3 but fr larger, 8mm, with narrow wings. P.

4 LOVAGE *Levisticum officinale.* Tall strong-smelling branched per. Lvs 2-3-pinnate, *glossy* with broad toothed lobes. Fls greenish-yellow; rays 12-20, grooved; bracts numerous, upper ones *joined* at base. Fr oblong-egg-shaped, 5-7mm, yellow or brown. Widely cultivated and naturalised in grassy places, to 1200m. June-Aug. T–except F.

5 SOUTHERN MASTERWORT *Peucedanum venetum.* Med/tall hairless, *purplish,* per; stems much branched. Lvs 2-4-pinnate, the lobes usually pinnately-lobed; upper lvs much smaller with broad *sheathing stalks.* Fls white in *broad umbels,* petals oval; rays 10-25; bracts narrow lance-shaped. Fr oblong, 5·5-6mm, winged. Grassy and stony places, to 1500m. June-Aug. eP,sA,Ap.

6 MASTERWORT *Peucedanum ostruthium.* Med/tall almost hairless per, stems with *alternate* branches. Lvs 2-trefoil, the lobes oval, toothed; upper lvs small with *inflated sheathing* stalks. Fls white or pinkish, no lower bracts; rays 30 or more. Fr almost rounded, 4-5mm, broadly winged. Meadows, woods, rocky places and stream banks, 1400-2800m. June-Aug. P,A,Ap,(B,S). **6a** *P. verticillare* is taller with *opposite or whorled* branches; fr 7-9mm, broadly winged. c,eA,cAp.

7 AUSTRIAN HOGWEED *Heracleum austriacum.* Short/med hairy per, stems slender, branched. Lvs *pinnate,* the lobes oval toothed, hairy. Fls white, petals *deeply notched,* or pink in flattish umbels, unequal; rays 6-13; bracts usually present. Fr broad oval, 7-11mm, broadly winged. Fields and scrub, to 2100m. July-Sept. c,e,seA. **7a** *Hogweed* H. *sphondylium* is a much larger plant with thick, often bristly stems, umbels with 15-45 rays. T. **7b** *H. minimum* is like 7 but smaller with small *hairless* lf lobes; umbel rays 3-6. Limestone screes, to 2500m. July-Aug. swA.

8 BROAD-LEAVED SERMOUNTAIN *Laserpitium latifolium.* Med/tall almost hairless per. Lvs large 2-pinnate, lobes oval-heart-shaped, *toothed.* Fls white with notched petals; rays 25-40, bracts numerous. Fr oval 5-10mm, narrowly-winged. Open woods and stony places to 2000m. July-Aug. P,A,Ap. **8a** *L. siler* has narrower, *untoothed* lflets, and 20-50 rays. P,A,Ap. **8b** *L. nestleri* like 8 but umbels with few bracts and these *soon falling.* Limestone rocks. P,Cev. **8c** *L. krapfii* like 8b but fls *greenish-yellow* or *pinkish;* fr wings of different sizes. s,e,seA. **8d** *L. peucedanoides* like 8a but umbels *only* 2-15 rayed, lower lvs 2-3-trifoliate. s,seA. **8e** *L. nitidum* like 8 but bracts with *hairy margins.* sA–Italy. **8f** *L. halleri* like 8 but smaller, lflobes *cut into 5-7 linear lobes;* fls white or pink. A-not France.

Wintergreen Family   Pyrolaceae

Attractive evergreen creeping, hairless, perennials with basal or whorled, stalked lvs. Fls 5-petalled drooping, closed or open cups in spike-like racemes or umbels. Fr a dry capsule.

1 LESSER WINTERGREEN *Pyrola minor.* Low per. Lvs in basal rosettes, broadly elliptical, finely toothed, pale green; stalk *shorter* than lfblade. Fls white or lilac-pink, 7mm, globular-cups in stalked spikes; style *not* protruding. Woods, moors and heaths, to 2700m. June-Aug. T.

2 INTERMEDIATE WINTERGREEN *Pyrola media.* Low/short per. Lvs in basal rosettes, rounded to oval, finely toothed, dark green; stalk as long or shorter than lfblade. Fls white or pale pink, 7-10mm, globular-cups in stalked spikes; style straight, *protruding.* Woods, moors and heaths, to 2200m. June-Aug. B, A, AA, nAp.

3 ROUND-LEAVED WINTERGREEN *Pyrola rotundifolia.* Like 2 but lf stalk *longer* than lfblade. Fls pure white, 8-12mm, globular-cups; style *S-shaped,* protruding. Woods, heaths and marshes, usually over limestone, to 2300m. June-Sept. T—except c, sAp. **3a** *P. norvegica* is shorter, with more rounded lvs, the veins *forming a network* around the edge. To 1450m. July-Aug. S.

4 PALE-GREEN WINTERGREEN *Pyrola chlorantha.* Low per. Lvs oblong to rounded, finely toothed, pale green above, darker beneath; stalk *longer* than lfblade. Fls pale *yellowish-green,* 8-12mm, in stalked spikes; style *S-shaped,* protruding. Coniferous woods and grassy places, to 2200m. June-Aug. S, P, A.

5 NODDING WINTERGREEN *Orthilia secunda* (= *Pyrola secunda*). Low per. Lvs in basal rosettes, oval, elliptical or rounded, toothed, pale green; stalk shorter than lfblade. Fls greenish-white, 5-6mm, oval-cups in *one-sided spikes;* style straight, protruding. Woods, moors and rocky places, to 2200m. July-Aug. T.

6 ONE-FLOWERED WINTERGREEN *Moneses uniflora.* Low per. Lvs *opposite,* rounded or oval, toothed, tapered into the stalk, pale green. Fls *solitary,* white, 13-20mm, wide open cups; style straight, protruding. Woods, especially coniferous, to 2100m. May-Aug. T.

7 UMBELLATE WINTERGREEN *Chimaphila umbellata* (= *Pyrola umbellata*). Low per. Lvs in *whorls* or opposite, narrowly-oval, broadest above middle, toothed, narrowed into the stalk. Fls pale pink, 7-12mm, globular-cups in *umbel-like* clusters; style not protruding. Woods, especially coniferous and rocky places, to 500m, not reaching alpine levels in south. June-July. G, S, A.

Birdsnest Family   Monotropaceae

8 DUTCHMAN'S PIPE or YELLOW BIRDSNEST *Monotropa hypopitys.* Low greenless saprophytic per, with erect yellowish stems, unbranched. Lvs *scale-like,* oval-elliptic, yellowish, turning brown eventually. Fls pale yellowish or ivory white, long bells with 4-5 petals, in drooping clusters; becoming erect in fr. Damp woods, especially beech and pine, to 1800m. June-Sept. T—except Ap.

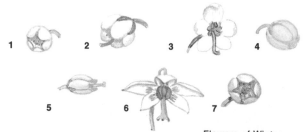

Flowers of Wintergreens

1

2

3

4

5

6

7

8

Diapensia Family   Diapensiaceae

1 DIAPENSIA *Diapensia lapponica.* Low evergreen, *cushion-forming* subshrub, hairless. Lvs narrow-spoon-shaped, shiny, leathery, untoothed. Fls white, 10 mm, 5-petalled, solitary on short stalks; stamens 5, stigma 3-lobed. Bare mountain ridges, to 1600 m. May-July. B*, S.

Crowberry Family   Empetraceae

2 CROWBERRY *Empetrum nigrum.* Low, heath-like, evergreen shrub, usually mat-forming; young twigs reddish. Lvs *alternate,* narrow-oblong, deep shiny green, margins *rolled under.* Fls tiny, pink or purplish, 1-2 mm, at base of lvs, 6-petalled; male and female on separate plants. Fr a rounded *black* berry, green at first. Moors and bogs, to 3050 m. B, P, A*, n, cAp. **2a** *E.n. hermaphroditum* is a more upright plant with broader lvs, fls with both sexes, the stamens often *persisting* round the fr. S, A.

Heather Family   Ericaceae

Small evergreen or deciduous shrubs or subshrubs with usually untoothed, deep green, lvs. Fls generally 4-5-parted, bell-, cup- or funnel-shaped, clusters or solitary. Fr a dry capsule or a fleshy berry, edible in *Vaccinium* and *Arctostaphylos.*

3 MATTED CASSIOPE *Cassiope* (= *Harrimanella*) *hypnoides.* Prostrate mat-forming subshrub. Lvs *alternate,* scale-like, narrow-oblong, pointed, *overlapping* up stem. Fls white with crimson sepals, rounded-bells, 4-5 mm, solitary, drooping on long stalks. Damp, mossy tundra, by streams and snow patches, to 1900 m. July-Aug. S–Arctic.

4 CASSIOPE *Cassiope tetragona.* More upright than 3 with *opposite* lvs, forming four ranks. Fls creamy-white with yellowish sepals, drooping bells, 6-8 mm. Dry stony heaths, usually on limestone, to 1650 m. June-Aug. S–Arctic.

5 LAPLAND RHODODENDRON *Rhododendron lapponicum.* Low, *mat-forming,* evergreen subshrub. Lvs oblong, leathery, dark green above, rusty-scaly beneath, margins *rolled under.* Fls violet-purple open bells, 8 mm long, in clusters of 3-6. Dry heaths and stony places, in calcareous soils, to 1350 m. May-June. S.

6 CRANBERRY *Vaccinium oxycoccus.* Low creeping evergreen subshrub. Lvs oval, dark green above, bluish-white beneath; margin rolled under. Fls pink, with 4 spreading or *reflexed* petals and prominent stamens, stalks long, downy, solitary or up to four in a cluster. Fr a rounded or pear-shaped berry, red or brownish, often speckled. Peat bogs, to 2000 m. May-Aug. B, S, A. **6a** *V. microcarpum* has more triangular lvs and smaller fls with *hairless* stalks.

7 COWBERRY *Vaccinium vitis-idaea.* More or less prostrate, creeping, *evergreen* subshrub. Lvs elliptical to oblong, dark green above, paler with dark dots beneath, margins rolled under. Fls white tinged pink, open bells, 5-6 mm long, in drooping clusters of 3-6. Fr a rounded *red berry.* Moors, heaths, coniferous woods and sub-alpine pastures, to 3050 m. May-Aug. T.

8 BOG WHORTLEBERRY *Vaccinium uliginosum.* Upright *deciduous* shrub to 0·75 m. Lvs oval, bluish-green, with netted veins. Fls pale pink bells, 4-6 mm, in clusters of 1-3. Fr a rounded, *bluish-black berry.* Moors, heaths, coniferous woods and sub-alpine pastures, to 3000 m. May-June. T. **8a** *V.u. microphyllum* is lower and smaller, often *mat-forming.* S–Arctic.

9 BILBERRY or WHORTLEBERRY *Vaccinium myrtillus.* Rather like 8 but lvs bright green and *toothed.* Fls pale green tinged pink, 4-6 mm, usually solitary. Fr a bluish-black berry. Heaths, moors and open woods, to 2800 m. Apr-July. T.

Heather Family (contd.)

1 ALPENROSE *Rhododendron ferrugineum*. Evergreen shrub to 1 m, often forming dense thickets. Lvs elliptic-oblong, shiny, deep-green above, *reddish-scaly* beneath, margins *rolled under*. Fls pale to deep pinkish-red, bell-shaped, in small clusters. Fr a dry capsule. Mountain slopes, open woods or scrubland, to 3200 m. May-Aug. P,A,nAp. **1a Hairy Alpenrose** *R. hirsutum* □ has bright green, hairy-edged lvs and bright pink fls, smaller than 1. Open woods, scrub and screes, on limestone, to 2600 m. May-July. c,eA.

2 MARSH ANDROMEDA or BOG ROSEMARY *Andromeda polifolia*. Creeping hairless undershrub. Lvs narrow-elliptical, shiny greyish-green above, whitish beneath. Fls bright pink, turning white, drooping rounded-bells, in small clusters. Sphagnum bogs and wet heaths, to 1150 m. June-Aug. T-except Ap.

3 BEARBERRY *Arctostaphylos uva-ursi*. Prostrate *mat-forming,* evergreen shrub. Lvs oval, broadest above middle, dark green, *untoothed*. Fls greenish-white to pink, drooping bells, in small clusters. Fr a shiny red berry, edible. Heaths, open woods and rocky places, to 2800 m. June-Sept. T. **3a Alpine Bearberry** *A. alpinus* (= *Arctous alpinus*) □ is deciduous with bright-green, *toothed* lvs; fr a black berry. To 2700 m. May-July. T-except n,sAp.

4 SPRING HEATH *Erica herbacea* (= *E. carnea*) has lvs in *fours;* fls bright or flesh-pink, urn-shaped, with *projecting* dark-purple anthers. Coniferous woods and stony places, to 2700 m. Mar-June. A-especially in the east, ncAp. **4a Bell Heather** *E. cinerea*. Thickly-branched dwarf hairless undershrub. Lvs in *threes,* linear, deep green, often bronzed. Fls bright red-purple bells, in branched clusters. Heaths, woods, dry moors and rocky places, to 1500 m. May-Sept. T-except Ap. **4b Cornish Heath** *E. vagans* □ has lvs in fours or fives; fls lilac-pink or white, with projecting purple-brown anthers, in long *dense leafy spikes*. Heaths and woodland over acid rocks, to 1800 m. May-Aug. B*,P,wA.

5 CROSS-LEAVED HEATH *Erica tetralix*. Short *greyish downy* undershrub. Lvs in *fours,* linear, pointed. Fls pale pink, drooping rounded-bells, in compact heads; anthers *not* protruding. Bogs, wet heaths and pine woods on acid soils, to 2200 m. June-Oct. B,S,P,c,wA.

6 HEATHER or LING *Calluna vulgaris*. Short/med carpeting subshrub. Lvs tiny, opposite and in rows, *scale-like*. Fls pinkish-lilac or pale purple, sepals and petals coloured, *not joined*, in slender lfy spikes. Moors, heaths and open woods on acid soils, to 2700 m. July-Oct. T.

7 DWARF ALPENROSE *Rhodothamnus chamaecistus*. Dwarf hairy shrub. Lvs elliptical-oblong, bright green, *hairy-margined*. Fls pale pink, large saucer-shaped, with *separate petals,* solitary or two to three together. Dry rocky slopes and screes on limestone, to 2400 m. May-July. eA.

8 CREEPING AZALEA *Loiseleuria procumbens*. Prostrate *mat-forming,* evergreen undershrub, hairless. Lvs oblong, shiny, deep-green; margins rolled under. Fls tiny, pale pink, in clusters of two to five. Dry stony or peaty places, often exposed, on acid soils, 1500-3000 m. May-July. T-except Ap.

9 MOUNTAIN HEATH *Phyllodoce caerulea*. Low short undershrub, heath-like. Lvs alternate, narrow-oblong, *rough-edged*. Fls lilac to pinkish-purple, drooping bells, *larger* than 4-6, in clusters of 2-6 at top of shoots. Heaths and rocky moors, 2000-2600 m. June-Aug. B*,S,cP.

1 Alpenrose

Primrose Family   Primulaceae

Perennials, sometimes annuals, with undivided lvs. Fls with 5 petals (except Chickweed Wintergreen, p. 174) often joined into a short/long tube, sepal 5. Fr a small capsule.

PRIMULAS have basal rosettes of lvs and open fls in long-stalked umbel-like clusters, or directly from the lfy rosettes; petals lobes notched, joined in a tube at base, sepal-tube with short teeth.

1 PRIMROSE *Primula vulgaris.* Low hairy per. Lvs oblong, *crinkled, tapering* to stalk. Fls pale yellow with orange central marks, 20-30mm, *solitary* on long hairy stalks, fragrant. Woods, scrubland and grassy banks, to 1500m. Mar-June. T. *Hybridises* with 2 and 3.

2 OXLIP *Primula elatior.* Low hairy per. Lvs like 1 but *more abruptly* narrowed to stalk. Fls pale yellow with orange central marks, 15-25mm, 1-20 in a *nodding, one-sided cluster; not* fragrant. Meadows and woods, to 2700m. Mar-Aug. T. **2a** *P.e.* intricata has *less* abruptly narrowed lvs and markedly *tapered* fr capsules. A.

3 COWSLIP *Primula veris.* Low/short hairy per. Lvs *very abruptly* narrowed at base into stalk, crinkled. Fls deeper yellow than 1-2 with orange spots in centre, 10-15mm, up to 30 in a nodding, one-sided cluster, *fragrant.* Meadows and pastures, to 2200m. Apr-June. T.

4 BIRDSEYE PRIMROSE *Primula farinosa.* Low per, *mealy white* on stems and underneath lvs. Lvs spoon-shaped or elliptical, toothed. Fls lilac-pink, purple, rarely white, with *yellow eye,* 8-16mm, 2 to many in an umbel. Marshes and damp pastures, usually on acid soils, to 3000m. May-Aug. nB, S, P, A.

5 LONG-FLOWERED PRIMROSE *Primula halleri* (= *P. longiflora*). Low/short per. Lvs pale green, oblong or spoon-shaped, slightly toothed, *mealy yellow* beneath, Fls lilac or violet with a yellow eye, 15-20mm, each with a *long thin tube,* 2-12 in an umbel. Meadows and rock crevices, to 2900m. June-July. A–rare in w.

6 STICKY PRIMROSE *Primula glutinosa.* Low *sticky* per. Lvs deep green, narrow-spoon-shaped, blunt-toothed. Fls deep violet, 12-18mm, 2-8 in an umbel; petal lobes *deeply notched,* fragrant. Meadows and granitic rocks, 1800-3300m. June-Aug. c, eA.

7 SPECTACULAR PRIMROSE *Primula spectabilis.* Low/short, somewhat sticky, per. Lvs bright *glossy green,* oblong to rhombic, tending to curl backwards, with a *horny-white edge.* Fls pinkish-red or violet, 20-30mm, 2-5 in an umbel; petal lobes *rather wavy,* deeply notched. Rocky and stony slopes on limestone, to 2500m. May-Aug. seA. **7a Glaucous Primrose** *P. glaucescens* has glossy, *bluish-green* lvs which tend to curl inwards, *without* darkish dots above. To 2400m. sA–Italy only excluding Dolomites.

8 WULFEN'S PRIMROSE *Primula wulfeniana.* Like a smaller version of 7. Lvs *glossy blue-green* with a membranous edge. Fls pinkish-red or lilac with a *white eye,* 25-30mm, 1-2 in an umbel. Limestone rocks and meadows to 2200m. May-Aug. seA–Carnic, Karawanken and Julian Alps. **8a** *P. clusiana* has green lvs with a *horny and hairy* edge; fls pinkish-red or violet, with a white eye, 1-4 in an umbel. May-July. n, neA–Germany, e, n Austrian Alps. *Not* overlapping in distribution with 8.

9 LEAST PRIMROSE *Primula minima.* Very low per with tiny lfy rosettes. Lvs glossy deep green, wedge-shaped, the *apex deeply toothed.* Fls bright pink with a white eye, 15-30mm, 1-2 in a *short-stalked* umbel; petal lobes deeply notched. Grassy places and rocky ledges on limestone and granite, often by snow patches, 2000-3000m. June-July. eA, seA.

1

2

3

4

5

6

7

8

9

Primrose Family (contd.)

1 ENTIRE-LEAVED PRIMROSE Primula integrifolia. Low, slightly sticky per. Lvs bright green, lance-shaped or oval, with hairy, untoothed edges. Fls rosy-purple or pinkish-lilac, 15-25 mm, 1-3 in an umbel, stalks often reddish. Stony and grassy places, often by snow patches, on acid rocks, 1900-2700 m. May-Aug. c,eP,cA. **1a** P. tyrolensis □ has more rounded, finely toothed lvs and lilac-pink fls with a white centre, solitary or sometimes 2 on a short stalk. Dolomitic rocks, to 2300 m. May-June. Dolomites*.

2 ALLIONI'S PRIMROSE Primula allionii. Low, very sticky per with small lf rosettes. Lvs greyish-green, oblong or rounded, narrowed at base, toothed or untoothed, gland dotted. Fls rose-pink to purplish, rarely white, 5-20 mm, solitary or up to 5, scarcely stalked. Shady limestone cliffs, to 1900 m. Mar-May. MA.

3 VISCID PRIMROSE Primula latifolia (= P. viscosa). Low/short sticky, rather fleshy, per. Lvs pale green, broad-oblong, narrowed at base into stalk, sometimes toothed at top; Fls purplish or dark violet with a mealy-white throat, 15 mm, 2-20 in a one-sided umbel, stalk long, sticky. Acid rock crevices, 1800-3050 m. June-July. eP,sw,c,sA,-not Austria.

4 MARGINATE PRIMROSE Primula marginata. Low/short, mealy-white per. Lvs lance-shaped, deeply toothed. Fls bluish- or pinkish-lilac with a mealy throat, 18-28 mm, 3-15 in a mealy-stalked umbel. Rock crevices and screes, to 2600 m. June-July. swA – France and Italy.

5 AURICULA or BEARS-EAR Primula auricula Low/short per. Lvs smooth or mealy-white, lance-shaped or rounded, toothed or not. Fls yellow with a mealy-white throat, 15-25 mm, 2-30 in a long-stalked umbel. Damp grassy places and rock crevices, to 2900 m. May-July. A, Ap.

6 PIEDMONT PRIMROSE Primula pedemontana. Low/short per. Lvs smooth shiny-green, oblong or lance-shaped, tapered into the stalk, edge normally toothed and with tiny red dots. Fls purple or deep pink with a white eye, 20-25 mm, 2-15 in an umbel. Acid rocks and stony pastures, 1400-3000 m. June-July. swA – not Switzerland. **6a** P. apennina is smaller with the lvs covered with yellowish-brown dots; fls pale to rose pink. Sandstone cliff ledges. May-Aug. nAp*. **6b** P. daonensis like 6 but lvs narrower, long-stalked, very sticky, covered with red dots; fls paler. Acid rocks and stony pastures, 1600-2800 m. June-July. c,sA – Rhaetian Alps to the Dolomites.

7 VILLOUS PRIMROSE Primula villosa. Low/short, very sticky, fleshy per. Lvs oval, usually toothed, tapered to a thin stalk, covered with tiny red dots. Fls rose-pink to lilac with a white eye, 20-30 mm, 2-5 on a sticky stalk much longer than lvs. Acid rocks and stony pastures, to 2200 m. Apr-June. sw*,s,seA. **7a** P. hirsuta □ is smaller with more rounded, coarse-toothed lvs. Fls lilac to deep purplish-red, 1-15 on stalks as long as lvs. To 3600 m. Apr-July. P,A.

8 SCOTTISH PRIMROSE Primula scotica. Like a small Birdseye Primrose. Low ann/bien with mealy-white stems and lvs. Lvs elliptical, untoothed. Fls dark purple with a pale yellow eye, 5-8 mm, in short-stalked clusters. Short coastal turf, at low altitudes. May-June. nB. **8a Northern Primrose** P. scandinavica has lvs mealy only beneath, and long-stalked fl clusters. Damp meadows on calcareous soils to 1500 m. S. **8b** P. stricta is a small version of 8a with narrower stalked lvs, with scarcely any meal. Fls pale lilac or violet. To 1150 m. June-July. S.

9 VITALIANA Vitaliana primuliflora (= Androsace vitaliana, Gregoria vitaliana). Variable low mat or tuft-forming per. Lvs in tiny rosettes, oblong-lance-shaped to linear, untoothed, green or grey-green. Fls yellow, 9-11 mm, with oblong petal lobes, 1-5 together, almost stalkless. Screes, rocky and stony places or short turf, often on acid rocks, 1700-3100 m. May-July. P,A,cAp.

1

1a

3

2

4

5

7a

8

6

7

9

Primrose Family (contd.)

ROCK-JASMINES – **see p. 346** for leaf identification

1 MILKWHITE ROCK-JASMINE *Androsace lactea.* Low, slightly hairy per, forming loose mats of small rosettes. *Lvs linear,* pointed, *untoothed.* Fls milk-white with a yellow eye, 8-12mm, solitary or up to six in a long-stemmed umbel; petal lobes *notched.* Rocks and screes and turf, to 2400m. May-Aug. A–more common in the east.

2 BLUNT-LEAVED ROCK-JASMINE *Androsace obtusifolia.* Low, slightly hairy per, forming loose tufts, Lvs spoon-shaped, *blunt-ended,* with a few star-shaped hairs along edge. Fls white or pale pink, 6-9mm, solitary or up to seven in long-stemmed umbels; petals *not notched.* Acid rocks and screes, 1500-3500m. June-Aug. A, nAp.

3 PINK ROCK-JASMINE *Androsace carnea.* Low, tufted, slightly hairy per, with closely-packed, bright green, lf rosettes. Lvs linear, pointed, 10-15mm long, *hairy, margin bristly.* Fls pink or white with a yellow eye (both illustrated), 5-8 mm, two to eight in short or long-stemmed umbels; petals rounded. Acid rocks and screes or short turf, 1400-3100 m. July-Aug. eP,w,cA. **3a, 3b, 3c,** p.337.

4 CILIATE ROCK-JASMINE *Androsace chamaejasme.* Low, lax, per forming flattish cushions or tufts. Lvs oblong, 5-16mm, with *an edge* of long silky-white hairs. Fls white or pale pink with a yellow eye, 5-7mm, two to seven in short or long-stemmed umbels. Rocks and short turf, on limestone or acid rocks, to 3000m. June-July. cP*, A. **4a** *A. villosa* ☐ is very variable but with more rounded lf rosettes. Lvs covered *all over* by long silky hairs, especially towards the top. Limestone rocks and screes, short turf, 1200-3000m. P, A, Ap.

5 PYRENEAN ROCK-JASMINE *Androsace pyrenaica.* Low per forming dense, *deep-green,* rounded-cushions. Lvs narrow-oblong, 3-7mm, overlapping, covered in *straight hairs.* Fls white with a yellow eye, 4-5mm, solitary. Granitic rocks and screes, 2000-3000m. June-Oct. c,eP.

6 SWISS ROCK-JASMINE *Androsace helvetica.* Low per forming dense, *grey,* rounded-cushions. Lvs oblong to spoon-shaped, 2-6mm, *covered* in short straight hairs. Fls white with a yellow eye, 4-6mm, solitary. Limestone screes and rocks, 2000-3500m. May-Aug. A. **6a** *A. vandelii* ☐ (= *A. imbricata*) has downy, *starry-haired,* lvs and fls on *short stalks.* 2000-3000m. P,A–especially in the Pennine Alps.

7 CYLINDRIC ROCK-JASMINE *Androsace cylindrica* (= *A. hirtella*). Low per forming dense grey-green, rounded cushions. Lvs oblong, 5-8mm, *downy with* straight or starry-hairs. Fls white with a yellow eye, or sometimes pink, 7-9mm, solitary, on *short slender stalks.* Limestone rocks and cliffs, 2000-3500m. July-Aug. cP*.

8 MATHILDA'S ROCK-JASMINE *Androsace mathildae.* Low cushion-forming per; lf rosettes large. Lvs linear, 10-15mm, *shiny-green,* hairless. Fls white or pink, 5mm, solitary, on short slender stalks. Rock crevices, 2000-2800m. July-Aug. cAp–Abruzzi*.

9 HAIRY ROCK-JASMINE *Androsace pubescens.* Low per, forming grey-green cushions; lf rosettes rather loose. Lvs oblong or spoon-shaped, 4-10mm, covered in short, *straight or branched, hairs.* Fls white or pink with a yellow eye, 4-6mm, solitary, short-stalked. Rock cliffs and screes, limestone or granite, 2000-3800m. June-July. P,sw,cA–not Austria. **9a** *A. ciliata* ☐ is larger with bristle-edged lvs and fls 5-8mm, pink or violet with an orange or yellow eye. 2800-3400m. cP*. **9b** *A. hausmannii* ☐ has narrow lance-shaped fleshy lvs, slightly recurved, *starry-haired*; fls white or pink with a yellow eye. Limestone crevices, 1900-3100m. July-Aug. eA–centred on the Dolomites.

10 ALPINE ROCK-JASMINE *Androsace alpina.* Low per, rather flat, mat-like, cushions. Lvs oblong-lance-shaped, 5-10mm, covered in *short starry hairs.* Fls white or pink with a yellow eye, 7-9mm, solitary, short-stalked. Granite screes and rocks, 2000-4000m. July-Aug. A. **10a** *A. wulfeniana* ☐ denser cushions, lvs 3-5mm, fls larger, deep-pink, 10-12mm, petals straight-ended. Sandstone and shaly cliffs and rocks, 2000-2600m. June-July. eA–Tauern and Carnic Alps. **10b,** p.337.

Primrose Family *(contd.)*

1 ANNUAL ANDROSACE *Androsace maxima.* Low hairy ann. Lvs round or oblong, toothed, all in a basal rosette. Fls *tiny*, white or pale pink *surrounded* by larger green sepals, in an umbel. Dry fields and waste places, to 1600m. Apr-May. S,P,A,cAp.

2 NORTHERN ANDROSACE *Androsace septentrionalis.* Low downy ann/bien. Lvs oblong to elliptical, toothed, all in a basal rosette. Fls white or pale pink 4-5mm across, *long-stalked*, in 5-30 fld umbels. Dry meadows and sandy places, to 2200m. Apr-July. S,wA*. **2a** *A. chaixii* is similar but with *laxer* 5-8 fld umbels and *larger* fls. Open woodland, short turf and rocky ground, to 1800m. Apr-July. French A, MA*.

3 ELONGATED ANDROSACE *Androsace elongata.* Similar to 2 but *almost hairless* and the lvs often toothless. Fls tiny white, with a yellow throat, 3mm across, concealed amongst the sepal teeth. Dry, grassy places, to 2200m. May-July. P, nA.

4 ALPINE BELLS *Cortusa matthiola.* Short hairy per. Lvs *rounded,* lobed and toothed, deep green, in basal tufts. Fls rosy-purple nodding bells, 9-11mm, in clusters at top of long lfless stems. Woods and damp places, over limestone, to 2200m. June-July. A.

5 DWARF SNOWBELL *Soldanella pusilla.* Low hairless per forming small mats of deep green, rounded or kidney-shaped, lvs. Fls rosy violet, *narrow nodding-bells* with a *shallow* fringe, solitary. Wet alpine soils and turf on acid rocks, often by snow patches, to 3100m. May-Aug. c,eA,nAp.

6 LEAST SNOWBELL *Soldanella minima.* Similar to 5 but lvs smaller and stems and lfstalks *covered* in short glandular hairs. Fls pale violet or white, *conical bells* with a *shallow* fringe, solitary. Damp soils and turf over limestone, to 2500m. May-July. eA. **6a** *S.m. samnitica* has *narrow* bell fls. cAp. **6b Austrian Snowbell** *S. austriaca* □ similar to 6 but lvs more rounded, slightly *heart-shaped* at base. Fls whitish. eA–Austrian.

7 ALPINE SNOWBELL *Soldanella alpina.* Low hairless, mat-forming per. Lvs deep green, round-heart-shaped or kidney-shaped, stalked. Fls violet or violet-blue, wide nodding bells, *deeply-fringed,* in clusters of two to four. Wet pastures and stony places, particularly on limestone, to 3000m. Apr-Aug. c,eP,A,Ap. **7a Hungarian Snowbell** *S. hungarica major* has lfstalks and stems *covered* in short glandular hairs; lvs often violet beneath. Coniferous woodland and wet moors, to 1700m. May-July. eA–Austrian. **7b Mountain Snowbell** *S. montana* similar to 7a but *more* hairy and fls in clusters of *six to eight.* eA– Austrian and Italian.

8 PYRENEAN SNOWBELL *Soldanella villosa* is more robust than 7a and 7b, the lfstalks and stems *very* hairy. Lvs *green below.* Fls violet, deeply fringed bells in clusters of three to four. Damp shady places on limestone, to 1700m. May-July. wP*.

9 WATER VIOLET *Hottonia palustris.* Pale green aquatic hairless per with *submerged* pinnate lvs. Fls pale lilac with yellow eye, in whorled racemes above the water. Still fresh water, ditches and ponds, to 1500m. May-July. T–but local.

10 SOWBREAD *Cyclamen purpurascens* (= *C. europaeum*). Low tuberous rooted per, almost hairless. Lvs *rounded-heart-shaped,* stalked, shiny deep green, often mottled, purplish beneath. Fls carmine-pink, deeper at mouth, with *reflexed* petals, throat rounded, fragrant. Flstalk, *coiling* in fruit. Stony woods and scrub, particularly on limestone, to 1800m. June-Oct. A,especially e,se. **10a** *C. hederifolium* (= *C. neapolitanum*) □ has lobed, *ivy-like,* lvs; fls pale pink or white with an angular, *5-sided,* throat, often before lvs. Scrubland and stony places, to 1200m. Aug-Nov.sA-Italian and Julian Alps,Ap,(B).

Primrose Family (contd.)

1 YELLOW PIMPERNEL *Lysimachia nemorum.* Evergreen creeping hairless per. Lvs pale green, rounded or lance-shaped, pointed, opposite. Fls bright yellow, *saucer-like,* 6-8·5mm, with *narrow* sepal-teeth. Damp or shaded places, to 1800m. May-July. T–except Ap.

2 CREEPING JENNY *Lysimachia nummularia.* Similar to 1 but lvs rounded-heart-shaped, blunt. Fls yellow, *cup-shaped,* 8-16mm, with *broad* sepal-teeth. Ditches, damp grassland and other moist places, to 1800m. May-July. T–except Ap.

3 BROOKWEED *Samolus valerandi.* Low/short hairless per, stems branched or unbranched. Lvs oval or spoon-shaped, pale green, in a *basal rosette* and alternate up the stem. Fls white *cup-shaped,* 2-3mm, in lax racemes. Damp or shady places often on limestone, to 1200m. June-Aug. T.

4 CHICKWEED WINTERGREEN *Trientalis europaea.* Low hairless per with a creeping rootstock. Lvs lance-shaped, shiny green, mostly in a *single whorl* near the top of the stem. Fls white starry with 5-9 petals, usually solitary. Damp grassy places, coniferous woods and other acid places, to 2000m. June-July. B, S, A–very local.

5 CHAFFWEED *Anagallis minima.* Tiny hairless erect ann. Lvs oval, *alternate.* Fls white or pink hidden at base of lvs, the *petals shorter* than the sepal teeth. Damp open woods, heaths and sandy places, to 1150m. June-Aug. T–except Ap.

6 BOG PIMPERNEL *Anagallis tenella.* Slender mat-forming hairless per. Lvs oval or rounded, short-stalked, *opposite.* Fls pink, rarely white, open cups 6-10mm. Damp turf, bogs and marshy places, to 1200m. May-Sept. B, P, A.

Thrift Family   Plumbaginaceae

Alpine species with basal tufts of lvs. Fls 5-parted in rounded clusters with papery bracts on slender lfless stalks.

7 PLANTAIN-LEAVED THRIFT *Armeria alliacea* (= *A. plantaginea*). Short/med per. Lvs *spear-shaped, pointed, several veined,* grey-green. Flheads purple to white, 10-20mm. Dry grassland, stony meadows and screes, to 3000m. May-Sept. P, wA, nAp. **7a**, p.337.

8 HALLER'S THRIFT *Armeria maritima halleri* (= *A. halleri*). Low/short downy per. Lvs grassy, *hairy-margined, 1-veined.* Flheads bright pink or reddish, 10-15mm. Stony meadows particularly on serpentine rocks, 1800-2900m. June-Aug. eP. **8a Mountain Thrift** *A. m. alpina* (= *Statice montana*) has *hairless* lvs. Screes and damp meadows, 1400-3100 m. July-Aug. P,A,Ap.

Phlox Family   Polemoniaceae

9 JACOB'S LADDER *Polemonium caeruleum.* The only European alpine member of the family *Polemoniaceae.* Med/tall tufted hairless per. Lvs *pinnate,* alternate up the stem. Fls blue or white open-cups, 18-25mm, 5 petals *joined* halfway. Rocks, damp meadows and woodland, to 2300m. May-Aug. P, A–uncommon.

Bindweed Family   Convolvulaceae

Twining annuals or perennials, sometimes parasitic. Lvs alternate, no stipules. Fls 5-parted funnel or bell-shaped. Fr a capsule.

10 FIELD BINDWEED *Convolvulus arvensis.* Per with twining or creeping stems, slightly downy when young. Lvs arrow-shaped, stalked. Fls pink and/or white, trumpet-shaped, 15-30mm. Cultivated fields, meadows and waste places, to 1850m. June-Sept. T.

11 COMMON DODDER *Cuscuta epithymum.* Slender climbing ann, stems *thread-like,* reddish. Lvs tiny *scale-like.* Fls tiny, pale pink, cupshaped, in tight clusters, petals and sepals *pointed.* Parasitic on gorse, heather and many other plants. Grassy places and heaths, to 2200m. June-Oct. T. **11a**, p.337.

Bogbean Family   Menyanthaceae

Aquatic perennials with five-parted fls and alternate lvs. Petals joined at base into a short tube. Fr a capsule. Closely related to the Gentian Family.

1 BOGBEAN *Menyanthes trifoliata.* Creeping hairless per, with short erect shoots. Lvs *trifoliate,* lflets oblong or rhombic, slightly toothed, held *above* water. Fls pink and white, 15mm, *fringed* with long white hairs, in loose racemes. Marshes, bogs, ponds and ditches, to 1800m. May-July. T.

2 FRINGED WATERLILY *Nymphoides peltata.* Creeping hairless per with floating lvs and stems. Lvs *rounded-heart-shaped,* green, sometimes purple spotted. Fls yellow, 30-40mm, with *serrated* edge to petals. Lakes, ponds and slow-flowing rivers, to 1200m. June-Sept. T−except S. Often local.

Gentian Family   Gentianaceae

Hairless annuals or perennials with opposite untoothed lvs. Fls solitary or in whorled or branched clusters, generally 4-5-parted, but sometimes with as many as 12 petals or lobes; petals joined at base or with a long tube or trumpet. Fr a many-seeded dry capsule, splitting into two.

3 GREAT YELLOW GENTIAN *Gentiana lutea.* Med/tall per. Lvs large oval, pointed, *ribbed,* bluish-green, the upper clasping the stem. Fls yellow, 5-9 petal-led, *starry,* in whorled clusters at base of upper lvs. Anthers *not* joined. Meadows, marshes and rocky slopes, to 2500m. June-Aug. P, A, Ap. **3a** *G.l. symphyandra* has its anthers *joined together. seA.*

4 SPOTTED GENTIAN *Gentiana punctata.* Short/med per. Lvs elliptical or lance-shaped, grey-green, ribbed; the lower stalked. Fls pale greenish-yellow, purple-spotted, upright bells, 15-35mm long in clusters at base of upper lvs. Sepal tube with *5-8 teeth.* Meadows, open woods and rocky places, to 3050m. July-Sept. A. **4a** *G. burseri* has yellow, brown-spotted fls; sepal tube papery, *split down one side.* To 2700m. July-Aug. P, swA−France and Italy.

5 PURPLE GENTIAN *Gentiana purpurea.* Short/med per. Lvs lance-shaped or oval, strongly ribbed, the lower stalked. Fls *reddish-purple* with dark-purple spots, upright bells, 15-25mm long, in clusters at base of upper lvs. Sepal tube papery, *split down one side.* Meadows and open woods, 1600-2750m. July-Oct. S, A, AA, nAp. **5a Brown Gentian** *G. pannonica,* has purple fls with *red-black* spots; sepal tube with *5-8 teeth.* To 2500m. July-Sept. n, e, seA.

6 WILLOW-LEAVED GENTIAN *Gentiana asclepiadea.* Short/med per with slender erect or arching stems. Lvs lance-shaped, *long-pointed,* stalkless. Fls blue *trumpets* with paler bands, 35-50mm long, at base of lvs. Woods, damp meadows and rocks, often on limestone, to 2200m. Aug-Oct. A, nAp.

7 MARSH GENTIAN *Gentiana pneumonanthe.* Low/short per with slender, more or less erect, stems. Lvs narrow-oblong or lance-shaped, l-veined, stalkless. Fls bright blue trumpets, *green-striped* outside, 25-50mm long, solitary or several. Marshy places and heaths on acid soils, to 1500m. July-Oct. T−except Ap.

3 Great Yellow Gentian

Gentian Family *(contd.)*

1 STYRIAN GENTIAN *Gentiana frigida.* Low tufted per. Lvs strap-shaped, pale green. Fls yellowish-white, blue-flushed and striped outside, long-bell-shaped, 20-35 mm, solitary or two to three. Meadows and stony places, on limestone, 2000-2500 m. July-Sept. c A–Styrian Alps*.

2 KARAWANKEN GENTIAN *Gentiana froelichii.* Low tufted per. Lvs linear-lance-shaped or oblong, *pointed.* Fls clear blue, *unspotted,* erect tubular-trumpets, 30-40 mm, solitary or two on short stems. Screes and turf, on limestone, 1400-2400 m. July-Sept. seA–Karawanken Alps. .

3 CROSS GENTIAN *Gentiana cruciata.* Med leafy per. Lvs oval-lance-shaped, shiny green. Fls dull blue, greenish outside, 20-25 mm, erect, petal tube with *four* lobes, in tight clusters at base of lvs. Meadows, woods and rocky places, to 2000 m. July-Oct. A, Ap.

4 PROSTRATE GENTIAN *Gentiana prostrata.* Low *annual* with thin prostrate or erect stems. Lvs small, narrow-oblong or spoon-shaped. Fls steel-blue, greenish at base, 10-20 mm long, solitary, petal tube with *four lobes.* Short turf and stony places, 2000-2800 m. July-Aug. cA–including E. Switzerland*.

5 PYRENEAN GENTIAN *Gentiana pyrenaica.* Low tufted per. Lvs small, narrow-lance-shaped, pointed, *overlapping.* Fls violet-blue, wide erect trumpets, 20-30 mm long, appearing *ten-lobed,* solitary. Stamens creamy-white. Damp meadows and boggy areas, to 2800 m. June-Sept. eP.

6 CLUSIUS'S GENTIAN *Gentiana clusii.* Low tufted per. Lvs leathery, elliptical to oblong-lance-shaped. Fls large trumpets, mid to dark blue, 40-60 mm, *not* or scarcely green spotted inside, solitary. Sepal teeth *triangular,* widest at base. Mountain pastures and stony places, generally on limestone, to 2800 m. Apr-Aug. A, nAp.

7 PYRENEAN TRUMPET GENTIAN *Gentiana occidentalis.* Similar to 6 but sepal-teeth *narrowed* at base, and petal-lobes pointed. Turf and stony places, on limestone, to 3000 m. May-Aug. wP. **7a** *G. ligustica* has fls with green spotting inside and petal-lobes drawn out into a *long fine point.* To 2800 m. May-Aug. MA, cAp.

8 TRUMPET GENTIAN *Gentiana acaulis* (= *G. kochiana*). Low tufted per. Lvs elliptical or lance-shaped, blunt, greyish-green. Fls large trumpets, deep blue with *green spotting* inside, sometimes purplish or pinkish, rarely white, 40-70 mm, petal-lobes pointed, solitary, stalked. Sepal-teeth narrow-lance-shaped, broadest *above* base. Turf and stony places and bogs, on acid rocks, 1400-3000 m. May-Aug. eP, A, Ap. **8a** *G. angustifolia* has narrow strap or spoon-shaped lvs. Turf and stony places, on limestone, to 2500 m. May-Aug. P, swA, J.

9 SOUTHERN GENTIAN *Gentiana alpina.* Low tufted per like a dwarf form of 8. Lvs *round-oval,* greyish-green. Fls medium trumpets, deep blue or sky blue, green-spotted inside, 40 mm, solitary, *scarcely stalked;* petal-lobes rounded. Turf and stony places, on acid rocks, 2000-2600 m. June-Aug. cP, sw, wcA.

8 Trumpet Gentians

Gentian Family (contd.)

1 SPRING GENTIAN Gentiana verna. Low tufted per. Lvs lance-shaped or ellipti-cal, *bright green*, mostly in a basal rosette. Fls solitary, pale to deep blue or purplish, rarely white, 15-25mm, petal tube with 5 oval lobes. Calyx *narrowly winged*. Meadows, heaths and marshy places on acid and alkaline rocks, to 3000m. Mar-Aug. B,P,A,nAp. **1a** G.v. tergestina (= G. tergestina) has narrower, longer, *lvs* and *broader wings* on calyx. Dry turf on limestone, to 2000m. Apr-June. P,sA–Italian,Ap.

2 SHORT-LEAVED GENTIAN Gentiana brachyphylla. Low tufted per. Lvs rounded to diamond-shaped, *bluish-green*, leathery, mostly in a basal rosette. Fls solitary, bright pale to mid blue, 15-25mm, with narrow oval lobes. Calyx very slender, *not winged*. Turf, stony places and alluvium on acid rocks, 1800-4200m. July-Aug. P,A. **2a** G.b. favratii has narrowly winged calyces and deep blue fls with *rounded petal lobes*. Turf and stony places on limestone, 2000-2800m. July-Sept. A. **2b** G. pumila □ like 2 but with *narrow-lance-shaped* lvs and sapphire-blue fls with triangular-pointed, petal lobes. Damp meadows on limestone, 1600-2800m. June-Aug. eA. **2c** G.p. delphinensis like 2b but with *blunt* petal lobes. eP,swA.

3 BAVARIAN GENTIAN Gentiana bavarica. Low mat-forming per. Lvs small, yellowish-green, oblong to spoon-shaped, blunt, *overlapping* along stem. Fls solitary, dark blue, sometimes violet or white, 16-20mm, petal lobes oblong-blunt. Damp meadows and marshy places, 1800-3600m. July-Sept. A. **3a** G. rostanii has bright-green *narrow-lance-shaped*, pointed lvs and sky blue fls. 2400-2900m. sw,scA*. **3b Triglav Gentian** G. terglouensis □ is like a small version of 3 but with sky blue fls, petal lobes *pointed;* lvs oval-lance-shaped. Meadows and stony places on limestone, 1900-2700m. seA*. **3c** G.t. schleicheri comes from w,sA,MA*.

4 SNOW GENTIAN Gentiana nivalis. Low, tiny, *slender ann*, erect; stems *branched*. Lvs opposite, oval or elliptical. Fls bright, deep blue, 8mm, petal tube with 5 pointed lobes. Meadows, marshes, heaths and stony places, 1600-3100m. June-Aug. T,B.*

5 BLADDER GENTIAN Gentiana utriculosa. Low/short, slender, erect ann; stems branched. Lvs opposite, basal ones in a rosette, lance-shaped or elliptical, pale green. Fls intense blue, 12-18mm, petal tube with 5 pointed lobes. Calyx slightly inflated, with *broad* wings. Damp meadows, bogs and stony places, to 2500m. May-Aug. A,Ap.

GENTIANELLA – like Gentians but fls with a frilly whitish throat.

6 SLENDER GENTIAN Gentianella tenella. Low unbranched ann. Lvs spoon-shaped or elliptical, mostly in a basal rosette. Fls all from base, sky-blue, violet, rarely white, 4-6mm, long stalked, with *4-petal lobes*. Damp pastures and stony places, usually on acid rocks, 1500-3100m. July-Sept. S,P,A. **6a** G. nana is small with *5-parted* fls. Rocky places and moraines on limestone, 2200-2800m. July-Sept. eA–Italy and Austria.

7 APENNEAN GENTIANELLA Gentianella columnae. Low branched per. Lvs lance-shaped. Fls purple or whitish, 10-15mm, with 4 petal lobes. Calyx with 2 broad outer lobes *not enclosing* the 3 narrow, inner ones. Meadows and stony places, to 1800m. July-Sept. cAp.

8 FIELD GENTIAN Gentianella campestris. Short branched ann/bien. Lvs oval, pointed on stem. Fls bluish-lilac or white, 15-25mm, with *4 petal lobes*. Calyx with 2 broad outer lobes *enclosing* the 3 narrow, inner ones. Grassland and heath, to 2750m. July-Oct. T. **8a** G. hypericifolia like 8 but sepal lobes broadest above, *not below,* the middle. w,cP.

Gentian Family *(contd.)*

1 FELWORT *Gentianella amarella.* Low/short ann/bien, stems branched *above* base. Lvs oval to narrow-lance-shaped. Fls reddish-violet, sometimes white or yellowish, 14-22mm, with 4 or 5 petal lobes. Calyx with 5 *equal lobes.* Meadows and sandy places, to 1800m. June-Oct. B,S,A.

2 GERMAN GENTIAN *Gentianella germanica* agg. (including *G. ramosa, G. austriaca* and *G. lutescens*). Like 1 but branched *from the base.* Fls *always* with 5 petal lobes, lavender, blue or violet, 20-35mm, sometimes white. Meadows, marshes and waste places, usually over limestone, to 2700m. B*,A,Ap.

3 COMMON CENTAURY *Centaurium erythraea.* Variable short/med bien, stems branched. Lvs elliptical to spoon-shaped, in a *basal rosette,* smaller up stems. Fls pink or purplish, rarely white, 10-14mm, in rather flat-topped clusters; petal tube with 5 lobes. Dry grassland, scrub and stony places, to 1400m. June-Sept. T−except Ap. **3a** *G. suffruticosum* like 3 but rosette lvs *broad* rather than narrowed at the base. June-Sept. P.

4 LESSER CENTAURY *Centaurium pulchellum.* Low/short branched or un-branched ann, *without* a basal lf-rosette. Lvs oval to lance-shaped. Fls pinkish-purple, rarely white, 5-9mm, in loose clusters, rarely 4, lobes. Open places and damp meadows, to 1200m. June-Sept. T−except Ap.

5 YELLOW-WORT *Blackstonia perfoliata.* Med erect greyish ann. Lvs oval or triangular, *joined round* stem; basal lvs oblong in a rosette. Fls yellow, 10-15mm, with *6-12 oblong petals.* Grassy, shaded and rocky areas, to 1300m. May-Sept. B,P,A.

6 LOMATOGONIUM *Lomatogonium carinthiacum.* Short rather thin stemmed ann. Lvs mostly basal, oval to oblong, pale green. Fls pale blue, saucer-shaped, 10-25mm, solitary *on long lfless stems;* petals 4 or 5. Meadows and grassy stream banks, 1400-2700m. Aug-Oct. c,eA.

7 MARSH FELWORT *Swertia perennis.* Med erect, unbranched per. Lvs oval to elliptic, yellowish-green, the upper clasping the stem. Fls blue or violet-red, rarely yellowish-green or white, *starry,* 20-30mm, in branched clusters; petals 4 or 5, pointed. Marshes and wet meadows, to 2500m. July-Oct. P,A,nAp.

Periwinkle Family Apocynaceae

8 LESSER PERIWINKLE *Vinca minor.* Trailing *evergreen per,* almost hairless, stems thin rooting at nodes. Lvs opposite, lance-shaped or elliptical, deep shiny-green. Fls solitary, blue-violet, 25-30mm, *propeller-shaped,* with 5 joined petals. Fr linear, *forked* at base. Woodlands and shaded banks and rocks, to 1320m. Feb-May. P,A,(B).

2 German Gentians

Milkweed Family   Asclepiadaceae

A mainly tropical family with one alpine species.

1 SWALLOW-WORT *Vincetoxicum hirundinaria.* Med erect, clump-forming, per. Lvs *opposite,* oval to lance-shaped, untoothed. Fls white or yellowish-green, purplish in bud, 3-10mm, 5-parted, in clusters at base of lvs. Fr a *double-pod* usually. Woods, scrubland and stony places, to 1800m. May-Aug. P, A. **Poisonous.**

Borage Family   Boraginaceae

Annuals or perennials, often rough or bristly, with uncut alternate lvs and fls borne in spiralled clusters. Fls 5-parted; petals joined into a short or long tube. Fr, consisting of 4-nutlets, hidden at base of persisting sepal tube. A difficult family with many species looking rather alike.

2 GROMWELL *Lithospermum officinale.* Med/tall bristly per, tufted. Lvs broad to narrowly lance-shaped, pointed. Fls yellowish or greenish-white, 4-6mm, in dense spirals. Nutlets *shiny* white. Hedges, scrub and woodland margins, to 1600m. May-Aug. T.

3 BLUE GROMWELL *Buglossoides* (= *Lithospermum*) *purpurocaerulea.* Med hairy, tufted, per. Lvs lance-shaped to narrowly elliptical, pointed. Fls at first reddish-purple *but soon turning* bright blue, 14-19mm, in small cluster at ends of stems. Nutlets shiny white. Scrub and woodland margins, to 1200m. Apr-June. T−except S. **3a** *B. gastonii* has lvs *clasping* the stem and smaller fls, 12-14mm; nutlets yellowish. Woods and rocky slopes, to 2000m. wP.

4 CORN GROMWELL *Buglossoides* (= *Lithospermum*) *arvensis.* Short/med, rough-hairy ann. Lvs oblong to spoon-shaped, pointed. Fls *small,* dull white, purplish or blue, 4-9mm, in small clusters. Nutlets brownish. Wood margins, waste and rocky places, to 2300m. Apr-Sept. T.

5 SHRUBBY GROMWELL *Lithodora* (= *Lithospermum*) *oleifolia.* Laxly branched undershrub to 0·5m, stiff-hairy. Lvs oblong to spoon-shaped, blunt, dull green above, *whitish beneath.* Fls pale pink becoming blue, 6-7mm, in small clusters; petal lobes rounded. Rocky places, to 1100m. May-July. eP−Spanish.

GOLDEN DROPS *Onosma.* Tufted bristly pers with tubular yellow fls.

6 PYRENEAN GOLDEN DROP *Onosma bubanii.* Short grey-green per. Lvs narrow-oblong covered in *straight bristles.* Fls pale-yellow, 16-20mm long, smooth; flstems *unbranched.* Dry stony places, mainly on limestone, to 1700m. May-June. P−Spanish. **6a** *O. vaudensis* is more robust, the lower lvs *broadest above* the middle and the flstems *branched.* wA−Rhone Valley.

7 GOLDEN DROP *Onosma arenaria.* Per or bien with a single stem and basal rosettes of oblong-spoon-shaped lvs. Lvs covered with *straight and star-shaped hairs.* Fls pale-yellow, 12-19mm, in branched clusters. Rocky and stony places to 1700m. May-July. eA−Austrian. **7a** *O. austriaca* has several branched fl stems and larger fls, 18-22 mm. eA−Austrian. **7b Swiss Golden Drop** *Onosma helvetica* □ is a tufted per. Lvs like 7. Fls pale-yellow, 20-24 mm, *finely hairy.* Stony and sandy places, to 2500 m. May-June. swA.

8 BORAGE *Borago officinalis.* Med ann, roughly hairy. Lvs oval, pointed, wavy-margined, the lower stalked. Fls bright blue, 20-25mm, starry with pointed rather reflexed petal lobes and a *protruding* column of stamens. Dry, often waste places to 1800m. May-Sept. P, s, seA, Ap.

8 Borage   1 Swallow-wort, fr.

Borage Family *(contd.)*

1 LESSER HONEYWORT *Cerinthe minor.* Low/med hairless ann/bien/per. Lvs oblong to oval, greyish-green rough and often *white spotted;* upper lvs stalkless. Fls pale yellow, sometimes with 5 violet spots in the throat, 10-12mm, bell-shaped with *pointed lobes.* Meadows, fields and waste places, over limestone, to 2200m. May-Sept. A, Ap. **1a** *C.m. auriculata* is per with more swollen fls and flstalks with *short bristles. A, Ap.*

2 SMOOTH HONEYWORT *Cerinthe glabra.* Short/med hairless bien/per. Lvs oblong to oval, *smooth, unspotted;* upper lvs heart-shaped at base, unstalked. Fls yellow with 5 dark reddish-purple spots in the throat, 8-13mm, tubular-bells with *rounded lobes.* Meadows and damp woods, on limestone, to 2600m. May-July. P, A, Ap.

3 MOLTKIA *Moltkia suffruticosa.* Dwarf cushion-forming densely-lfy subshrub. Lvs linear, pointed, rough-hairy, green above and *whitish beneath.* Fls blue, 13-16mm, narrow upturned bells with rounded lobes, in dense clusters. Stony places, on limestone, to 1200m. May-July. AA, seA, nAp.

4 VIPER'S BUGLOSS *Echium vulgare.* Med/tall erect bristly bien. Lvs elliptical to lance-shaped, the upper narrower. Fls blue, pink in bud, 10-19mm, funnel-shaped with rounded lobes, in large branched clusters; stamens *protruding.* Meadows and dry rocky places, to 1800m. July-Sept. T.

5 COMMON LUNGWORT *Pulmonaria officinalis.* Short tufted hairy per. Lvs *heart-shaped,* green, spotted white. Fls pink turning blue or bluish-violet, 13-18mm, funnel-shaped, in small clusters. Damp open woods, mainly on limestone, to 1900m. Mar-May. A, J. **5a** *P. obscura* has *unspotted* or faintly green-spotted lvs. A, J.

6 STYRIAN LUNGWORT *Pulmonaria stiriaca.* Short tufted hairy per. Lvs *oblong,* pointed, narrowed at the base, green, *spotted white.* Fls bright blue, 14-18mm, funnel-shaped. Open woods and damp places, to 1300m. Apr-June. eA. **6a** *P. saccharata* has reddish-violet or bluish-violet fls. MA, n, cAp.

7 MOUNTAIN LUNGWORT *Pulmonaria montana.* Short/med tufted per with *sticky stems.* Lvs long-lance-shaped, shining above with soft hairs, unspotted or green-spotted; upper lvs heart-shaped at base. Fls pink turning bright blue, 15-20mm, funnel-shaped. Meadows and damp woods, to 1900m. Apr-May. P, A.

8 NARROW-LEAVED LUNGWORT *Pulmonaria angustifolia* (= *P. azurea*). Short/med rough, tufted, per. Lvs *narrow lance-shaped,* tapered into a short stalk, unspotted. Fls red or purplish turning bright blue, 12-20mm, funnel-shaped. Sepal tube narrow in fr. Woods and meadows, on acid soils, to 2600m. Apr-July. swA. **8a** *P. visianii* has shorter lvs and a broad sepal tube in fr. To 2600m. Apr-July. c, e, seA. **8b** *P. kerneri* like 8a but with *white spotted* lvs. Apr-July. neA.

9 LONG-LEAVED LUNGWORT *Pulmonaria longifolia.* Rather like 8 but with *white-spotted* lvs, sometimes green spotted. Fls red turning violet or blue-violet, 8-12mm, funnel-shaped, in dense clusters. Woods and shady places, to 2000m. Apr-July. B, P, Cev.

4 Viper's Bugloss

Borage Family *(contd.)*

1 COMMON COMFREY *Symphytum officinale.* Med/tall rough-hairy per, stems *winged.* Lvs broad lance-shaped, lowest very large, stalked. Fls creamy-white, pinkish or purple-violet, tubular-bells, 12-18mm, in *forked, spiralled* clusters. Damp meadows and ditches, to 1600m. May-June. T. **1a Tuberous Comfrey** *S. tuberosum* is a smaller plant, with pale *yellowish-white* fls, the basal lvs *withered* at flg time. Woods and damp places, to 1600m. T—except S.

2 ALKANET *Anchusa officinalis.* Short/tall, rough-hairy per/bien, well branched. Lvs *long* lance-shaped, the lower ones stalked. Fls violet or reddish, sometimes white, 7-15mm, fl-tube 5-7mm long, in spiralled clusters. Meadows, banks and rocky places, often over limestone, to 1800m. May-Sept. T,(B). **2a Bugloss** *A. arvensis* (= *Lycopsis arvensis*) is annual with small blue fls, 4-6mm. T. **2b** *A. barrelieri* is similar to 2, but fls blue or bluish-violet, 7-10mm, fl-tube *only* 1·5mm long. Woods and fields, to 2300m. May-July. MA—Italian,nAp.

3 MADWORT *Asperugo procumbens.* Low sprawling ann, bristly. Lvs *opposite* usually, lance-shaped. Fls purplish, 2-3mm, 1-3 on short *drooping stalks* at base of lvs. Fr surrounded by enlarged calyx. Cultivated and waste ground, to 2600m. May-Nov. B*,S,A,nAp.

FORGET-ME-NOTS *Myosotis.* Annuals or perennials, usually softly-hairy, with oblong or lance-shaped lvs, generally stalkless. Fls small, blue, often pinkish in bud, in forked, loose spiralled clusters. Seeds shiny, brown or black.

4 FIELD FORGET-ME-NOT *Myosotis arvensis.* Low/med bien. Basal lvs broadest above the middle, in a rosette. Fls bright blue, 3-4mm; sepal tube with many *hooked* hairs. Dry and waste places, shady areas, to 2000m. Apr-Oct. T. **4a** *M. ramosissima* is smaller, ann, often only a few cm tall, the sepal teeth *spreading* in fr. T.

5 CHANGING FORGET-ME-NOT *Myosotis discolor.* Slender low/short ann. Lvs oval-lance-shaped. Fls *pale cream* at first but turning blue, 3-4mm; sepal teeth *incurved* in fr. Bare and waste places, to 2000m. May-June. T. **5a** *M. stricta* has lvs with *hooked* hairs and tiny pale or deep blue fls, 1-2mm. T. **5b** *M. speluncicola* is similar to 5a but fls *white.* S,swA—France,cAp.

6 WOOD FORGET-ME-NOT *Myosotis sylvatica* agg. Short per, well branched. Lvs oval or elliptic. Fls *sky-blue*, 6-8mm, flat; sepal tube with short spreading hooked hairs. Woods and grassy places, to 2000m. Apr-July. T. **6a** *M. decumbens* has fl tubes *longer* than the calyx. P,A.

7 ALPINE WOOD FORGET-ME-NOT *Myosotis alpestris.* Low/short per, often tufted. Leaves oval to spoon-shaped, hairy beneath, the basal sometimes stalked. Fls bright or deep blue, sometimes whitish, 6-9mm, flat; sepal tube with *hooked* hairs, teeth incurved in fr; seeds black. Damp woods and meadows, 1500-2800m. Apr-Sept. B*,P,A,Ap.

8 ALPINE FORGET-ME-NOT *Myosotis alpina* (= *M. pyrenaica*) like 7 but lvs hairless beneath and *fl stalks* as well as sepal tube with hooked hairs. Rocks, screes and mountain pastures, 1500-2800m. Apr-Sept. P. **8a** *M. stenophylla* is like 7 and 8 but with *no* hooked hairs. eA.

9 TUFTED FORGET-ME-NOT *Myosotis laxa* (incl. *M. caespitosa*). Short/med ann or bien, branched from base. Fls bright blue, 2-4mm; sepal tube long-stalked, with *only* straight hairs. Fr seeds dark brown. Wet stony and grassy places, to 1600m. May-Sept. T.

10 WATER FORGET-ME-NOT *Myosotis scorpioides* agg. Low/med creeping per; hairs closely *pressed* to stems, lvs and sepal tube; *no* hooked hairs. Fls sky-blue, sometimes pink or whitish, 4-8mm, petals slightly *notched.* Wet places, ditches, river margins, to 2000m. June-Sept. B,S,A.

11 LAMOTTE'S FORGET-ME-NOT *Myosotis lamottiana.* Like 10 but the stems *roughly* hairy, shining, and the fls rather smaller. Wet meadows, 1500-2000m. June-Sept. P, Cev.

Borage Family *(contd.)*

1 KING OF THE ALPS *Eritrichium nanum.* Low dense, cushion-forming, hairy per; like a dwarf forget-me-not. Lvs in rosettes, narrow-oblong to spoon-shaped. Fls pale to brilliant blue, rarely whitish, 7-9mm, in short-stalked clusters of 3-7. Acid rocks and screes, sometimes dolomite, 2500-3600m. July-Aug. c,s,seA.

2 BUR FORGET-ME-NOT *Lappula squarrosa* (= *L. myosotis*). Med branched hairy ann/bien. Lvs greyish, oblong to lance-shaped, unstalked. Fls pale blue, 2-4mm, in loose lfy clusters. Fr erect, egg-shaped with two rows of hooked bristles forming a *mitre-like* erection. Dry slopes and waste places, to 2500m. June-Aug. T–except B. **2a** *L. deflexa* (= *Hackelia deflexa*) has larger fls, 3-6mm; fr *drooping*. S,eP,A.

3 BLUE-EYED MARY *Omphalodes verna.* Low/short creeping per, slightly hairy. Lvs mostly clustered at base, oval to *heart-shaped,* pointed, long-stalked. Fls sky blue, 8-10mm, in loose racemes; petals rounded. Damp mountain woods, to 1200m. Mar-May. e,seA,nAp,(B). **3a** *O. scorpioides* is bien with lance or spoon-shaped lvs; fls small, blue, 3-4mm. eA.

4 HOUND'S TONGUE *Cynoglossum officinale.* Med greyish, softly hairy bien. Lvs oblong to lance-shaped, the upper clasping the stem. Fls dull *maroon-purple,* funnel-shaped, 5-6mm long. Fr flattened, covered with *hooked spines.* Dry and stony places, to 2400m. May-Aug. T. **4a** *C. magellense* (= *C. apenninum*) is a lower *per* with narrow-lance-shaped lvs and reddish fls, 8mm long. Meadows, to 2000m. June-Aug. c,sAp. **4b** *C. nebrodense* is smaller than 4 with reddish-violet fls and *rounded fr.* Woods to 2150m. c,sAp.

5 SOLENANTHUS *Solenanthus apenninus.* Tall bien. Basal lvs large, elliptical or broad lance-shaped, the upper smaller, *clasping stem.* Fls purple, funnel-shaped, 7-9mm long, in *dense* branched spirals. Woods and pastures, to 1800m. c,sA.

Verbena Family   Verbenaceae

6 VERVAIN *Verbena officinalis.* Med roughly-hairy per, stems stiff, *square.* Lvs opposite, diamond-shaped, the lower pinnately-lobed, stalked. Fls lilac-blue, 2-5mm, in slender lfless spikes; petal tube 5-lobed, more or less *2-lipped.* Waste places and rocks, to 1500m. June-Sept. T–except Norway.

Mint Family   Labiatae

Aromatic *square-stemmed* annuals or perennials, sometimes subshrubs. Lvs usually undivided *opposite.* Fls clustered or whorled, rarely solitary; calyx tubular, 5-toothed, often 2-lipped; corolla *2-lipped* and open-mouthed. Stamens 4 or 2. Fr 4 nutlets hidden inside calyx.

7 ALPINE SKULLCAP *Scutellaria alpina.* Short slightly hairy per. Lvs oval, toothed, the lower stalked. Fls purplish, lower lip often *whitish,* 20-25mm long, in *quadrangular* clusters. Bracts often purplish. Limestone rocks and screes, grassy slopes, to 2500m. June-Aug. P,w,sA, Ap.

8 MOUNTAIN GERMANDER *Teucrium montanum.* Low/short mat-forming subshrub. Lvs leathery, elliptical, untoothed, whitish beneath, margins *rolled under.* Fls in flattish clusters, yellowish-white, 12-15mm long. Rocks, screes and dry pastures, usually on limestone, to 2400m. May-Aug. P,A,Ap.

9 PYRENEAN GERMANDER *Teucrium pyrenaicum.* Low, slender, creeping, hairy per. Lvs rounded to oval, *toothed.* Fls in rounded clusters, white or purple with a white lower lip, 14-16mm long. Rocks, screes and dry pastures, usually on limestone, to 2000m. June-Aug. P.

10 WALL GERMANDER *Teucrium chamaedrys.* Short tufted per, slightly hairy. Lvs oblong or broadly-oval, toothed, *dark green,* shiny. Fls pale to deep purple, 9-16mm long, in *lfy spikes.* Dry places, banks and open woodland, to 1800m. May-Sept. P,A,Ap. **10a** *T. lucidum* is larger and more or less *hairless.* June-Aug. swA.

Labiate Family *(contd.)*

BUGLES *Ajuga*. Tufted perennials, sometimes annuals. Upper lip of corolla very short, the lower 3-lobed. Calyx with 5 equal teeth.

1 PYRAMIDAL BUGLE *Ajuga pyramidalis*. Low/short creeping per, stems *hairy all round*. Lvs oval, slightly toothed or untoothed, the lower stalked. Fls pale violet-blue or deep violet, rarely pink or white, 10-18mm, in dense lfy pyramidal spikes. Meadows, stony places and scrub, often on acid soils, 1300-2800m. Apr-Aug. T. **1a** *A. genevensis* has more distinctly toothed, *often shallowly lobed,* lvs and bright blue, rarely pink, fls. Meadows and woodland clearings, on calcareous soils. P, A, Ap, (B).

2 COMMON BUGLE *Ajuga reptans*. Low short, creeping per with *rooting runners;* stems hairy on two sides. Fls blue, rarely pink or white, 14-17mm, in lfy spike, the lf-like bracts often purplish. Damp grassy places and woods to 2000m. Apr-July. T.

3 GROUND PINE *Ajuga chamaepitys*. Low greyish-hairy ann, *smelling* faintly of pine when crushed. Lvs with *3-narrow lobes,* the lobes often further lobed or toothed. Fls yellow, red spotted, 10-15mm, partly hidden amongst lvs. Stony and grassy places, often on calcareous soils, to 1600m. May-Sept. B, A.

4 TENORE'S BUGLE *Ajuga tenorii*. Low, often stemless slightly hairy or hairless per. Lvs oblong to spoon-shaped, toothed. Fls bright blue, 16-25mm; upper bracts *shorter* than the fls. Siliceous soils, to 1400m. May-June. c, sAp.

5 SIDERITIS *Sideritis hyssopifolia*. Short/med hairy or hairless sub-shrubby per. Lvs linear to oval or spoon-shaped, slightly toothed or untoothed, scarcely stalked. Fls *pale yellow,* sometimes purple tinged, 10mm, in dense oblong clusters. Pastures, woods and rocky places, often on limestone, to 1800m. July-Aug. P, swA, Jura. **5a** *S. endressii* is similar, the fls in loose clusters, yellow outside but brown or purple tinged *inside,* 9-20mm. To 1800m. c, eP.

HEMP-NETTLES *Galeopsis*. Annuals with fls in terminal or branched clusters or whorls. Fls with the upper lip hooded and the lower 3-lobed. Calyx bell-shaped, 5-toothed.

6 LARGE PINK HEMP-NETTLE *Galeopsis ladanum* (= *G. intermedia*). Short/med ann, stems square hairy almost *to the base.* Lvs oval to lance-shaped, narrowed at base, toothed. Fls deep pink with yellow blotches on the lip, 15-28mm; calyx green. Fields and stony places, often on acid soils, to 2400m. July-Oct. T. **6a Red Hemp-nettle** *G. angustifolia* □ has *narrower*, linear or linear-lance-shaped lvs and white-hairy calyces. To 2000m. T. **6b** *G. reuteri* is taller than 6 with wiry rounded stems, *hairless in the lower half.* Rocks and screes, to 1600m. MA.

7 PYRENEAN HEMP-NETTLE *Galeopsis pyrenaica*. Short/med hairy ann, stems square. Lvs oval to triangular, toothed, short-stalked, *velvety-hairy.* Fls pinkish-purple with darker blotches, 17-25mm. Acid sands and gravels by streams, to 2200m. Aug-Sept. eP. **7a Downy Hemp-nettle** *G. segetum* has larger pale yellow fls, the lower lip sometimes purple blotched. July-Sept. eP, swA, (B).

8 LARGE-FLOWERED HEMP-NETTLE *Galeopsis speciosa*. Med/tall branched, hairy ann. Lvs oval to lance-shaped, pointed, toothed, short-stalked. Fls yellow with a large *purple blotch* on the lower lip, 27-34mm. Cultivated land and waste places, to 1740m. July-Sept. T−except Ap.

6a Red Hemp-nettle

Labiate Family (contd.)

1 HAIRY HEMP-NETTLE *Galeopsis pubescens.* Med *white-hairy* ann. Lvs oval, pointed, square or heart-shaped at the base, toothed, stalked. Fls bright pinkish-red with yellow blotches on the lower lip, 20-25mm. Open woods, fields, banks and hedgerows, to 1600m. July-Sept. A,n,cAp.

2 COMMON HEMP-NETTLE *Galeopsis tetrahit.* Short/med rough-hairy ann, branched. Lvs lance-shaped to broadly oval, pointed, *narrowed* at the base, toothed, stalked. Fls pale pinkish-purple with darker markings, rarely yellowish or whitish, 15-20mm. Open woods, heaths, paths and waste places, to 2400m. July-Oct. T. **2a** *G.t. glaucocerata* has *hairless stems.* French Alps. **2b** *G. bifida* has smaller fls than 2, rarely exceeding 15mm long, the middle lobe of the lower lip *distinctly* 2-lobed. T.

SALVIAS. Perennials with opposite or whorled fls forming long spikes. Upper lip of corolla straight or curved, the lower 3-lobed. Calyx 2-lipped. Stamens 2; style protruding.

3 JUPITER'S DISTAFF or STICKY SAGE *Salvia glutinosa.* Med/tall branched, *stickily-hairy,* per. Lvs oval to heart-shaped, toothed, stalked. Fls *yellow* with reddish-brown markings, 30-40mm, in whorls of 2-6. Woods, copses and clearings, generally on limestone, to 1800m. June-Sept. c,eP,A,Ap.

4 MEADOW CLARY *Salvia pratensis.* Med/tall hairy, slightly aromatic, per. Lvs oval or oblong, blunt-toothed, stalked, the upper lvs smaller, unstalked. Fls *violet-blue,* 20-30mm, in whorls of 3-6. Dry grassland, usually on limestone, 1920m. May-Aug. T.

5 WHORLED CLARY *Salvia verticillata.* Med/tall hairy, rather unpleasant smelling, often purple tinged, per. Lvs oval-heart-shaped, with *one or two lobes* at base, toothed, stalked. Fls lilac-blue or purplish, 8-15mm, in tight whorls of 15-30. Dry grassland, paths and stony places, to 2400m. May-Aug. P,c,e,sA,Ap,(B,S).

6 WIND SALVIA *Salvia nemorosa.* Med hairy per. Lvs oblong-heart-shaped, blunt toothed, stalked, the upper lvs smaller, unstalked. Fls *small,* violet-blue, 8-12mm, in whorls of 2-6. Meadows and grassy places, to 1450m. May-Aug. c,sA,n,cAp.

4 Meadow Clary

Labiate Family *(contd.)*

DEADNETTLES *Lamium*. Annuals or perennials with upright stems and whorls of fls at the lf-bases. Calyx tubular with 5-teeth. Corolla with the upper lip 'hood-like' and the lower generally heart-shaped.

1 BALM-LEAVED ARCHANGEL *Lamium orvala*. Med/tall hairless or slightly hairy per. Lvs triangular-oval, stalked and with coarse, irregular teeth. Fls pink to dark purple, or whitish, 25-45mm, the upper and lower lips with *serrated edges*. Woods, shady places and banks, to 1600m. May-July. s,seA.

2 WHITE DEADNETTLE *Lamium album*. Short/med hairy creeping per, stems erect; faintly aromatic. Lvs heart-shaped, toothed, stalked. Fls *white*, 20-25mm, hairy. Banks, hedgerows, paths and waste places, to 2300m. Apr-Nov. T.
2a Spotted Deadnettle *L. maculatum □ is more strongly aromatic and with pinkish-purple* fls. To 2000m. Apr-Oct. T,(B).

3 RED DEADNETTLE *Lamium purpureum*. Low/short hairy, often purplish, ann. Lvs oval or heart-shaped, blunt-toothed, stalked. Fls pinkish-purple, 10-18mm. Cultivated and waste ground, to 2500m. Mar-Dec. T. **3a** *L. hybridum* has more sharply toothed lvs and rather smaller fls. T.

4 HENBIT DEADNETTLE *Lamium amplexicaule*. Low/short hairy ann. Lvs rounded or oval, blunt-toothed, the lower stalked but the upper stalkless and *half-clasping* the stem. Fls pinkish-purple, 14-20mm. Cultivated and waste ground, to 2550m. Apr-Dec. T.

5 YELLOW ARCHANGEL *Lamiastrum galeobdolon*. Short/med hairy creeping per, stems erect. Lvs narrow to broadly oval, dark green, blunt-toothed, stalked. Fls *yellow* streaked with green, 17-25mm, in dense whorls of 8-16; lower lip *3-lobed*. Woods and shady places, to 2000m. Apr-July. T. **5a** *L.g. montanum* has more sharply toothed lvs and *whorls* of 20-30 fls. **5b** *L.g. flavidum* like 5a but plants not creeping and with smaller fls. A,Ap.

6 SHRUBBY HOREHOUND *Ballota frutescens*. Much branched, slightly pungent, shrub to 60cm. Lvs *small,* oval to oblong, untoothed or finely toothed, stalked. Fls white or lilac, 12-15mm, in small whorls of 8-12. Rocky places, to 1200m. swA–French and Italian.

7 BLACK HOREHOUND *Ballota nigra*. Med/tall, rather straggly, hairy per; strongly aromatic. Lvs oval or oblong, toothed, stalked. Fls pink or lilac, 12-14mm, in dense whorls at lf-bases; calyx with 5 *finely-pointed* teeth, curved back in fr. Banks, woodland margins and waste places, to 1530m. June-Sept. T.

CATMINTS *Nepeta*. Med/tall hairy pers with dense whorls of fls up lfy stems. Lower lip of corolla 3-lobed, the upper 2-lobed. Calyx with 5 almost equal teeth, 15-veined.

8 BROAD-LEAVED CATMINT *Nepeta latifolia*. Med/tall densely hairy, rather bluish-green, per. Lvs oval-oblong, heart-shaped at base, toothed, almost stalkless. Fls *blue,* 8-11mm, in whorls forming a long spike. Meadows and pinewood clearings, to 1700m. July-Sept. P.

9 COMMON CATMINT *Nepeta cataria*. Med/tall branched grey-woolly per with a *minty scent*. Lvs oval-heart-shaped, toothed, stalked. Fls *white* with purple spots, 7-10mm, in a spike with the lower whorls separated. Banks, hedgerows and rocky places, to 1500m. June-Sept. T.

2a Spotted Deadnettle

Labiate Family (contd.)

1 LESSER CATMINT *Nepeta nepetella.* Med/tall greyish-green branched per. Lvs lance-shaped to oblong, *grey-hairy beneath,* toothed, stalked. Fls white, pink or bluish-violet, 10-12mm, in loose, often branched spikes. Dry rocky and stony places and river gravels, to 1700m. July-Aug. P,sA,Ap.

BETONYS or WOUNDWORTS *Stachys.* Hairy perennials, sometimes annuals with erect stems. Fls in whorls forming dense or loose lfy spikes. Lower lip of corolla 3-lobed, the upper 1-2-lobed. Calyx with 5 equal teeth, 5-10-veined.

2 ALPINE BETONY *Stachys monieri* (= *S. densiflora*). Short/med bristly-hairy per with *basal lfy rosettes.* Lvs oblong, heart-shaped at the base, toothed, the upper short-stalked. Fls pink or purplish, 20-24mm, in dense short, blunt spikes. Dry meadows, generally on limestone, to 2400m. July-Aug. P,A. **2a Yellow Betony** *Stachys alopecuros* has more triangular-shaped lvs and *pale yellow* fls, 15-20mm. Meadows, scrub and screes, on limestone, to 2000m. June-Aug. c,eP,A.

3 BETONY *Stachys officinalis* (= *Betonica officinalis*). Variable short/tall *softly hairy* or almost hairless per with basal lfy rosettes. Lvs oblong to oval, heart-shaped at the base, toothed, the lower stalked. Fls reddish-purple, 12-18mm, in dense blunt spikes. Light woodland, grassy places and heaths, to 1800m. June-Oct. T.

4 DOWNY WOUNDWORT *Stachys germanica.* Med/tall *downy greyish-white* per. Lvs oblong to heart-shaped, finely toothed, the lower long-stalked but the upper stalkless, green above, greyish-white beneath. Fls pale pinkish-purple, 15-20mm, in whorled spikes. Woodland edges and clearings, rocky ground and screes, often on limestone, to 1750m. June-Sept. B*,P,A,Ap. **4a Alpine Woundwort** *S. alpina* has *glandular-hairy* stems and dull purple fls, 15-22mm. Shady and damp stony places, on limestone, to 2000m. B*,P,A,Ap.

5 HEDGE WOUNDWORT *Stachys sylvatica.* Med/tall, rough-hairy, *green* per, pungent. Lvs oval-heart-shaped, toothed, all stalked. Fls dull reddish-purple with white markings on the lower lip, 13-18mm, in loose spikes. Hedges, banks and shady places, to 1700m. June-Oct. T.

6 MARSH WOUNDWORT *Stachys palustris.* Short/med hairy per. Lvs oblong to lance-shaped, heart-shaped at the base, toothed, *only the lower* stalked. Fls pale purple, 12-15mm, in whorled spikes. Damp places, often by fresh water, to 1600m. June-Oct. T. Often hybridises with 5.

7 YELLOW WOUNDWORT *Stachys recta.* Variable short/tall sparsely hairy *aromatic* per. Lvs oblong or oval, green, finely toothed, the lower stalked, the upper narrower and stalkless. Fls *pale yellow* with purplish streaks, 15-20mm, in narrow whorled, branched spikes. Dry rocky and waste places, to 2250m. June-Sept. B*,P,A,Ap. **7a** *S. annua* is shorter and *annual* with all the lvs stalked. Fls white or pale yellow. Cultivated fields and waste places, generally on limestone. T.

8 BASTARD BALM *Melittis melissophyllum.* Short/med hairy, strong smelling, per, with erect stems. Lvs oblong or oval, heart-shaped at the base, toothed, stalked. Fls white with pink markings, pink or purple, 25-40mm, in *few fld whorls* at base of lvs. Woods, hedges, banks and other shady places, to 1400m. May-July. B,P,A,Ap.

8 Bastard Balm, colour forms

Labiate Family (contd.)

1 GROUND IVY Glechoma hederacea. Low creeping, softly-hairy per with long rooting runner. Lvs kidney-shaped, blunt toothed, long stalked. Fls pale violet or pinkish, 15-22mm, in loose whorls at lf-bases. Woods, grassy and waste places, to 1600m. Mar-June. T.

2 NORTHERN DRAGONHEAD Dracocephalum ruyschiana. Med per with erect stems. Lvs narrow-lance-shaped, hairless, untoothed, the margins rolled under. Fls blue-violet, rarely pink or white, 20-28mm, in terminal clusters of 2-6. Dry grassy places and open woods, to 2200m. June-Sept. S, P, A. **2a** D. austriacum □ has large fls and lvs divided into 3-5 narrow-lance-shaped segments. May-June. w*, cA.

3 SELF-HEAL Prunella vulgaris. Low creeping, slightly hairy per. Lvs oval to diamond-shaped, blunt-toothed or untoothed, the lower stalked. Fls deep violet-blue, 13-15mm, in dense oblong clusters with lvs at the base. Woods and dry meadows, particularly over limestone, to 2400m. June-Oct. T. **3a Large Self-heal** Prunella grandiflora □ is larger, the fls 20-25mm, violet-blue with a whitish tube and lower lip, without lvs at the base of the flhead. T−except B.

4 CUT-LEAVED SELF-HEAL Prunella laciniata. Low creeping, hairy per. Lvs oblong, pinnately lobed, the lower stalked. Fls creamy-white, rarely rose or purplish, 15-17mm, in oblong heads. Dry grassy and waste places, to 1320m. June-Oct. P, A, (B).

5 ALPINE CALAMINT Acinos alpinus (= Calamintha alpina). Short/med hairy per with slender stems. Lvs elliptical to oval, untoothed or with a few small teeth near the tip. Fls violet with white marks on the lower lip, 10-20mm, in small clusters at lf bases. Meadows, rocky places and screes, usually on limestone, to 2500m. June-Sept. P, A, Ap. **5a Basil-thyme** A. arvensis (= Calamintha acinos) □ is similar but annual usually, the fls only 7-10mm long. To 2000m. T.

6 LARGE-FLOWERED CALAMINT Calamintha grandiflora. Short/med slightly hairy, per. Lvs oval or oblong, pointed, toothed, stalked. Fls large, pink, 25-40mm, in loose lfy-heads. Loamy woodland, to 2100m. July-Sept. P, A.

7 WOOD CALAMINT Calamintha sylvatica. Med/tall hairy, per, mint-scented. Lvs oval or almost rounded, toothed, stalked, dark green. Fls pink or lilac, 10-22mm, in loose lfy spikes. Open woods and thickets and stony places, to 1600m. June-Oct. T−except S.

8 LESSER CALAMINT Calamintha nepeta. Med/tall hairy per. Lvs greyish, broadly-oval, blunt, with small teeth. Fls pale lilac or white, 10-15mm, in loose lfy spikes; sepal tube with protruding hairs in fr. Dry stony and scrubby places, to 1300m. July-Oct. P, w, sA, Ap, in eB but not at Alpine levels.

9 MICROMERIA Micromeria marginata (= M. piperella, Thymus marginatus). Dwarf spreading thyme-like subshrub with short erect branches. Lvs oval, blunt, untoothed. Fls purplish or violet, 12-16mm, in loose heads. Rocks and stony places, to 1600m. June-Aug. MA.

5a Basil-thyme

Labiate Family *(contd.)*

1 MARJORAM *Origanum vulgare.* Low spreading downy per, pleasantly aromatic. Lvs oval, untoothed or slightly so, stalked. Fls purplish or whitish, 4-7mm, in branched clusters, with *dark purple bracts* and protruding stamens. Dry grassland, screes and rocky places, to 2000m. July-Sept. T.

THYMES *Thymus.* Aromatic creeping or subshrubby pers. Lvs small, thick, blunt, untoothed, bristly on the edge towards the base. Fls in small heads or spikes, with protruding stamens.

2 WILD THYME *Thymus serpyllum.* Low matted per with *non-flowering* creeping branches, stems hairy all round. Lvs linear to elliptical with a *hairy margin,* unstalked. Fls pale pink to rosy-purple, 3-6mm long, in rounded heads. Dry grassy and stony places and scrub, to 3000m. Apr-Sept. T.

3 GLABRESCENT THYME *Thymus glabrescens.* Low subshrubby per with creeping branches ending in a fl cluster; stems hairy all round. Lvs elliptical-lance-shaped to narrow-oval, those of the fl stems at least 3mm broad. Fls pale pink to purplish, 3-6mm, in rounded heads, often with a separate cluster below. Dry grassy and stony places, to 1500m. May-Aug. A. **3a** *T.g. decipiens* has fl stem lvs *less than* 3mm broad. S,eA.

4 HAIRY THYME *Thymus praecox* subsp. *polytrichus.* Low creeping subshrub; stems hairy *on two sides only.* Lvs oval to spoon-shaped, short-stalked. Fls rosy-purple, 3-6mm, in rounded heads. Meadows and rocky places, to 3000m. May-Sept. c,eA. **4a** *Thymus nervosus* is similar but with *very narrow lvs,* not more than 1·5mm broad, spoon-shaped. c,eP.

5 LARGER WILD THYME *Thymus pulegioides.* Low/med tufted subshrubby per, *no* rooting runners. Stems hairy all round. Lvs oval to oblong, hairless above, stalked. Fls pinkish-purple, 6mm, in *whorled-spikes.* Grassy and waste places, to 2000m. July-Sept. T. **5a** *T. alpestris* has creeping, non-flg, branches as well. July-Sept. eA.

6 HYSSOP *Hyssopus officinalis.* Med aromatic subshrubby per. Lvs narrow-lance-shaped, hairy or hairless, untoothed. Fls violet or blue, rarely white, in *lfy whorled-spikes.* Dry hills and rocky places, to 2000m. July-Sept. P,A,Ap,(B). **6a** *H.o. aristatus* has a 1-2mm *whitish tip* to the upper lvs. c,eP.

MINTS *Mentha.* Aromatic pers with creeping stolens. Fls small, 4-lobed, lilac or pale purple, in dense spikes or spiked-whorls; stamens protruding. The species cross readily producing many hybrids.

7 CORN MINT *Mentha arvensis.* Low/med hairy per with a sickly scent. Lvs lance-shaped to broadly oval, toothed. Fls lilac or white, in dense whorls, the *stem tip lfy.* Damp places and ditches, to 1800m. July-Oct. T.

8 WATER MINT *Mentha aquatica.* Short/med hairy, often purplish per, pleasantly aromatic. Lvs oval, toothed, stalked. Fls lilac *in dense oblong-heads,* sometimes branched below. Swamps, ditches and damp places, to 1700m. July-Oct. T.

9 HORSE MINT *Mentha longifolia.* Med/tall downy per with a musty scent; stems *white* or *greyish.* Lvs oblong-elliptical, pointed, toothed, usually unstalked, whitish-downy. Fls lilac or white, in *narrow whorled spikes,* branched below. Fields, hedges and damp places, to 1900m. July-Oct. P,A,Ap. **9a Spearmint** *Mentha spicata* is strongly aromatic with *green,* almost hairless, lvs. T. Commonly used as a culinary herb.

10 PENNYROYAL *Mentha pulegium.* Short/med downy per, strongly aromatic; stems prostrate or erect. Lvs oval, scarcely toothed, short-stalked. Fls lilac, in dense whorled spikes but *without* a terminal head. Wet places and damp meadows, to 1800m. July-Oct. T−except S.

Labiate Family *(contd.)*

1 LAVENDER *Lavandula angustifolia.* Much branched aromatic shrub to 50 cm. Lvs lance-shaped to linear, grey-green. Fls lavender-blue or purplish, in dense oblong clusters. Warm rocky slopes, usually on limestone, to 1800 m. June-July. P, swA. **1a** *L.a. pyrenaica* has bracts longer than the calyces. eP.

2 DRAGONMOUTH *Horminum pyrenaicum.* Low/med tufted, slightly hairy per. Lvs in *basal clusters,* oval to rounded, blunt-toothed, long stalked, deep green. Fls dark violet-blue, 17-21 mm long, in small clusters forming *one-sided* spikes. Meadows, rocky places, open woods, on limestone, to 2500 m. June-Aug. P, A.

Figwort Family   Scrophulariaceae

Annuals, biennials or perennials with rounded or ridged (not square usually) stems and alternate or paired lvs. Fls flat, cup-shaped or tubular with 4-5 lobes, sometimes 2-lipped; stamens 2 or 4. Fr a capsule containing many small seeds. **For Figworts** *(Scrophularia* spp.) **see p.318.**

3 CREEPING SNAPDRAGON *Asarina procumbens* (= *Antirrhinum asarina*). Low stickily-hairy per with thin *sprawling stems.* Lvs *all opposite,* oval-heart-shaped, toothed, stalked. Fls solitary, pale whitish-yellow, a yellow patch on the lower lip, 30-35 mm long. Shaded rocky and stony places, to 1800 m. Apr-Sept. c, eP.

4 CHAENORHINUM *Chaenorhinum origanifolium* (= *Linaria origanifolia*). Low/short slightly-hairy, usually rather sprawling per. Lvs *green,* lance-shaped or oval, untoothed, the lower opposite. Fls snapdragon-like, purple with an orange patch on the lower lip, open-mouthed, 9-20 mm, in loose lfy racemes; spur short, 2-5 mm. Rocky and stony places and screes, to 1500 m. Apr-July. P.

5 SOFT SNAPDRAGON *Antirrhinum molle.* Dwarf shrubby per, softly downy, with thin rather sprawling stems. Lvs oval to elliptical, the lower opposite. Fls white or pale pink with a yellow mouth, 25-35 mm, spurless, in lax lfy spikes. Calcareous rocks and walls, stony places, to 1800 m. May-Aug. e, cP–Spanish and Andorra. **5a Rock Snapdragon** *A. sempervirens* □ is a more slender brittle plant with *leathery* deep green oblong or elliptical lvs covered in short hairs. Fls *smaller* 15-25 mm, white with yellow or violet mouth coloration. Rocky places, to 2000 m. cP–French and Spanish. **5b Common Snapdragon** *A. majus* □ is a tall *erect* per with large 33-45 mm yellow, pink or purple fls. To 1600 m. May-Oct. Cev,P.

6 COMMON TOADFLAX *Linaria vulgaris.* Variable med/tall erect hairless per. Lvs linear, alternate. Fls pale to mid-yellow, 25-33 mm, in dense spikes; spur long, 10-13 mm. Fields, rocky and waste places, to 1600 m. May-Sept. T–except Ap. **6a** *L. angustissima* has *smaller* pale yellow fls with a 7-10 mm long spur. A. **6b Pyrenean Toadflax** *L. supina* (= *L. pyrenaica*) □ is a low/short grey-green ann/per, often stickily hairy above, at least the lower lvs *in whorls.* Fls pale yellow, often violet tinged, 13-20 mm, in short spikes; spur long, 10-15 mm. To 2000 m. P,MA. **6c** *L. tonzigii* □ like 6b but lvs *broader* oblong-elliptical and fls larger, in rounded clusters. Dolomitic screes, 2000-2500 m. sA–Bergamasque Alps.

7 ALPINE TOADFLAX *Linaria alpina.* Low hairless ann/per with sprawling stems. Lvs linear to oblong-lance-shaped, *whorled,* grey-green. Fls *violet* with a yellow patch on the lower lip, or white, 13-22 mm, in rounded clusters; spur long 8-10 mm. Screes, rocks and river gravels, 1500-3800 m. May-Aug. T–except B, S.

8 STRIPED TOADFLAX *Linaria repens.* Short/med hairless, greyish per with *erect stems.* Lvs linear to elliptical, pointed, in whorls, the upper alternate. Fls white or pale lilac, deeper veined, 8-15 mm long, in loose spikes; *spur short,* 3-5 mm. Dry places and rocks, to 2300 m. June-Sept. Hybridises with 6. T-except Ap.

9 IVY-LEAVED TOADFLAX *Cymbalaria muralis.* Trailing, thin stemmed, hairless per. Lvs palmately-lobed, mostly alternate. Fls lilac or violet with a yellow patch on the lower lip, 9-15 mm long, spur short, 1.5-3 mm, solitary on long stalks at lf bases. Shady rocks and woods, to 2000 m. Apr-Oct. P, A, Ap(B). **9a**, p.337.

10 FAIRY FOXGLOVE *Erinus alpinus.* Low tufted hairy per. Lvs oval, broadest at the tip, *toothed,* clustered towards the base. Fls bright purple, rarely white, 6-9 mm across, with 5-notched petal lobes, unspurred. Rocks, screes and stony grassland, to 2400 m. May-Oct. P, A, cAp(B).

Figwort Family *(contd.)*

MULLEINS *Verbascum*. Tall robust biennials forming a large lfy rosette in the first year. Lvs alternate up the stems. Fls flattish, 5-petalled, in long tapered, sometimes branched spikes; stamens 5.

1 ORANGE MULLEIN *Verbascum phlomoides*. Greyish-white or yellowish mealy bien. Basal lvs oblong-elliptical, toothed or untoothed; upper lvs oval to lance-shaped, sharply pointed. Fls yellow or orange-yellow, 20-55mm, stamens with white or yellowish hairs; fl-spikes unbranched. Dry stony places and scrub, to 1400m. July-Sept. P,A,Ap,(B). **1a** *V. densiflorum* is similar but the stem lvs held closely against the stem. P,A,Ap.

2 AARON'S ROD *Verbascum thapsus*. Greyish or whitish mealy bien. Basal lvs elliptical to oblong, broadest towards the tip; upper lvs held close to the *winged stem*. Fls yellow, 12-35mm, petal lobes oval, stamens with white hairs; fl-spikes, unbranched. Dry stony or gravelly places and scrub, to 1850m. July-Sept. T. **2a** *V.t. crassifolium* is similar but the lower lvs are stalked and the petal lobes oblong. P,A. **2b** *V. argenteum* has *branched* fl-spikes, the stamens with violet-coloured hairs. sAp.

3 HOARY MULLEIN *Verbascum pulverulentum*. Densely white mealy bien. Basal lvs oval to oblong, broadest towards the tip; stem lvs held away from the *rounded stems.* Fls yellow, 18-25mm, stamens with white hairs; fl-spikes *branched.* Stony and waste places, to 1200m. July-Aug. B*,P,wA,Ap. **3a** *V. lychnitis* has *angled stems* and lvs almost hairless above, dark green. Fls yellow or white. T.

4 DARK MULLEIN *Verbascum nigrum*. Green, *not mealy,* hairy per with ridged stems. Basal lvs oval to oblong, *heart-shaped* at base, blunt-toothed, stalked. Bracts and calyces *hairy.* Fls yellow or cream with purple spots, stamens with *violet hairs;* fl-spikes unbranched or with a few short branches. Dry open places, banks and rocks, to 1800m. July-Sept. T−except Ap. **4a** *V. chaixii* has *unridged* stems and grey-green lvs. Fls yellow with a *purple throat,* in branched spikes. June-Sept. P,A.

5 LANATE MULLEIN *Verbascum lanatum*. Greenish bien with ridged stems. Lvs oval or oblong, heart-shaped at base, blunt toothed, stalked, green above, greyish beneath. Bracts and calyces *hairless.* Fls yellow, 16-28mm, stamens with violet-hairs; fl-spikes dense, unbranched. Rather similar to 4, but with *almost hairless bracts.* Mountain woods, to 2000m. June-Sept. e,seA.

2 Aaron's Rod   **4** Dark Mullein

1

2

3

4

5

Figwort Family *(contd.)*

1 LARGE YELLOW FOXGLOVE *Digitalis grandiflora.* Med/tall bien/per. Lvs oval-lance-shaped or almost triangular, finely toothed, stalkless, shiny green above, greyish-hairy beneath. Fls pale yellow outside, cream netted maroon inside, 40-50mm long, tubular-bells, in long loose spikes. Woods and stony places, to 2000m. June-Aug. P, A. Poisonous. **1a Small Yellow Foxglove** *D. lutea* has dense spikes of *small,* 9-25mm, pale yellow or whitish fls. P, wA, Ap.

2 WULFENIA *Wulfenia carinthiaca.* Short/med, almost hairless, per. Lvs in *basal rosettes,* rounded to oval, broadest above the middle, blunt toothed, stalked, deep green. Fls dark violet-blue, 12-15mm long, tubular-bells, in dense spikes. Moist humus rich grassland over acid rocks, 1500-2000m. July-Aug. seA*.

SPEEDWELLS *Veronica.* Annuals or perennials with opposite lvs. Fls solitary or in spike-like racemes with 4 joined-petals and 4 sepals, the lowermost petal often smaller; stamens 2. Fr a small flattened, heart-shaped, capsule. (See also p.320).

3 VIOLET SPEEDWELL *Veronica bellidioides.* Low/short, tufted, hairy per. Lvs oblong or spoon-shaped, usually finely toothed, short-stalked, the basal ones rosetted. Fls deep violet-blue, 9-10mm, in small clusters. Dry pastures, often on acid soils, 1400-3000m. July-Aug. P, A. **3a** *V.b. lilacina* has lilac fls with a *white centre.* P, swA.

4 THYME-LEAVED SPEEDWELL *Veronica serpyllifolia.* Low/short *creeping and rooting,* hairy per with erect fl stems. Lvs oval, untoothed or finely toothed, short-stalked, pale green. Fls white or pale blue with darker lines, 6-10mm, short stalked, in loose spikes of 20-40. Grassy shady places, to 2500m. May-Aug. T. **4a** *V.s. humifusa* has spikes of 8-15, long-stalked fls.

5 ALPINE SPEEDWELL *Veronica alpina.* Low hairy or hairless per. Lvs bluish-green, oval to elliptical, untoothed or finely toothed,*stalkless.* Fls deep blue, 7mm, in small clusters. Meadows and stony places, to 2000m. July-Aug. T.

6 ROCK SPEEDWELL *Veronica fruticans.* Low/short sprawling, hairy per. Lvs narrow-oblong to oval, usually broadest above the middle, untoothed or finely toothed, short-stalked. Fls deep blue with a *reddish centre,* 11-15mm, in small clusters. Stony grassland and rocky places, to 3000m. July-Sept. T. **6a** *V. fruticulosa* □ᵢ has smaller fls marked with red. Limestone rocks and screes, to 2800m. P, A.

7 PYRENEAN SPEEDWELL *Veronica nummularia.* Low matted hairy per with short upright stems, woody at base. Lvs broadly-elliptical to oval or rounded, *untoothed,* short-stalked. Fls blue or pink, 6mm, in small clusters. Damp rocks and screes, often on schists, to 1800m. June-Aug. P.

8 SPIKED PYRENEAN SPEEDWELL *Veronica ponae.* Short/med, creeping, hairy per; flstems erect. Lvs oblong-lance-shaped to oval, toothed, almost stalkless. Fls bluish-lilac, 10mm, in *loose spikes.* Damp rocks, woods and shady places, 1200-2500m. June-Aug. P.

9 SPANISH SPEEDWELL *Veronica aragonensis.* Low/short rather weak, sprawling hairy per. Lvs oblong to elliptical, untoothed or blunt-toothed. Fls pale blue, 8-10mm, in loose spikes. Limestone rocks and screes, to 2100m. June-July. cP–Spain.

10 LARGE SPEEDWELL *Veronica austriaca.* Variable short/med erect or somewhat sprawling per. Lvs broadly-oval to narrow lance-shaped or pinnately-lobed, toothed, short-stalked. Fls bright blue, 10-13mm, in *paired spikes;* calyx hairy. Grassy places, to 1800m. June-Aug. e,sA. **10a** *V.a. teucrium* (= *V. teucrium*) is taller with *untoothed* or sharply toothed lvs. P, A, Ap. **10b** *V. prostrata* □ᵢis more sprawling than 9 with smaller fls, 6-8mm, calyx *hairless.* P, A, n, cAp.

1

2

3

4

5

6

6a

7

8

9

10

10b

Figwort Family *(contd.)*

1 LEAFLESS-STEMMED SPEEDWELL *Veronica aphylla.* Low *stemless* or short-stemmed, matted, hairy per. Lvs in rosettes, elliptical to spoon-shaped, sometimes toothed, short-stalked. Fls deep blue or lilac, 6-8mm, in clusters of 2-6. Rocks and stony pastures, usually on limestone, 1200-3000m. July-Sept. P, A, Ap.

2 GERMANDER SPEEDWELL *Veronica chamaedrys.* Low/short sprawling hairy per. Lvs oblong to oval, toothed or lobed, short-stalked or unstalked. Fls bright blue with a *white eye,* 10mm, in opposite spikes towards tops of stems. Grassy places, hedges, stony and waste places, to 2250m. Mar-July. T.

3 BROOKLIME *Veronica beccabunga.* Short/med creeping and rooting, rather fleshy, *hairless* per. Lvs *rounded to oblong,* toothed, short-stalked. Fls pale to dark blue, 5-7mm, in opposite spikes. Streams, pools and wet places, to 2500m. May-Sept. T.

4 SPIKED SPEEDWELL *Veronica spicata.* Low/med hairy creeping per, with erect stems. Lvs linear to lance-shaped or oval, blunt-toothed, short-stalked. Fls blue, 4-8mm, in dense *terminal spikes.* Grassland, wood margins, dry and rocky places, to 2050m. July-Nov. T−B*.

PAEDEROTAS. Like Speedwells but the fls 2-lipped and with a short tube.

5 BLUISH PAEDEROTA *Paederota bonarota* (= *Veronica paederota*). Low/short hairy, greenish or bluish per. Lvs oval to rounded, toothed, short-stalked, the upper lvs narrower. Fls violet, blue or pinkish, 10-13mm long, in short terminal spikes. Limestone rock crevices, to 2500m. June-Aug. s, eA.

6 YELLOW VERONICA *Paederota lutea* (= *Veronica lutea*). Low/short, dark green, hairy per. Lvs oval to lance-shaped, toothed, unstalked or short-stalked. Fls *pale yellow,* 10-13mm long, in short terminal racemes. Limestone rock crevices, to 2100m. June-Aug. e, seA.

COW-WHEATS *Melampyrum.* Variable semi-parasitic annuals with opposite, usually toothed, lvs. Floral lvs (bracts) toothed or not. Fls 2-lipped with a short tube, the upper lip rounded, the lower 3-lobed. Calyx 4-toothed; stamens 4.

7 CRESTED COW-WHEAT *Melampyrum cristatum.* Short/med, slightly downy per. Lvs narrow lance-shaped, unstalked; bracts purple, *serrately-toothed.* Fls yellow and purple, 12-16mm, mouth almost closed, in short squarish spikes. Dry grassy and rocky places, woodland margins, to 1500m. June-Sept. T. **7a** *M. velebiticum* has *yellow fls,* 20mm long, *open-mouthed* and only the uppermost bracts toothed if at all. sw, s, seA. **7b** *M. subalpinum* like 7a but bracts *deeply toothed.* eA. **7c** *M. vaudense* like 7 but *fls yellow* 16-20mm long; bracts deeply toothed. swA, J. **7d** *M. italicum* like 7c but bracts only *shallowly-toothed.* Fls 14-16mm. Ap.

8 FIELD COW-WHEAT *Melampyrum arvense.* Short/med per. Lvs lance-shaped, toothed or untoothed; bracts green, whitish or reddish-pink, *deeply toothed.* Fls purplish-pink marked with yellow, 20-25mm, mouth closed, in loose spikes. Dry grassy and rocky places to 1500m. June-Sept. T−B*.

9 COMMON COW-WHEAT *Melampyrum pratense.* Very variable low/med hairless or bristly ann. Lvs linear to oval, untoothed; *bracts green,* untoothed or toothed. Fls whitish to bright yellow, sometimes purple tinged, 10-18mm, mouth usually closed, in loose lfy *one-sided spikes.* Open woods, clearings and peaty ground, to 2250m. June-Sept. T. **9a Wood Cow-wheat** *M. sylvaticum* is shorter *with smaller fls,* 8-10mm, often purple spotted, *open-mouthed,* to 2500m. July-Sept. T−B*. **9b** *M. nemorosum* has stalked lvs and at least the *uppermost bracts purple.* To 1500m. S, A.

10 TOZZIA *Tozzia alpina.* Short/med hairy, semi-parasitic, per; stems *4-angled.* Lvs opposite, oval, toothed, unstalked. Fls golden-yellow, purple-spotted inside, 4-10mm long, funnel-shaped, with *5-lobes.* Damp meadows and streamsides, often amongst coarse herbs, on limestone, to 2250m. June-July. P, A, nAp.

Figwort Family (contd.)

EYEBRIGHTS *Euphrasia*. Small annuals with small opposite, toothed and unstalked lvs. Fls in small clusters or spikes, open mouthed, the upper lip 2-lobed, the lower larger and 3-lobed.

1 COMMON EYEBRIGHT *Euphrasia rostkoviana*. Short hairy, often red-tinted, ann. Lvs oval or oblong, toothed. Fls white, yellow-throated, the upper lip often lilac, 8-12mm long. Meadows, open woods and stony places, to 3000m. July-Oct. T−except S. **1a** *E. hirtella* is shorter with fls only 4-7mm long. Meadows, to 2300m. P,A,Ap. **1b** *E. drosocalyx* is like 1a but the lvs are *narrowed* at the base and only slightly hairy. A.

2 EASTERN EYEBRIGHT *Euphrasia picta*. Short ann. Lvs rounded to triangular or oblong, toothed. Fls white or lilac, with violet veins, yellow-throated, 7-10mm long. Meadows and pastures, 1500-2500m. July-Sept. c,eA*.

3 WIND EYEBRIGHT *Euphrasia nemorosa*. Low/short hairy ann. Lvs elliptical to oblong or triangular, toothed, *usually densely hairy*. Fls small, white to lilac, 5-7mm. Meadows, to 2300m. June-Sept. T−except Ap.

4 GLOSSY EYEBRIGHT *Euphrasia stricta*. Short erect, branched ann, usually strongly tinged with purple. Lvs *glossy green,* hairless, oval or lance-shaped, toothed. Fls lilac or white veined blue, yellow-throated, 7-10mm long. Meadows, dry grassy places and scrub, to 2600m. June-Sept. T−except B. **4a** *E. pectinata* is rarely purple tinged and with slightly smaller white or sometimes lilac fls. P,sw,s,seA,Ap.

5 ALPINE EYEBRIGHT *Euphrasia alpina*. Low/short, purple tinged ann, few branched. Lvs oval or lance-shaped, toothed, hairy or hairless. Fls *usually deep lilac,* 8-14mm. Meadows and pastures on acid rocks, 1500-2500m. May-Sept. P,A,n,cAp. **5a** *E. christii* has *pale yellow* fls. csA−Switzerland and Italy. **5b Dwarf Eyebright** *E. minima* is smaller than 5 and *not* purple tinged. Fls 4-6mm long, whitish or pale yellow with a lilac upper lip. To 3250m. P,A,Ap. **5c** *E. micrantha* like 5 but the lvs always hairless and narrower and the fls lilac or purple, 4-7mm. Heaths, often with *Calluna*. T−except Ap.

6 IRISH EYEBRIGHT *Euphrasia salisburgensis* agg. Low/short purple or bronze tinged ann. Lvs narrow oblong or oval, few toothed, narrowed at the base, hairless. Fls white, sometimes purplish, yellow-throated, 5-7·5mm long. Meadows, scrub and screes, on limestone, to 2600m. July-Sept. T. **6a** *E. portae* has *green lvs* usually; fls 7·5-9mm with a white lower and a lilac upper lip. To 2300m, sA−Italy.

ODONTITES. Semi-parasitic annuals with opposite stalkless lvs. Fls small, in one-sided terminal spikes, 2-lipped, the lower lip glandular-hairy 3-lobed; stamens protruding.

7 STICKY ODONTITES *Odontites viscosa*. Short/med *glandular-hairy* branched ann. Lvs linear to oblong untoothed or almost so. Fls yellow, 5-6mm long. Woods and dry slopes, to 1600m. July-Sept. P,swA.

8 YELLOW ODONTITES *Odontites lutea*. Short/med *almost hairless,* well branched, ann. Lvs linear, toothed or not. Fls bright yellow, 5-8mm long, in loose spikes. Dry meadows and scrubland, on limestone, to 1800m. July-Sept. P,A,Ap.

9 RED BARTSIA *Odontites verna*. Short *often purplish-hairy,* well branched ann. Lvs lance-shaped, toothed. Fls *pink,* 8-10mm long, in dense lfy spikes. Meadows, cultivated fields, waste places and pathways, to 1800m. Aug-Oct. T.

10 ALPINE BARTSIA *Bartsia alpina*. Low/short semi-parasitic downy per, *unbranched.* Lvs opposite oval, toothed, stalkless. Fls dark purple, 15-20mm long, 2-lipped, open mouthed, in short *purple-bracted* spikes. Damp meadows and snowbeds, to 2700m. June-Aug. T−except Ap. **10a** *B. spicata* often has slightly branched stems, the upper bracts narrower and untoothed. cP.

Figwort Family (contd.)

LOUSEWORTS *Pedicularis*. Semi-parasitic pers with pinnately-lobed lvs, the lobes toothed or further lobed; stem lvs alternate or in groups. Fls 2-lipped, open-mouthed and deadnettle-like, the upper lip curved and hood-like, often beaked, the lower 3-lobed; calyx 5-lobed; stamens 4. **For Key, see p.351.**

1 STEMLESS LOUSEWORT *Pedicularis acaulis*. Low *stemless* per. Lvs in tufts, pinnately-lobed. Fls rose-pink, 18-35mm, clustered at lf-bases. Damp or shady meadows, to 1500m. July-Aug. s,seA–Italy and Yugoslavia.

2 LEAFY LOUSEWORT *Pedicularis foliosa*. Short/med hairy per; stems lfy mainly in the *upper half*. Lvs 2-3-pinnately-lobed. Fls pale yellow, 15-25mm, in dense short spikes, the upper lip not beaked. Bracts like lvs, *longer* than the fls. Meadows, streamsides and scrub, often on limestone, to 2500m. July-Aug. P,A,Ap. **2a** *P. hacquetii* is taller, the calyces *densely woolly*. To 1700m. seA,cAp.

3 BEAKLESS RED LOUSEWORT *Pedicularis recutita*. Short/med almost hairless per. Lvs pinnately-lobed. Fls greenish-yellow tinged with dull crimson, 12-15mm, in dense spikes; upper lip *not beaked*. Bracts shorter than fls. Meadows and damp or shady places, 1500-2500m. July-Aug. A.

4 VERTICILLATE LOUSEWORT *Pedicularis verticillata*. Low/short hairy or hairless per. *Lvs in whorls* of 3-4, pinnately-lobed; bracts often purplish. Fls purplish-red, 12-18mm, in dense blunt spikes, the upper lip not beaked. Damp mountain pastures, 1500-3100m. June-Aug. P,Cev,A,Ap. **4a Pink Lousewort** *Pedicularis rosea*. Low/short hairy per, often purple tinged above. Lvs pinnately-lobed, *mostly basal*, not whorled. Fls pink to lilac, 12-18mm, in tight rounded clusters, the upper lip darker, not beaked. Bracts *longer* than fls. Screes and stony grassland to 2700m. July-Aug. s,eA-not Switzerland. **4b** *P.r. allionii* has the *upper bracts* 2-lobed. P*,swA. **4c** *P. hirsuta* like 4a but with woolly upper stems and calyces. To 1400m. June. S.

5 CRIMSON-TIPPED LOUSEWORT *Pedicularis oederi*. Low/short hairy per; stems few lvd or lfless. Lvs pinnately-lobed. Fls yellow with a *crimson tip* to the upper lip, 12-20mm, in dense clusters. Bracts shorter than fls. Damp meadows and rocky places, generally on limestone, to 1950m. July-Aug. S,A*. **5a** *P. flammea* is shorter. To 1300m. S*.

6 RED RATTLE or MARSH LOUSEWORT *Pedicularis palustris*. Low/med hairy or hairless ann/bien, branched from the base, often purplish. Lvs 1-2-pinnately-lobed. Fls reddish-pink or pale pink, 15-25mm, in loose spikes, the upper lip with two tiny lobes, not beaked; calyx 2-lipped, *inflating* in fr. Bracts as long as fls. Damp meadows and marshes on acid soils, to 1800m. May-Sept. S,A. **6a Common Lousewort** *P. sylvatica* is generally per, the upper lip of the fls *without* lobes; calyx *not* 2-lipped. T–except Ap. **6b** *P. asparagoides* like 6a but stems *unbranched*. P,A.

7 CRESTED LOUSEWORT *Pedicularis comosa*. Short/tall hairy per with lfy stems. Lvs 2-pinnately-lobed. Fls pale yellow, 15-25mm, in dense blunt spikes, the upper lip with a *short beak*. Bracts shorter or longer than the fls. Meadows and stony slopes, 1400-2250m. June-Aug. P,sw,wA,Ap.

8 LONG BEAKED YELLOW LOUSEWORT *Pedicularis tuberosa*. Short hairy per with lfy stems. Lvs 2-pinnately-lobed. Fls pale yellow, 15-20mm, in dense heads, the upper lip *long-beaked;* calyx hairy *all over*. Bracts shorter than the fls. Meadows, open woods and screes on acid rocks, 1200-2900m. June-Aug. P,A,Ap. **8a** *P. elongata* is taller with *2 lines* of hairs down the stems and smaller fls; calyx hairy only along teeth margins. Dry meadows and screes, on limestone, to 2300m. seA. **8b** *P. julica* like 8a, but calyx hairy *all over*. seA. **8c Lapland Lousewort** *P. lapponica* has less deeply lobed lvs than 8 and smaller fl clusters, fls held *almost horizontally;* calyx bell-shaped. To 1700m. S.

Figwort Family *(contd.)*

1 BEAKED LOUSEWORT *Pedicularis rostratocapitata.* Low/short per; stems with *two opposite lines* of hairs. Lvs hairless, green tinged purple, 2-pinnate, with oblong, toothed or untoothed, segments; *bracts pinnate.* Fls pink or purplish-red, 15-25mm, long-beaked, in terminal clusters; sepal tube usually hairless. Pastures and screes, usually on limestone, to 2800m. June-Aug. c,e,seA. **1a** *P. kerneri* has a hairless lower lip to the fl. Usually on acid pastures and screes. P,A.

2 ASCENDING LOUSEWORT *Pedicularis ascendens.* Short more or less hairless per. Lvs pinnate with lobed segments, bracts pinnate. Fls *dull yellow,* 12-16mm, beaked, in dense spikes; sepal tube hairless. Meadows on limestone, to 2000m. sw,wcA,Ap.

3 FLESH-PINK LOUSEWORT *Pedicularis rostratospicata* (= *P. incarnata*). Short/med hairy stemmed per. Lvs 2-pinnate, hairless; *bracts untoothed.* Fls pink to purplish-red, 10-13mm, long-beaked, in loose spikes; sepal tube hairy. Damp calcareous meadows, 1500-2700m. July-Aug. eP*—Mt. Canigou,A.

4 PYRENEAN LOUSEWORT *Pedicularis pyrenaica.* Short per; stems with two opposite lines of hairs. Lvs 2-pinnate, hairless; *bracts as lvs* but shorter. Fls pink, 15-20mm, long-beaked, in rounded clusters; sepal tube hairless. Meadows and screes, 1500-2800m. June-Aug. P. **4a** *P. mixta* has *hairless* and more densely lfy stems; fls pink with a deep crimson beak. P. **4b Mt. Cenis Lousewort** *P. cenisia* □ like 4 but the stems, bracts and calyces *white-hairy;* fls as 4a. Meadows 1500-2600m. July-Aug. swA,AA,Ap.

5 TUFTED LOUSEWORT *Pedicularis gyroflexa.* Short downy per. Lvs 2-pinnate; bracts as lvs but shorter. Fls rose-red, 18-25mm, *short-beaked,* in dense clusters; sepal tube downy. Meadows and screes on limestone, 1600-2800m. July-Aug. P*,sw,sA. **5a** *P.g. praetutiana* has *hairless lvs.* c,sAp. **5b** *P. elegans* like 5a but with smaller, stalked fls; sepal tube only *slightly hairy.* c,sAp.

6 TAUERN LOUSEWORT *Pedicularis portenschlagii.* Low per; stems with two opposite lines of hairs. Lvs hairless, pinnate, with oval, toothed, segments; bracts as lvs but shorter. Fls rose-pink or purplish, 20-25mm, *solitary or 2-3-clustered;* sepal tube hairless or almost so. Meadows and screes on acid rocks, 1700-2600m. June-Aug. ec,neA.

7 FERN-LEAVED LOUSEWORT *Pedicularis asplenifolia.* Low per; stems *red-hairy in the upper half.* Lvs hairless, narrow pinnate; bracts as lvs but shorter, red-woolly. Fls rose-red, 15-17mm, short-beaked, in clusters of 2-5; sepal tube red-woolly. Meadows and screes, on acid rocks, 1900-2800m. July-Aug. ec,eA.

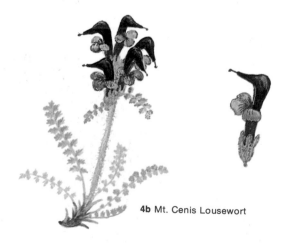

4b Mt. Cenis Lousewort

Figwort Family *(contd.)*

RATTLES *Rhinanthus.* Erect, semi-parasitic annuals. Fls yellow in terminal spikes, mixed with lfy bracts, 2-lipped, the upper lip hooded or beak-like; calyx larger, rounded, rather inflated. A very difficult group with variable species.

1 YELLOW RATTLE *Rhinanthus minor.* Short/med almost hairless ann, stems often spotted. Lvs oval-oblong to narrow lance-shaped, toothed, dark green; bracts triangular, finely toothed, hairless. Fls bright yellow with a violet tip, 13-15 mm long, straight, open-mouthed. Grassy meadows, to 2000 m. May-Sept. T.

2 APENNINE RATTLE *Rhinanthus wettsteinii.* Short ann. Lvs lance-shaped with a few teeth towards the tip; bracts longer than calyx, triangular, *hairy.* Fls yellow, 18 mm long, *mouth closed.* Grassy meadows, to 1500 m. June-Sept. c, sAp. **2a** *R. antiquus* has broader lvs and hairless bracts. sA.

3 ARISTATE YELLOW RATTLE *Rhinanthus aristatus.* Low/med ann, *stems black streaked.* Lvs narrow lance-shaped, toothed, bracts narrow triangular, hairless, the lower teeth *long and slender.* Fls yellow, tipped with violet, 15-18 mm long, curved, open-mouthed. Meadows and thickets, to 2500 m. June-Sept. A. **3a** *R. alpinus* has bracts without long slender teeth at the base and fls yellow tinged purple. eA–Austrian. **3b** *R. carinthiacus* like 3a but bracts and calyces *hairy.* seA–Austrian.

4 BURNAT'S YELLOW RATTLE *Rhinanthus burnatii.* Med ann, stems unstreaked. Lvs oblong, toothed; bracts longer than calyx, both *hairy along edges.* Fls plain yellow, 20 mm, mouth closed. Meadows, to 2000 m. June-Sept. sA–French and Italian **4a** *R. songeonii* has hairless bracts and calyces. sA–French and Italian.

5 SOUTHERN YELLOW RATTLE *Rhinanthus ovifugus.* Med ann, stems black streaked. Lvs narrow lance-shaped or oval, toothed; bracts slightly longer than calyx, hairless, the lower teeth long and slender. Fls yellow, 20 mm long, slightly curved, *mouth closed.* Meadows and thickets to 2000 m. June-Sept. w, sA, Ap. **5a** *R. borbasii* has bracts much longer than calyces. eA.

6 NARROW-LEAVED RATTLE *Rhinanthus angustifolius.* Med ann, stems black streaked. Lvs *linear to lance-shaped,* toothed; bracts longer than calyx, hairless, with small teeth. Fls pale yellow 16-20 mm, slightly curved, mouth closed. Meadows to 2500 m. June-Sept. P, A, Ap.

7 GREATER YELLOW RATTLE *Rhinanthus alectorolophus.* Med/tall ann, stems *unstreaked.* Lvs lance-shaped to oval, toothed, *pale green;* bracts broad-triangular, hairless, evenly toothed. Fls yellow, 20 mm long, slightly curved, mouth closed. Meadows and grassy places, to 2300 m. May-Sept. A.

8 TOOTHWORT *Lathraea squamaria.* Low/med, slightly downy parasitic per; stems white, pale pink or yellowish, *scaly.* Scale lvs alternate, oval-heart-shaped, untoothed. Fls white tinged with pale purple, 14-17 mm long, tubular bells, in dense one-sided spikes, drooping at first. On trees and shrubs, particularly Hazel, Beech and Alder, to 1600 m. Mar-May. T.

Gloxinia Family   Gesneriaceae

9 RAMONDA *Ramonda myconi* (= *R. pyrenaica, Verbascum myconi*). Low hairy per. Lvs deep green, forming flat rosettes, oval, toothed, corrugated, rusty-hairy beneath. Fls blue to violet, 20-30 mm, 5-lobed, solitary or in small clusters on long stems; anthers yellow. Shady rock crevices, to 1800 m. June-Aug. w, cP.

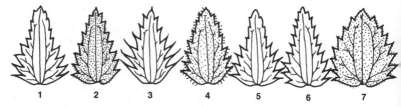

Rhinanthus bracts

Globularia Family   Globulariaceae

Perennials or dwarf subshrubs, often mat forming. Lvs usually untoothed, alternate or in rosettes, shiny. Fls small in dense globular heads; corolla 2-lipped, the upper lip 2-lobed, the lower 3-lobed; stamens 4.

1 APENNINE GLOBULARIA *Globularia incanescens.* Low *deciduous* creeping per. Lvs *grey-green,* rather mealy, rounded to lance-shaped, narrowing abruptly into the stalk, untoothed. Flheads pale blue, 15mm, on lfy stalks 3-6cm long. Rocky and stony slopes, to 2000m. May-Aug. AA, nAp.

2 COMMON GLOBULARIA *Globularia punctata* (= *G. aphyllanthes, G. willkommii*). Short/med tufted evergreen per, stems erect. Basal lvs oval to spoon-shaped, sometimes 3-toothed or notched at tip, long-stalked, green; stem lvs *pointed, stalkless.* Flheads blue, 10-15mm. Meadows, rocks and open woods, often on limestone, to 1650m. May-June. sS, P, A, Ap.

3 MATTED GLOBULARIA *Globularia cordifolia.* Low *creeping evergreen* subshrub, forming dense mats. Lvs all basal, deep green, spoon-shaped, notched or slightly 3-lobed at tip. Flheads lilac-blue or grey-blue, 10-20mm, on erect stalks with 1-2 scales sometimes. Rocks and screes, to 2600m. May-July. eP, A, Ap. **3a** *G. repens* ☐ is smaller with lance-shaped or elliptical *pointed* lvs, 10-20mm long. P, swA. **3b** *G. meridionalis* ☐ like 3 but *more robust,* lvs 20-90mm long. seA, c, sAp.

4 LEAFLESS-STEMMED GLOBULARIA *Globularia nudicaulis.* Short/med tufted evergreen per; stems erect *lfless* but with several small scales. Lvs dark green, oval to spoon-shaped, rounded or notched at tip, stalked. Flheads blue, 15-30mm. Rocky places, dry meadows and open woods, to 2700m. June-Aug. P, A. **4a** *G. gracilis* is smaller and with *short runners;* flheads not more than 15mm. P.

Butterwort Family   Lentibulariaceae

Insectivorous perennials of bogs and wet places. Lvs in sticky rosettes in the Butterworts – margins inrolled; finely divided with tiny bladders and submerged below water in Bladderworts. Fls 2-lipped, spurred; lower lip 3-lobed.

5 ALPINE BUTTERWORT *Pinguicula alpina.* Low per. Lvs elliptical-oblong or lance-shaped, pale yellowish-green. Fls *white* with 1-2 yellow spots in throat, 8-10mm long; spur short, blunt. Bogs, stream banks and damp rocks, to 2600m. June-Aug. S, P, A.

6 SOUTHERN BUTTERWORT *Pinguicula leptoceras.* Low per. Lvs oblong to oval, yellowish-green. Fls blue with a *white-hairy throat patch,* 16-25mm long; spur short, 4-6mm long. Marshes and wet meadows, to 2500m. May-July. A, AA, nAp. **6a** *P. villosa* is smaller with *brownish* lvs; fls *very small,* 7-8mm, pale violet with 2 yellow spots in throat. Sphagnum bogs usually, to 1100m. July. S.

7 LONG-LEAVED BUTTERWORT *Pinguicula longifolia.* Low per. Lvs *narrow-*elliptical to narrow-lance-shaped. Fls large, lilac to pale blue with a large *white-hairy patch* in throat, 30-40mm long, petal *lobes rounded; spur long,* curved, 10-16mm. Wet rocks, to 1600m. July. cP*. **7a** *P.l. reichenbachiana* has very narrow sepals and smaller lvs. AA, cAp*.

8 LARGE-FLOWERED BUTTERWORT *Pinguicula grandiflora.* Low/short per. Lvs oblong to oval, pale green. Fls violet to pale lilac, white-throated, 25-35mm, petal *lobes oblong;* spur 10-12mm long. Bogs and wet rocks, to 2500m. Apr-July. B–sw Ireland, P, wA, Jura.

9 COMMON BUTTERWORT *Pinguicula vulgaris.* Low per. Lvs oblong to oblong-oval, yellowish-green. Fls violet 15-22mm long with a white-throat patch; *spur short* 3-6mm. Bogs, wet heaths and rocks, to 2300m. May-July. T.

10 LESSER BLADDERWORT *Utricularia minor.* Slender aquatic per. Lvs submerged, cut into *thread-like lobes,* bearing tiny *translucent bladders.* Stems slender, erect, carrying 2-6 fls above the water. Fls pale-yellow, 6-8mm long, with a short spur. Ponds, ditches and bogs, to 1850m. June-Sept. Probably T.

Broomrape Family   Orobanchaceae

Hairy unbranched perennials, sometimes annuals, parasitic on the roots of other plants and *without* green pigment. Stems erect with scale-like lvs and dense spikes of tubular 2-lipped fls; upper lip slightly 2-lobed, lower lip 3-lobed. Fr an egg-shaped, many-seeded capsule. Often very local. *Identification of the host plant* often helps in identification of these difficult species.

1 SAND BROOMRAPE *Orobanche arenaria.* Short/med glandular-hairy, bluish or violet tinted per. Scale-lvs lance-shaped. Fls bluish-violet, 25-35 mm long, with *hairy anthers* and a white stigma. Bracts 3 to each fl. Parasitic on Wormworts, Dog Daisies and other herbs. Alluvial river flats, to 1800 m. June-July. P, A, Ap. **1a** *O. purpurea* has smaller bluish-violet fls *veined with deep violet* and hairless anthers. Parasitic on Sneezeworts, Wormwoods and other Composites. Waste places, to 1800 m. T–except S.

2 THYME BROOMRAPE *Orobanche alba.* Short, glandular-hairy, purplish-red tinted, *fragrant,* per. Scale lvs lance-shaped. Fls purplish-red, yellow or whitish, 15-25 mm long, middle lobe of lower lip *larger* than adjacent lobes, stamen-stalks *hairy at base.* Bracts short, one to each fl. Parasitic on Thymes and other Labiates. Woods and grassy places, to 1800 m. Apr-Aug. T–except S. **2a Thistle Broomrape** *O. reticulata* is taller, brownish-violet or yellowish-purple tinged fls, stamens hairless usually; lower lip with *3-equal lobes.* On Thistles. Fields and stony places, to 2500 m. B*, P, A, Ap.

3 AMETHYST BROOMRAPE *Orobanche amethystea.* Short/med slightly glandular-hairy, brownish, pink or bluish tinted ann/per. Scale lvs narrow-lance-shaped to oval, pointed. Fls *cream or white* tinged with violet, pink or brown, 15-25 mm long, stamen stalks hairy at base; upper lip *deeply two-lobed.* Bracts one to each fl. Parasitic on Eryngos, Carrot, Woundworts and various Composites. Fields and stony places, to 2200 m. June-Aug. P, wA. **3a Mugwort Broomrape** *O. loricata* has white or pale yellow fls tinged and veined violet, upper lip only *slightly two-lobed.* On Wormwoods and other Composites, or occasionally on Umbellifers. May-July. eP, A. **3b Common Broomrape** *O. minor* is smaller than 3a. Parasitic mainly on Clovers. P, A, Ap.

4 CLOVE-SCENTED BROOMRAPE *Orobanche caryophyllacea.* Short/med yellowish or purplish ann. Scale lvs triangular-lance-shaped, pointed. Fls *fragrant,* pink or pale yellow tinged dull purple, 20-32 mm long, stamen stalks hairy at base, stigma purple. Bract short, one to each fl. Parasitic on Bedstraws. To 1600 m. T–except S. **4a Germander Broomrape** *O. teucrii* is smaller and parasitic on Germanders. P, A. **4b** *O. lutea* like 4 but with yellowish or reddish-brown fls and a yellow or white stigma. On Clovers, Medicks and other Legumes. P, A, Ap.

5 KNAPWEED BROOMRAPE *Orobanche elatior.* Short/med reddish or honey-brown per. Fls yellow, often tinged pink, 18-25 mm long, stamens stalks hairy at base, stigma yellow. Bract one, *as long as fls.* On Knapweeds, Globe Thistles and other Composites. To 1600 m. June-July. B, eP, A. **5a Alsace Broomrape** *O. alsatica* has its *stem swollen* at the base and with smaller fls and bracts, fls often tinged purple or brown. On Umbellifers. A.

6 SERMOUNTAIN BROOMRAPE *Orobanche laserpitii – sileris.* Rather like 5 but more robust and *stem swollen* at the base. Fls brownish-violet, yellowish at base, 22-30 mm long with one bract to each. Parasitic on Sermountain. To 1600 m. June-Aug. A.

7 YELLOW BROOMRAPE *Orobanche flava.* Short/med yellowish or brownish per, stem scarcely swollen at the base. Fls *yellow,* slightly red tinged, 15-22 mm long, stamen stalks hairy at base, stigma pale yellow. Parasitic on Butterburs, Coltsfoot and Adenostyles. Stony places, to 1700 m. June-July. A. **7a Sage Broomrape** *O. salviae* is shorter with a hairy style and yellow stigma turning orange-brown. On Salvias, to 1200 m. A. **7b** *O. lucorum* like 7a but the stigma turning reddish or purplish with age. On Barberry. eA.

8 SLENDER BROOMRAPE *Orobanche gracilis.* Short/med yellow or reddish per. Scale lvs oval to lance-shaped. Fls yellow tinged red outside, *shining dark red inside,* 15-25 mm long, stigma yellow. Bract short, one to each fl. On various Legumes, to 1350 m. June-Aug. P, A, Ap.

1

2

3

4

5

6

7

8

Bedstraw Family   Rubiaceae

Perennials, rarely annuals, with weak square stems. Lvs in whorls. Fls small, funnel-shaped, 4-lobed; sepals absent. Fr a 2-lobed nutlet. (See also p.312.)

1 LADY'S BEDSTRAW *Galium verum*. Short/tall slightly hairy per; stem slender. Lvs linear, deep shiny green above. Fls *golden-yellow,* 2-3mm, short-tubed, fragrant. Meadows and hedgebanks to 2000m. June-Sept. T.

2 SWISS BEDSTRAW *Galium helveticum*. Low, sprawling, almost hairless per. Lvs oval, in whorls of 4-8, the edges with hooked prickles. Fls yellowish-white, 2mm, in branched clusters. Meadows and screes, to 3200m. June-Aug. A.

3 SIX-LEAVED WOODRUFF *Asperula hexaphylla*. Low/short sprawling hairless per. Lvs lance-shaped, generally in *whorls of 6.* Fls rose pink, 5-6mm long, in terminal clusters. Limestone rocks, to 2000 m. June-July. swA. **3a**□, p.337.

4 PYRENEAN WOODRUFF *Asperula pyrenaica* agg. Variable loose cushion-forming per; stems weak. Lvs linear to elliptical, in groups of 2-4, lower broader. Fls pale pink or purplish, 3-5mm long. Rocky and grassy places, on limestone, to 2000 m. June-Aug. P,A,Ap. **4a**, p.337.

5 SOUTHERN WOODRUFF *Asperula taurina*. Short/med hairy per, creeping underground. Lvs lance-shaped or oval, in whorls of 4, *3-veined.* Fls white or pale yellowish, 7-11mm long, in dense clusters surrounded by a ruff of lvs. Deciduous woods, scrub and shady rocks, to 1300m. c,sA,Ap.

6 BLUE WOODRUFF *Asperula arvensis*. Short/med slender hairless ann. Lvs lance-shaped, blunt, in whorls of 6-8. Fls *bluish-violet,* 3-6mm long, in dense branched clusters. Fields and waste places, to 1500m. Apr-July. B*,P,A,Ap.

Honeysuckle Family   Caprifoliaceae

Shrubs or subshrubs with opposite lvs. Fls 5-lobed, with a short or long tube, sometimes 2-lipped; ovary below fl, calyx of 5 tiny teeth. Fr a fleshy berry.

7 TWIN FLOWER *Linnaea borealis*. Evergreen creeping subshrub. Lvs oval, toothed. Fls pinkish-white, bell-shaped, 5-9mm, drooping, in pairs on long stalks, fragrant. Coniferous woods and heaths, 1200-2200m. June-Aug. B*,S,A*.

8 BLUE-BERRIED HONEYSUCKLE *Lonicera caerulea*. Deciduous shrub to 2m, slightly hairy. Lvs elliptical to oblong, hairless. Fls yellowish-white, bell-shaped with 5-equal lobes, 12-16mm, hairy outside, in pairs. *Berry blue-black.* Woods and scrub on acid soils, 1300-2600 m. May-July. S,P,A. **8a**, p.337.

9 PYRENEAN HONEYSUCKLE *Lonicera pyrenaica*. Deciduous shrub to 1m, *hairless.* Lvs bluish-green, oval, broadest above the middle, pointed. Fls white, often red tinged, bell-shaped with 5-equal lobes, 12-20mm, in pairs. Berry bright red. Woods and rocks, mainly limestone, to 1500m. May-July. P.

10 ALPINE HONEYSUCKLE *Lonicera alpigena*. Deciduous shrub to 3m, slightly hairy. Lvs oblong to elliptical, pointed, *hairy on edges* when young. Fls yellowish or greenish-yellow, tinged reddish-brown, *2-lipped,* 12-20mm long, in pairs. Berry scarlet. Woods, scrub and rocks, on limestone, to 2300m. May-July. P,A,Ap.

11 BLACK-BERRIED HONEYSUCKLE *Lonicera nigra*. Deciduous shrub to 2m, hairy or hairless. Lvs narrow elliptical to oblong, *bluish-green beneath.* Fls pale pink, 2-lipped, 6-10mm, hairy outside, in pairs, faintly fragrant. Berry black. Woods, scrub, stony places, to 1800m. May-July. P,A.

12 FLY HONEYSUCKLE *Lonicera xylosteum*. Deciduous shrub to 3m, hairy. Lvs almost rounded or oval, stalked, grey-green, *hairy above and beneath.* Fls yellowish-white, 2-lipped, 8-12mm long, hairy outside, in pairs. Berry bright red. Woods and scrub, generally on limestone, to 1800m. May-June. B*,S,P,A.

13 COMMON HONEYSUCKLE *Lonicera periclymenum*. Deciduous *woody climber.* Lvs oblong to elliptical, dark green above, bluish-green beneath. Fls creamy-white, often red tinged, 2-lipped, 35-50mm, in terminal clusters, *highly fragrant.* Berry red. Woods, scrub and hedgerows, to 1800m. June-Oct. T.

Honeysuckle Family *(contd.)*

1 WAYFARING TREE *Viburnum lantana.* Downy deciduous shrub to 6 m. Lvs *oval,* finely toothed, stalked, green and wrinkled above, whitish beneath. Fls creamy-white, cup-shaped, 5-9 mm, in dense flat-topped clusters, fragrant. Fr a black berry, at first red. Open woods and scrub, on limestone, to 1600 m. Apr-June. T. **1a Guelder Rose** *V. opulus* has *3-5-lobed lvs;* outer sterile fls of clusters *much larger* than the inner ones. Berry red when ripe. T−except Ap.

2 ALPINE or RED-BERRIED ELDER *Sambucus racemosa.* Deciduous shrub to 4 m; bark grey. Lvs *pinnate;* lflets oval or elliptical, toothed. Fls creamy-white, cup-shaped, in dense branched *pyramidal clusters.* Fr a shiny red berry. Woods and shady rocky places, to 2050 m. Apr-June. T,(B). **2a Common Elder.** *S. nigra* has *flat-topped* fl clusters and black berries. To 1600 m. T. **2b Dwarf Elder** *S. ebulus* is a spreading per to 2 m with pinnate lvs and lfy stipules. Fls white with *purple anthers;* berry black. T,(B).

Moschatel Family   Adoxaceae

3 MOSCHATEL *Adoxa moschatellina.* Low carpeting hairless, pale green per. Lvs *2-trifoliate,* long stalked; lflets lobed. Fls pale green, tiny, 5-lobed, in a tight box-like cluster. Fr a green berry, drooping. Damp woods, hedgerows and shady rocks, to 2400 m. Mar-June. T.

Valerian Family   Valerianaceae

Perennials with opposite lvs, whorled at base of stems; no stipules. Fls small with 5 petals joined into a tube, often spurred or pouched at the base, in clusters; stamens 3 (1 in *Centranthus*). Fr small, one-seeded, with a rather feathery persisting calyx.

4 COMMON VALERIAN *Valeriana officinalis.* Variable short/tall per with *furrowed hairless stems.* Lvs pinnate, the lower stalked. Fls pink or white, 2·5-5 mm long, the tube pouched at the base, in rounded branched clusters. Damp or dry meadows and stony places, to 2400 m. May-Aug. T−except S. **4a** *V.o. collina* has stems densely hairy below and smaller fls, 2-2·5 mm long. P,swA.

5 PYRENEAN VALERIAN *Valeriana pyrenaica.* Tall dark green per with solitary furrowed stem, *hairy at the nodes.* Basal lvs oval, heart-shaped at base, toothed, long stalked; stem lvs with 1-2 pairs of small lflets at base, stalkless. Fls pink, 2·5-3 mm long, in dense clustered heads. Damp meadows and woods, to 2400 m. June-Aug. P.

6 GLOBULARIA-LEAVED VALERIAN *Valeriana globulariifolia.* Low/short creeping per, with erect hairless stems. Basal lvs oblong or spoon-shaped, *untoothed,* stalked; stem lvs with 3-5 narrow lflets, stalked. Fls pink, 4-5 mm long, in small dense clusters. Limestone rocks, to 2200 m. June-Aug. P*. **6a Marsh Valerian** *V. dioica* is taller, the stems slightly *hairy at the nodes* and the fls 1·5-2·5 mm long. Marshes and wet places, to 1800 m. May-June. T.

7 THREE-LEAVED VALERIAN *Valeriana tripteris.* Short/med per, stems hairy at the nodes. Basal lvs oval, heart-shaped at the base, toothed, long stalked; stem lvs *trifoliate,* usually short-stalked. Fls pink or white, 2-4 mm long, in stalked clusters. Woods, scrub and rocky places, usually on limestone, to 2600 m. June-Aug. P,A,Ap. **7a** *V. montana* has *untoothed* lower lvs and oval toothed upper lvs, *not divided.* Apr-July. P,A,Ap.

8 ENTIRE-LEAVED VALERIAN *Valeriana saliunca.* Low/short tufted per, *stems hairless.* Basal lvs lance to spoon-shaped, untoothed, short-stalked; stem lvs *one pair only,* lance-shaped, sometimes 3-lobed, stalkless. Fls deep pink, 3·5-4·5 mm long, in small terminal clusters. Rocks, screes and stony slopes, usually on limestone, 1800-2700 m. July-Aug. A, cAp*.

9 CELTIC SPIKENARD *Valeriana celtica.* Low/short hairless per. *All lvs oval or oblong,* untoothed, stalked, the upper narrower. Fls yellowish or brownish, 1-2 mm long, in small whorled spikes; male and female on separate plants. Acid alpine pastures, 1800-2800 m. July-Aug. A.

1

2

3

4

5

6

7

8

9

227

Valerian Family *(contd.)*

1 ROCK VALERIAN *Valeriana saxatilis.* Low/short per, stems hairless. Basal lvs *elliptical-oblong or lance-shaped,* long-stalked, toothed or not; stem lvs absent or one pair. Fls white, 1-2mm, in loose branched clusters; male and female on separate plants. Limestone rocks, to 2500m. June-Aug. e,ec,sA,nAp.

2 ELONGATED VALERIAN *Valeriana elongata.* Low/short per with slender hairless stems. Basal lvs oval or oblong, scarcely toothed, long-stalked; stem lvs 1-2 pairs, oval or triangular, short-stalked. Fls *brownish or greenish,* 1-2mm, in long loose clusters; male and female on separate plants. Limestone rocks and screes, 1400-2200m. June-Aug. e,seA.

3 DWARF VALERIAN *Valeriana supina.* Low per with *hairy stems.* Basal lvs spoon-shaped or rounded, blunt toothed or untoothed, stalked; stem lvs 1-2 pairs, narrower, unstalked. Fls deep pink, 3-4mm, in dense clusters; male and female on separate plants. Limestone rocks and screes, often by snow patches, 1800-2900m. July-Aug. s,e,seA*.

4 NARROW-LEAVED VALERIAN *Centranthus* (= *Valeriana*) *angustifolius.* Med/tall branched, hairless, blue-green per. Lvs linear, blunt, untoothed with clusters of smaller lvs at the nodes. Fls pink, 7-9mm long, with a spur at the base of each, in loose, branched, clusters. Rocky slopes and screes, usually on limestone, to 2000m. May-Aug. swA,J,Ap. **4a** *C. lecoqii* is shorter, the stems scarcely branched and the lvs lance-shaped, pointed, *without* clusters of smaller lvs at each node. sP.

Bellflower Family   Campanulaceae

Perennials or annuals with alternate undivided lvs, often with a milky latex when broken; no stipules. Fls 5-parted, solitary or clustered in heads or spikes; corolla bell-shaped or starry, the petals joined into a short or long tube. Fr a many seeded capsule.

5 MOUNTAIN SHEEPSBIT *Jasione montana.* Low/med, hairy ann/bien, stems lfless in the upper half. Lvs narrow-oblong to narrow lance-shaped, usually untoothed but often *wavy-edged;* bracts oval or triangular, shorter than fls. Fls blue, rarely pink or white, in globular heads. Dry grassy places and heaths on acid soils, to 1700m. May-Sept. T.

6 DWARF SHEEPSBIT *Jasione crispa* (= *J. humilis*). Low densely tufted, hairy per, often with short non-flg shoots. Lvs oblong or lance-shaped, toothed or not; bracts oval, toothed, green or purplish. Fls blue in globular heads. Meadows and screes on acid soils, to 2500 m. July-Aug. c,eP,Cev. **6a**, p.337.

RAMPIONS *Phyteuma.* Fls in dense globular heads or spikes; corolla with 5 long strap-shaped lobes joined near the base; stigma long, protruding.

7 DEVIL'S CLAW *Physoplexis comosa* (= *Phyteuma comosum*). Low tufted hairless per. Lvs kidney-shaped to oblong, coarsely toothed, stalked, bright shiny-green. Fls pinkish lilac, tipped with blackish-violet, 16-20mm long, the lobes *not separating* at top, in globular heads. Limestone and dolomitic rock crevices, to 2000m. July-Aug. sA.

8 SPIKED RAMPION *Phyteuma spicatum.* Med/tall hairless per. Lvs oval-heart-shaped, toothed, long-stalked, the upper lvs linear; bracts linear, short. Fls *yellowish or greenish-white,* in cylindric spikes up to 6cm long. Meadows and woods, to 2100 m. May-July. B*,S*,P,A. **8a, 8b**, p.337.

9 BETONY-LEAVED RAMPION *Phyteuma betonicifolium.* Short/med per, stems often slightly hairy at the base. Lvs oval-lance-shaped, *truncated at base,* toothed, long-stalked; the upper lvs linear; bracts narrow, inconspicuous. Fls reddish-blue to reddish-violet in cylindric spikes up to 4cm long. Meadows and woods, to 2650m. May-Aug. A.

10 SCORZONERA-LEAVED RAMPION *Phyteuma scorzonerifolium.* Med/tall almost hairless per. Lvs *narrow-lance-shaped,* toothed, the upper smaller; bracts linear, inconspicuous. Fls pale bluish-lilac in cylindric spikes to 5cm long. Meadows and open woods, to 2200 m. June-July. sw,scA,n,cAp. **10a, 10b**, p.337.

Bellflower Family (contd.)

1 BLACK RAMPION *Phyteuma nigrum.* Short/med hairless per. Lvs oblong-elliptical, heart-shaped at base, blunt-toothed, stalked, the uppermost much narrower; bracts linear, *longer than fls.* Fls blackish-violet, rarely blue or white, in oval or cylindric heads. Meadows and woods, to 1200 m. July-Sept. neA.

2 MARITIME RAMPION *Phyteuma balbisii* (= *P. cordatum*). Short slender hairy or hairless per. Lvs rounded or kidney-shaped, toothed, long-stalked, the upper lvs heart-shaped, short-stalked; *bracts small.* Fls bluish-white in globular or oblong heads. Limestone rocks, to 2000 m. July-Aug. MA*.

3 ROUND-HEADED RAMPION *Phyteuma orbiculare.* Short/med sometimes slightly hairy per. Lvs narrow lance-shaped to heart-shaped, toothed, stalked; uppermost narrower, scarcely stalked; *bracts oval,* toothed or not, variable in size. Fls dark blue or violet-blue, rarely white, in globular heads. Dry meadows and rocky ground, often on limestone, to 2600 m. May-Oct. P, A, Ap. **3a** *P. sieberi* has *broader,* oval-lance-shaped, stalkless, upper lvs and broader bracts. Limestone rocks, 1600-2600 m. July-Sept. seA.

4 HORNED RAMPION *Phyteuma scheuchzeri.* Short/med hairless per. Lvs lance-shaped to heart-shaped, thick, bluish-green, toothed, long-stalked, the upper narrow lance-shaped, stalkless; bracts linear, *much longer* than fls. Fls deep blue in globular heads. Rocky slopes, to 2600 m. June-July. sA. **4a** *P. charmelii* has thinner *bright green lvs,* the basal ones usually withered at flg time. To 1900 m. July-Aug. P, sA, nAp*.

5 RHAETIAN RAMPION *Phyteuma hedraianthifolium.* Low/short, more or less erect, hairless per. *All lvs linear,* broadest in the middle, finely toothed; bracts linear, longer than fls. Fls dark blue-violet, straight in bud, in globular heads. Rocky and stony places, 1800-3100 m. July-Aug. cA–nent Austria.

6 GLOBE-HEADED RAMPION *Phyteuma hemisphaericum.* Low/short hairless per. Lvs lance-shaped to linear, *usually untoothed,* the upper linear; bracts oval, sometimes toothed at the base. Fls dark violet-blue, curved in bud, in globular heads. Meadows, stony slopes and screes, on acid rocks, to 2900 m. July-Aug. P, A, Ap.

7 DWARF RAMPION *Phyteuma humile.* Low/short hairless per. Lvs linear, *broadest above* the middle, scarcely toothed, crowded towards the base; bracts oval, toothed, as long as fls. Fls dark blue-violet, in globular heads. Stony meadows, acid rocks, screes and moraines, 1800-3250 m. July-Aug. swA.

8 ROSETTE-LEAVED RAMPION *Phyteuma globulariifolium.* Low almost hairless per, stems almost lfless. Lvs in a *basal rosette,* oblong or spoon-shaped, blunt-toothed, stalked; bracts oval, generally shorter than fls, with a *hairy margin.* Fls deep violet-blue, curved in bud, in globular heads. Acid rocks and screes, 2000-3000 m. July-Sept. eA. **8a** *P.g. pedemontanum* (= *P. pedemontanum*) is taller with pointed lvs and lance-shaped bracts, 1300-2600 m. P, w, cA. **8b** *P. rupicola* like 8 but the *basal lvs rounded* and the heads only 4-6 fld. wP. **8c** *P. nanum* like 8 but taller with broader lvs, the bracts *without* a hairy margin.

9 EDRAIANTHUS *Edraianthus graminifolius* (= *Wahlenbergia graminifolia*). Low/short tufted per, stems hairy. Lvs narrow lance-shaped or linear, untoothed, the upper lvs smaller; bracts oval, pointed, generally shorter than the fls. Fls blue or bluish-violet, 12-20 mm long, upright bells *in close clusters* of 3-6. Seed pods opening at top. Rocky and stony places, 1500-1800 m. May-Aug. c, sAp.

9 Edraianthus

Bellflower Family (contd.)

BELLFLOWERS *Campanula.* Perennials, sometimes biennial or annual, with solitary, clustered or racemed, nodding or upright bell-flowers, blue or purple, sometimes white (yellow in *C. thyrsoides*).

1 MT. CENIS BELLFLOWER *Campanula cenisia.* Low, rather sprawling hairy per with numerous non-flg runners forming lfy-rosettes. Lvs blue-green, oval, broadest above the middle, untoothed, stalkless. Fls blue, 10-15 mm, *starry open bells,* erect, solitary. Moraines, screes and rocky ledges, seldom on limestone, 2000-3100 m. July-Sept. A*.

2 CRIMPED BELLFLOWER *Campanula zoysii.* Low delicate, tufted, hairless per. Lvs oval to oblong, untoothed. Fls pale blue, 15-20 mm long, solitary or up to five, narrow bells *crimped in* at the end. Limestone rock crevices and screes, to 2300 m. July-Aug. seA–Julian and Karawanken.

3 SPREADING BELLFLOWER *Campanula patula.* Med hairy or hairless rather *rough* per; stems slender, erect. Lvs oval, broadest above the middle, *toothed,* the upper narrower and unstalked. Fls violet-blue, sometimes white, 20-25 mm long, erect wide bells in *spreading clusters;* sepal teeth linear. Grassy places, woods and scrub, to 1600 m. June-July. T. **3a** *C.p. costae* is more robust with long, *toothed, sepal teeth.* eP–Val d'Aran. **3b Rampion Bellflower** *C. rapunculus* has *smaller* pale blue or white fls, 10-12 mm, often in branched clusters. To 1600 m. T.

4 CREEPING BELLFLOWER *Campanula rapunculoides.* Med/tall slender, hairy or almost hairless, per. Lvs oval, heart-shaped, the lower long-stalked, the upper narrower and stalkless. As deep purple or violet, 20-30 mm, *drooping,* in long spikes; sepal teeth *curved backwards.* Fields and woods, to 2000 m. July-Aug. T-except B.

5 PEACH-LEAVED BELLFLOWER *Campanula persicifolia.* Med/tall hairless per. Lvs lance-shaped or oval, *peach-like,* pointed, finely toothed, unstalked. Fls violet-blue, 30-40 mm long, *half-nodding* broad bells in loose racemes. Meadows, woods and scrub, to 2000 m. May-Aug. T.

6 BEARDED BELLFLOWER *Campanula barbata.* Short hairy per, unbranched. Lvs mainly in basal rosettes, lance-shaped or oblong, bristly, wavy-edged, untoothed. Fls pale blue, 20-30 mm long, *white-hairy inside,* nodding in loose one-sided racemes. Meadows, open woods and stony places, to 3000 m. June-Aug. S, Cev, A.

7 ALPINE BELLFLOWER *Campanula alpina.* Low/short, hairy, erect per. Lvs mostly in basal rosettes, narrow-lance-shaped, *finely toothed.* Fls lilac to lavendar-blue, 15-20 mm long, half-nodding bells in loose clusters. Rocky and stony places, 1250-2400 m. July-Aug. eA.

8 PYRENEAN BELLFLOWER *Campanula speciosa.* Short/med erect, un-branched, rather bristly, bien/per. Lvs narrow-lance-shaped, 5-10 cm, lightly toothed, crowded towards base of the stem. Fls blue-violet, 28-32 mm long, almost erect bells in pyramidal clusters; epicalyx segments short, broad, bent backwards. Limestone rocks and screes, to 1600 m. July-Aug. P, Cev. **8a** *C.s. oliveri* is *very short,* the stems only 1-2 fld. eP. **8b** *C.s. corbariensis* like 8 but short and *branched from the base;* fls 15-20 mm long. neP–Corbières. **8c** *Campanula affinis* subsp. *bolosii.* Short/med bristly, unbranched, bien; lvs narrow-lance-shaped, 10-15 cm, stalkless; fls violet 20-40 mm long, rather inflated bells, petal lobes hairy-edged. eP.

9 LARGE-FLOWERED BELLFLOWER *Campanula alpestris* (= *C. allionii*). Low creeping, rather sprawling, hairy per. Basal lvs in rosettes, narrow-lance-shaped, unstalked, scarcely toothed; upper lvs linear. Fls pale to deep blue, 30-45 mm long, *usually solitary,* slightly nodding bells. Limestone rocks and screes, 1400-2800 m. July-Aug. swA, not Switzerland.

BELLFLOWERS

233

Bellflower Family *(contd.)*

1 ROCK BELLFLOWER *Campanula petraea.* Short/med hairy per. Lvs oval-lance-shaped, toothed, stalked, *white-hairy beneath.* Fls pale yellow, 12mm long, in tight heads. Limestone rocks, to 1300m. sA−French and Italian.

2 CLUSTERED BELLFLOWER *Campanula glomerata.* Variable short/med *rough-hairy,* tufted per. Lower lvs lance-shaped to oblong or elliptical, heart-shaped or rounded at base, toothed, long-stalked; upper lvs narrower, unstalked. Fls deep violet or blue-purple, 15-25mm long, in *tight heads;* sepal teeth lance-shaped. Meadows, woodland margins and scrub, to 1700m. June-Aug. T−except Ap. **2a** *C. foliosa* is larger with *slightly winged* lf stalks; sepal teeth very narrow. To 1800m. c,sA.

3 SPIKED BELLFLOWER *Campanula spicata.* Med/tall hairy per. Basal lvs tufted, narrow-lance-shaped, untoothed; stem lvs narrower, pointed. Fls lilac-purple or blue, 17-22mm long, in *long lfy spikes.* Meadows, rocky and stony places, 1500-2400m. July-Aug. A,n,cAp.

4 YELLOW BELLFLOWER *Campanula thyrsoides.* Short/med bristly-hairy bien. Lvs oblong-lance-shaped with wavy untoothed margins; upper lvs narrower, clasping. Fls *pale yellow,* 17-22mm long, in dense blunt spikes. Meadows and stony places, on limestone and schist, 1500-2700m. July-Sept. A,J.

5 RAINER'S BELLFLOWER *Campanula raineri.* Low tufted hairy per. Lvs greyish-green, oval or oblong, lightly toothed, almost stalkless. Fls usually solitary, pale blue, 30-40mm long, *broad erect bells.* Limestone crevices, 1300-2200m. Aug-Sept. sA−Bergamasque Alps. **5a** *C. morettiana* has more rounded, deeper-toothed, lvs and *smaller narrower* deep blue fls. 1500-2300m. Aug-Sept. sA−Dolomites.

6 GIANT BELLFLOWER *Campanula latifolia.* Med/tall hairy per; stems slightly angled. Lvs oval, heart-shaped at base, toothed, long-stalked, the uppermost stalkless. Fls large, blue, 40-55mm, in loose racemes; sepal teeth erect. Meadows, woods and riverbanks, to 1600m. July-Sept. T. **6a Nettle-leaved Bellflower** *C. trachelium* has more angular, *bristly stems* and rough-hairy lvs; fls violet-blue or pale blue, 30-40mm long. T. **6b** *C. bononiensis* like 6b but lvs *white-hairy beneath* and fls smaller, 10-20mm, in one-sided racemes. To 1500m. A.

7 HAREBELL *Campanula rotundifolia* agg. Variable short/med hairless per; stems slender, erect from a lfy mat. Lvs rounded to kidney-shaped, blunt-toothed, stalked, mostly crowded at base of stems; upper lvs narrow-lance-shaped, stalkless. Fls pale to mid-blue, 12-20mm long, nodding bells, solitary or in loosely branched clusters, erect in bud. Meadows, open woods, shrub and banks, to 2200m. May-Nov. T. **7a Flax-leaved Bellflower** *C. carnica,* has fls *drooping in bud;* fls larger, 22-26mm. Limestone crevices, to 2000m. s,seA. **7b** *C. scheuchzeri* like 7 but *all lvs* narrow lance-shaped or linear, unstalked. To 3400m. S,P,A,Ap. **7c** *C. rhomboidalis* like 7 but with *oval* stem lvs. P,w,sA. **7d Arctic Bellflower** *C. uniflora* ☐ is shorter than 7 with small *solitary* fls, 5-10mm; basal lvs oval, *untoothed.* Heaths and grassy places, to 1600m. July. S.

5 Rainer's Bellflower

Bellflower Family *(contd.)*

1 SPANISH BELLFLOWER *Campanula hispanica catalanica.* Med/tall per; stems erect, hairy in the lower half. Basal lvs *heart-shaped,* toothed, stalked; stem lvs lance-shaped, crowded in the lower half of stem. Fls blue, 10-14mm long, in loose few-fld clusters, erect in bud; sepal teeth linear, *pressed* against petal tube. Fr drooping. Rocky, stony or sandy places, to 2500m. June-Aug. eP. **1a** *C. fritschii* has stems lfless in the upper half and larger fls, 18-22mm. swA–Provence Alps. **1b** *C. apennina* like 1 but with hairless stems which are densely lfy up to the fls; *fr erect.* cAp. **1c** *C. bertolae* like 1b but fls larger, 12-20mm, on hairy stems; fr drooping. swA–Piedmont. **1d** *C. pseudostenocodon* like 1b but short, stems sometimes hairy in the lower part. Ap.

2 FRENCH BELLFLOWER *Campanula recta.* Short/med hairy per. Basal lvs oval-lance-shaped, *untoothed,* stalked; stems with numerous narrower lvs, stalkless. Fls blue, 15-22mm, in few-fl close clusters, *drooping* in bud; sepal teeth spreading. Fr drooping. Grassy and rocky places, to 1800m. June-Aug. P,Cev. **2a** *C. precatoria* has slightly toothed lvs; stem lvs half-clasping, but *absent* just below the fls. Meadows and pastures. eP.

3 PANICULATE BELLFLOWER *Campanula witasekiana.* Short slightly hairy per; stems angular, lfy up to the fls. Basal lvs rounded or kidney-shaped, blunt-toothed, stalked; upper lvs lance-shaped to linear, untoothed, stalkless. Fls blue, 12-16mm long, in branched clusters, drooping in bud; sepal teeth narrow-triangular, spreading. Meadows and rocky places to 1800m. July-Aug. e,seA.

4 FAIRY'S THIMBLE *Campanula cochlearifolia* (= *C. pusilla*). Variable *low,* slender, creeping, hairy or hairless per. Basal lvs heart-shaped to rounded, toothed, stalked, present at flg time; stem lvs lance-shaped, toothed. Fls blue or violet, rarely white, 12-16mm long, solitary or few in a cluster; drooping in bud; sepal teeth linear, spreading. Rocky and stony places, screes, often on limestone, to 3400m. June-Aug. P,A,n,cAp. **4a** *C. cespitosa* is taller with oval or diamond-shaped basal lvs; fls pale blue bells *narrowed at the top.* Limestone rocks and screes, to 2100m. Aug-Sept. e,seA.

5 JAUBERT'S BELLFLOWER *Campanula jaubertiana.* Very low creeping per; stems *densely hairy.* Basal lvs rounded to elliptical, heart-shaped at base, blunt-toothed, stalked; stem lvs oval to elliptical, toothed. Fls blue, 8-12mm long, solitary or 2-4 clustered, narrow bells drooping in bud. Limestone rocks, to 2000m. July-Aug. c,eP.

6 COTTIAN BELLFLOWER *Campanula stenocodon.* Short per, hairy in the lower part. Basal lvs rounded or heart-shaped, sharp-toothed, stalked, absent at flg time; stem lvs linear-lance-shaped, sometimes untoothed. Fls blue-purple, 12-18mm long, *narrow pendulous bells* in few-fld clusters, drooping in bud. Rocky and stony places to 1800m. July-Aug. swA–Cottian, MA. **6a** *C. beckiana* is taller with *many fls* in branched clusters. Meadows and open woods. July-Sept. neA.

7 SOLITARY HAREBELL *Campanula pulla.* Low/short slender per; stems hairy on the angles or hairless. Basal lvs oval or rounded, blunt-toothed, stalked; stem lvs oval or elliptical. Fls blue-purple, 18-24mm, *solitary* pendulous bells, drooping in bud. Grassy and rocky places and screes, often on limestone, 1500-2200m. July-Aug. neA.

8 PERFORATE BELLFLOWER *Campanula excisa.* Low creeping per; stems hairy. Basal lvs heart-shaped to rounded, sharp-toothed, stalked; stem lvs narrow-lance-shaped, untoothed. Fls blue or lilac-blue, 10-16mm long, narrow, rather pleated, bells, the lobes narrowed at the mouth and *apparently perforated.* Acid rocks and screes, 1400-2350m. June-Sept. sw,scA, not France.

4 Fairy's Thimbles

Scabious Family Dipsacaceae

Annuals or perennials with opposite lvs, the basal ones often rosetted. Fls in dense 'composite-like' heads, each floret with 4-5 petal lobes joined into a tube, a calyx, and 4 projecting stamens; outer florets often with 2-3 larger petal lobes. Fr small, one seeded.

1 ALPINE SCABIOUS *Cephalaria alpina.* Tall robust hairy per. Basal lvs pinnate or lyre-shaped with 3-8 pairs of lance-shaped lflets; stem lvs smaller with few lflets or undivided. *Flheads yellow,* 20-30mm, flattish, on longer slender stalks. Meadows, woods, scrub and screes, to 1800m. July-Aug. A, J.

2 SHINING SCABIOUS *Scabiosa lucida.* Short hairless per, stems unbranched, lfless in the upper half. Lvs *glossy-green,* oval-lance-shaped with shallow rounded-teeth, stalked; upper lvs pinnately-lobed with a *large end lobe.* Flheads reddish-purple, 10-20mm, flattish. Dry meadows and stony places, to 2700m. June-Sept. A, Ap. **2a** *S. stricta* has stems lfy for much of the way and mostly with undivided lvs. e, seA.

3 TYROLEAN SCABIOUS *Scabiosa vestina.* Short/med hairy per. Basal lvs narrow spoon-shaped, untoothed, stalked; stem lvs pinnately-lobed, the lflets linear or lance-shaped. Flheads purple, 20-30mm, flattish, the outer florets twice the size of the inner. Limestone rocks, to 1900m. June-Sept. sA–Italian. **3a** *S. graminifolia* is a denser more tufted plant with *silvery-grey* stems and lvs, all the lvs *undivided;* flheads blue-violet. To 1800m. P*, sw, s, seA.

4 PYRENEAN SCABIOUS *Scabiosa cinerea* (= *S. pyrenaica*). Short/med *whitish-hairy* per, stems lfless in the upper half. Basal lvs lance-shaped, toothed, stalked; stem lvs with a large end lflet and 1-2 pairs of smaller, linear, lflets. Flheads bluish-purple, 10-20mm, flattish. Dry meadows and stony places, to 2000m. June-Sept. P. **4a** *S.c. hladnikiana* has greyish lvs and stems lfy most of the way up. eA.

5 DEVILS-BIT SCABIOUS *Succisa pratensis* (= *Scabiosa succisa*). Med/tall hairy per. Basal lvs in a rosette, *elliptical,* untoothed, stalked, often purple-blotched; upper lvs narrower, sometimes toothed. Flheads rounded, mauve to dark purple, 15-25mm; florets all more or less equal in size, sometimes only female. Damp meadows and woods, to 2400m. July-Oct. T.

6 WOOD SCABIOUS *Knautia dipsacifolia* (= *Scabiosa sylvatica*). Med/tall hairy per. Lvs *bright green,* variable on same plant, oblong-oval, constricted or tapered to the base, toothed. Flheads lilac or purplish, 25-40mm, outermost florets *rather larger* than the inner ones. Shady places, woods and scrub, to 2000m. June-Sept. P, A.

2 Shining Scabious

Daisy Family Compositae

The Daisies or Composites form a very large and versatile family. Fls small, closely packed in compound heads surrounded by several to many sepal-like bracts – the flbracts. Petals joined in a tube ending in either 5 small teeth (forming the disc) or in a strap-shaped petal (the rays); flheads either all disc florets (thistles), ray florets (dandelions) or a central disc surrounded by rays (the Daisies). Fr tiny, surrounded by a feathery pappus which floats in the wind. See also pp. 324–34.

1 GOLDEN ROD *Solidago virgaurea.* Variable short/med hairy, or almost hairless, tufted per. Lvs lance-shaped or oblong, broadest above the middle, toothed, stalked, the upper narrower, stalkless. Flheads bright yellow, 15-20mm, *short rayed,* in branched racemes. Woods, clearings, stony places and heaths, to 2800m. July-Oct. T.

2 DAISY *Bellis perennis.* Low hairy per. Lvs *in rosettes,* spoon-shaped, slightly toothed. Flheads solitary, long-stalked, white with a yellow disc, 15-25mm. Grassy and waste places, to 2500m. Flg for most of the year. T.

ASTERS *Aster.* Flheads large with one row of long, blue, purple, lilac or white, spreading rays, surrounding a yellow disc.

3 ALPINE ASTER *Aster alpinus.* Low/short tufted hairy per. Lvs elliptical to spoon-shaped, untoothed, the upper narrower, stalkless. Flheads *solitary,* violet-blue, mauve or rarely white, 35-45mm. Dry meadows, rocky and stony places, to 3200m. July-Sept. T. **3a False Aster** *A. bellidiastrum* (= *Bellidiastrum michelii*) has broader often toothed lvs; flheads pink or white, 20-40mm, *on lfless stems.* To 2800m. A, MA, Ap.

4 EUROPEAN MICHELMAS DAISY *Aster amellus.* Short/med tufted hairy per. Lvs lance-shaped to oval, slightly toothed or untoothed. Flheads blue or purplish, 20-30mm, in *branched* flat-topped clusters, rarely solitary; outer flbracts *shorter* than inner. Meadows, open woods and rocky places, to 1400m. Aug-Sept. cP, A, n, cAp.

5 PYRENEAN ASTER *Aster pyrenaeus.* Med/tall hairy per. Lvs oblong-lance-shaped, lightly toothed, *half-clasping the stem.* Flheads lilac-blue, 20-35mm, long-stalked, solitary or 2-5 in a cluster; flbracts all the *same length.* Damp meadows and stony places, to 2000m. July-Sept. w, cP*.

6 THREE-VEINED ASTER *Aster sedifolius trinervis.* Med/tall slightly hairy per, sometimes ann. Lvs linear to lance-shaped or elliptical, the *lower 3-veined.* Flheads blue or lilac, few rayed, 15-20mm, in dense branched clusters. Fields, dry and waste places, to 1800m. June-Sept. Cev, P.

FLEABANES *Erigeron.* Like *Aster* but flheads small with several rows of narrow short rays and a pale yellow disc. A variable and difficult group.

7 GREEK FLEABANE *Erigeron atticus.* Short/med per; stems glandular-hairy. Lvs oval, broadest above the middle, the lower stalked. Flheads 25-35mm, in clusters, with *erect* violet or purplish rays. Meadows and rocky places, to 2200m. July-Sept. eP, A. **7a** *E. gaudinii* (= *E. glandulosus*) is shorter with stems *branched* at or below the middle. A. **7b Blue Fleabane** *E. acer* is a greyish ann/bien; stems hairy *not* glandular, often purplish; flheads, 10-15mm, rays dull purple, very short. T.

8 VARIABLE FLEABANE *Erigeron polymorphus* (= *E. glabratus*). Variable low/short slightly hairy per, stems slender. Lvs narrow-spoon-shaped or oblong, stalked. Flheads 15-20mm, with *spreading* lilac or white rays, the disc yellow or reddish-brown, solitary or in small clusters; flbracts *green* with a brown centre, slightly hairy. Short grassy and rocky places, to 3000m. July-Sept. P, Cev, A, Ap.

9 ALPINE FLEABANE *Erigeron alpinus.* Short per; stems unbranched, downy. Lvs narrow-elliptical to spoon-shaped, pointed, the lower stalked. Flheads solitary or few clustered, 20-30mm, with spreading lilac rays and a pale yellow disc; flbracts *lilac-tipped,* hairy. Meadows, rocky and stony places, 1500-3050m. July-Sept. T–B*. **9a** *E. epiroticus* is *shorter,* rarely exceeding 10cm; flheads solitary, rays purplish, flbracts *woolly.* cAp.

1

2

3

4

5

6

7

8

9

241

Daisy Family *(contd.)*

1 ONE-FLOWERED FLEABANE *Erigeron uniflorus.* Rather like a shorter version of the Variable Fleabane, p. 240. Flheads *small*, solitary, 10-15 mm, ray florets white or pale lilac, the disc yellow; flbracts *white-woolly*, lilac-tipped. Damp meadows, stony places and moraines, often around snow patches, 1200-3000 m. July-Sept. S, P, A, n, cAp. **1a** *E. humilis* has *black or dark violet* flbracts. To 1400 m. S.

2 NEGLECTED FLEABANE *Erigeron neglectus.* Low/short hairy per. Lvs spoon-shaped or lance-shaped, *hairy along edge only,* the upper stalkless. Flheads solitary, 12-18 mm, ray florets lilac, the disc yellow; flbracts hairy, lilac-tipped. Limestone rocks, 1800-2600 m. July-Sept. A—except extreme e, se.

3 CATSFOOT *Antennaria dioica* (= *Gnaphalium dioicum, Omalotheca dioica*). Low/short, mat-forming, downy per, with *rooting runners.* Lvs oval-spoon-shaped, blunt, untoothed, grey-green above, white-woolly beneath. Flheads in small clusters, rayless, white, pink or reddish; male 6 mm, female larger, on separate plants. Meadows, heaths and dry places, on acid soils, to 3000 m. May-July. T. **3a Alpine Catsfoot** *A. alpina* is shorter with lvs *almost hairless* and smaller flheads. To 2200 m. July-Aug. S.

4 CARPATHIAN CATSFOOT *Antennaria carpatica* (= *Gnaphalium carpaticum*). Low/short tufted, white-downy per; similar to 3 but *no runners.* Lvs oblong to linear, pointed. Flheads in small clusters, brown or blackish. Damp meadows, rocky and stony places, 1500-3100 m. July-Aug. S, P, A.

5 EDELWEISS *Leontopodium alpinum.* Low/short, white or greyish-woolly, tufted per. Lvs oblong, broadest above the middle, untoothed, the upper lvs narrower. Flheads small, yellowish-white, closely clustered, but *surrounded* by conspicuous large oblong, woolly-white, bracts. Grassy and rocky slopes, usually on limestone, 1700-3400 m. July-Sept. P, A, LA. **5a** *L. nivale* is a name sometimes applied to a dwarf form with short spoon-shaped lvs. cAp—Abruzzi.

6 DWARF CUDWEED *Omalotheca supina* (= *Gnaphalium supinum*). Low tufted, greyish-woolly, per. Lvs narrow lance-shaped, untoothed, 1-veined. Flheads small, rounded, rayless, reddish, hidden by brown bracts, *in clusters* amongst uppermost lvs. Damp meadows, stony places, moraines and by snow patches, usually on acid soils, 1400-3400 m. July-Sept. T. **6a** *O. hoppeanum* is not tuft-forming, taller and more slender, to 2650 m. A, Ap.

7 WOOD CUDWEED *Omalotheca sylvatica* (= *Gnaphalium sylvaticum*). Short/med tufted, grey-woolly, per. Lvs lance-shaped to linear, untoothed, usually 1-veined, green above, white-felted beneath. Flheads reddish or yellowish, hidden by brown sepal-like bracts, solitary or in long lfy spikes. Open woods and clearings, to 2500 m. July-Sept. T. **7a Highland Cudweed** *O. norvegica* (= *Gnaphalium norvegicum*) is shorter with broader, lance-shaped, *3-veined,* lvs, white-felted *on both sides.* Meadows, open woods and stony places, usually on acid soils, 1200-2800 m. S, P, A.

8 MOUNTAIN DOG-DAISY *Anthemis cretica* (= *A. montana*). Variable low/short downy to slightly hairy per. Lvs pinnately-lobed, stalked, whitish-hairy or green-hairless. Flheads solitary, white with a yellow disc, 25-40 mm, rays sometimes absent; flbracts woolly. Rocky places and screes, on acid rocks, to 2000 m. July-Sept. P, Cev, MA, nAp. **8a** *A.c. alpina* (= *Santolina alpina*) is usually taller, the lvs densely *yellowish-hairy;* rays generally absent. Limestone rocks. cAp—Abruzzi. **8b Carpathian Dog-daisy** *A. carpatica styriaca* ☐₁ Similar to 8 but lvs *pale green,* 1-2 pinnately-lobed. Flheads 30-50 mm; flbracts *hairless* or almost so. Grassy and stony places. eA. **8c** *A.c. petraea* similar to 8b cAp.

9 SOUTHERN DOG-DAISY *Anthemis triumfetti* (= *Cota triumfetti*). Med/tall slightly hairy, erect per; stems solitary, *branched above the middle.* Lvs pinnately-lobed, lobes toothed. Flheads solitary to each branch, white with a yellow disc, 30-50 mm. Woods and rocky places, to 1600 m. May-Aug. P, sw, csA, Ap. **9a Corn Camomile** *A. arvensis* is a branched *annual* with 1-3 pinnately-lobed greyish lvs. Cultivated areas and waste ground, to 1950 m. T.

Daisy Family *(contd.)*

MOON DAISIES *Leucanthemum*. Pers with toothed, or pinnately-lobed lvs. Flheads flat, daisy-like with white rays and a yellow disc.

1 MOON or OX-EYE DAISY *Leucanthemum vulgare* (= *Chrysanthemum leucanthemum*) agg. Variable short/tall grey or dark green, hairy or hairless per. Lower lvs oblong-spoon-shaped or oblong, pinnately-lobed, toothed or un-toothed, long-stalked; upper lvs stalkless, clasping the stem. Flheads usually solitary, 25-50mm; flbracts with a brown margin. Fields, meadows, open woods and pathways, to 2700m. June-Aug. T.

2 SAW-LEAVED MOON DAISY *Leucanthemum* (= *Chrysanthemum*) *atratum* agg. Variable short/med mat-forming per. Basal lvs spoon-shaped, blunt-toothed or lobed, long-stalked; stem lvs similar, stalkless. Flheads solitary, 20-50mm; flbracts with a dark brown margin and *a small scale-like appendage* at the top. Rocky and stony places, screes and gravels, usually on limestone, to 2850m. July-Sept. A, Ap.

3 ALPINE MOON DAISY *Leucanthemopsis* (= *Chrysanthemum* or *Pyrethrum*) *alpina*. Low tufted hairy per, stems rather weak, almost lfless. Lvs oval to spoon-shaped, toothed or pinnately-lobed, greenish or grey. Flheads solitary, 20-40mm, rays white but *often turning pink*. Short grass, rocky places, screes and moraines, 1800-2800m. July-Aug. T.

4 LAVENDER COTTON *Santolina chamaecyparissus*. Dwarf evergreen, *silvery-white-downy*, subshrub to 50cm, aromatic. Lvs linear, pinnately- lobed or toothed. Flheads *globular*, deep yellow, 10-15mm, rayless; flbracts white-downy. Dry rocky ground and banks, to 1200m. July-Sept. P.swA. **4a** *S.c. tomentosa* is taller with hairless lvs and *whitish or pale yellow* flheads. P, MA, LA, n, cAp.

MILFOILS or SNEEZEWORTS *Achillea*. Rather aromatic perennials with alternate, 1-3-pinnately-lobed lvs. Fls small, short-rayed, usually in flat-topped clusters, often branched.

5 ALPINE SNEEZEWORT *Achillea oxyloba*. Low/short hairy, creeping, per. Lvs mostly basal, pinnately-lobed. Flheads *solitary* or 2-3-clustered, white with a pale yellow disc, 20-30mm. Stony meadows and rocky places, on limestone, 1600-2800m. July-Sept. seA. **5a** *A.o. mucronulata* has 2-pinnately-lobed lvs, *mostly on the stems* and not basal. Ap. **5b** *A. barrelieri* like 5 but lvs and stems *silvery-hairy*. c, sAp.

6 DARK-STEMMED SNEEZEWORT *Achillea atrata*. Short *only slightly aromatic* per; stems *brown-hairy*, at least in the upper half. Lvs deep green, 2-pinnately-lobed, lobes narrow lance-shaped, almost hairless. Flheads white with a whitish disc, 12-18mm, in clusters of 2-10. Stony pastures, rocky places and screes, usually on limestone, 1700-3000m. July-Sept. A.

7 MUSK MILFOIL *Achillea moschata*. Low/short *almost hairless* per, strongly aromatic. Lvs bright green, pinnately-lobed, lobes lance-shaped. Flheads white with a whitish disc, 10-15mm, in loose clusters of 3-25. Stony pastures, rocks and screes, usually on acid rocks, 1450-3500m. July-Sept. cA, Ap.

8 DWARF MILFOIL *Achillea nana*. Low/short *greyish-downy* tufted per. Lvs 1-2-pinnately-lobed, lobes crowded, lance-shaped; lower lvs long-stalked. Flheads off-white, 9-11mm, in *tight clusters* of 6-8. Acid rocks and screes, 1700-3800m. July-Sept. A–except se, Ap.

3 Alpine Moon Daisy

245

Daisy Family (contd.)

1 SILVERY MILFOIL *Achillea clavennae.* Short creeping, silvery-hairy per. Lvs pinnately-lobed, the lobes lance-shaped, sometimes toothed; basal lvs long-stalked. Flheads white with an off-white disc, 10-20 mm, in loose clusters of 5-25. Rocks and screes, often on limestone, 1500-2500 m. July-Sept. c, e, seA.

2 SIMPLE-LEAVED MILFOIL *Achillea erba-rotta.* Short almost hairless per. Lvs *not* pinnately-lobed, spoon-shaped to lance-shaped, toothed, teeth at base often larger. Flheads white, 14-16 mm, in loose clusters of 10-30. Grassy meadows and rocky places, on acid soils usually, 2000-3200 m. July-Aug. swA. **2a** *A.e. rupestris* has mostly *untoothed* lvs. sAp.

3 LARGE-LEAVED SNEEZEWORT *Achillea macrophylla.* Med/tall more or less hairless per. Lvs *large,* pinnately-lobed, lobes 8-12, lance-shaped, toothed, pointed. Flheads white, 13-15 mm, in loose clusters of 5-40. Damp or shady places, on humus rich soils, to 2500 m. July-Sept. wA, nAp.

4 TANSY MILFOIL *Achillea tanacetifolia.* Med/tall slightly downy per. Lvs pinnately-lobed, lobes broad, toothed; lower lflets of upper lvs *clasping* the stem. Flheads pink, 5-6 mm across, rays 2·5-4 mm, many in *broad, branched clusters.* Pastures, wood margins and scrub, to 2500 m. July-Sept. sA. **4a** *A.t. distans* fls *usually white,* rays 1-2·5 mm long. A.

5 YARROW *Achillea millefolium* agg. Short/med downy per, aromatic. Lvs 2-3-pinnate, *feathery,* dark green, hairy above. Flheads white or pink with a creamish disc, 4-6 mm, in large flat-topped clusters. Grassy and waste places, to 2800 m. July-Sept. T. **5a** *A. collina* has lvs hairless above; fls white. s, e, seA. **5b** *A. pannonica* is like 5 but grey or silvery-hairy, fls white. Dry stony places. e, seA.

6 ANDORRAN MILFOIL *Achillea chamaemelifolia.* Short/med tufted, slightly hairy per; stems *branched at base.* Lvs pinnately-lobed, hairless, lobes linear, *untoothed.* Flheads white, 7-9 mm, in loose clusters. Rocky places, to 1700 m. June-Aug. eP.

7 CREAM-FLOWERED SNEEZEWORT *Achillea odorata.* Short hairy per; stems unbranched below flhead. Lvs 1-2-pinnately-lobed, segments ovate, toothed. Flheads *white to pale yellow,* 3-4 mm, in large dense clusters. Dry stony places, to 1450 m. May-July. P.

8 SNEEZEWORT *Achillea ptarmica.* Med/tall hairy per, *not* aromatic. Lvs not divided, lance-shaped, finely-toothed, stalkless, the upper half-clasping the stem, hairless. Flheads white with a creamish disc, 14-18 mm, in loose branched clusters. Damp grassy places on acid soils, to 1700 m. July-Sept. T—except Ap.

9 PYRENEAN SNEEZEWORT *Achillea pyrenaica.* Short/med slightly hairy per. Lvs deep green, not divided, lance-shaped, toothed, stalkless, *covered in* short glandular-hairs. Flheads white with a cream disc, 18-20 mm, in loose branched clusters. Damp grassland, to 1800 m. July-Sept. P, Cev.

8 Sneezewort

Daisy Family (contd.)

1 COLTSFOOT *Tussilago farfara*. Low/short creeping, downy per. Lvs all basal, round-heart-shaped with pointed teeth, stalked, *white-downy beneath*, green above. Flheads bright yellow, *short rayed*, solitary on stems covered with *purplish scales*, often before the lvs. Fr a white 'clock'. Bare and waste ground, fields and banks, to 2600 m. Feb-Aug. T.

2 ALPINE COLTSFOOT *Homogyne alpina*. Low/short creeping per with erect stems. Lvs mostly basal, *kidney-shaped*, blunt toothed, long-stalked, dark-shiny-green above, pale or purplish and hairy beneath; stem lvs bract-like. Flheads reddish-purple or violet, *rayless*, goblet-shaped, 10-15mm, solitary. Damp meadows, open woods, streamsides, to 3000 m. May-Aug. T–B*. **2a**, **2b**, p.337.

3 WHITE BUTTERBUR *Petasites albus* (= *Tussilago alba*). Low/short, patch-forming, downy per; stems scaly. Basal lvs round-heart-shaped, lobed and toothed, *white-downy beneath*, long-stalked. Flheads yellowish-white, rayless, brush-like, in dense rounded clusters; flbracts pale green. Damp meadows, woods, streamsides and gullies, to 2200 m. T−except B, sAp. **3a Alpine Butterbur** *P. paradoxus* (= *P. niveus*, *Tussilago nivea*) has triangular-heart-shaped lvs; flheads purplish-lilac with *reddish flbracts*. Damp places on limestone, to 2600 m. **3b Butterbur** *P. hybridus* like 3 but flheads pale lilac-pink or yellowish, *not fragrant*, flbracts purplish; lvs large, *grey-hairy beneath*, to 1800 m. B, S, A. **3c** *P. frigidus* like 3 but with smaller lvs, large teeth *only* and smaller flheads. To 1759 m. June. S.

4 ADENOSTYLES *Adenostyles alliariae*. Med/tall downy per. Lower lvs triangular-heart-shaped to kidney-shaped, coarsely toothed, long-stalked, *downy but green beneath*; upper lvs much smaller, *clasping stem*. Flheads small, reddish-purple, rayless, 6-8mm long, in dense branched clusters; 3-4 florets to each flhead. Woods, streamsides, scrub and damp rocky places, to 2700 m. July-Aug. P, A, Ap. **4a** *A.a. hybrida* (= *A. pyrenaica*) has larger flheads each with 12-15 florets. P. **4b** *A. alpina* (= *A. glabra*) is smaller with kidney-shaped lvs, *hairless or almost so* beneath. To 2500 m. A, J, Ap. **4c** *A. leucophylla* (= *A. tomentosa*) like 4a but shorter, stems and lvs *white-woolly*, upper lvs stalked. 1900-3100 m. sw, cA.

5 YELLOW OX-EYE *Buphthalmum salicifolium* (= *B. grandiflorum*). Short/med softly hairy per, *stems branched* in the upper half. Lvs oblong to narrow lance-shaped, slightly toothed or untoothed, the lower stalked. Flheads bright yellow, 30-50mm, long-rayed. Damp and stony places, woods, to 2050 m. June-July. A, nAp.

6 ARNICA *Arnica montana*. Short/med downy per, aromatic. Lvs mostly in *basal rosettes* elliptical to oblong, usually broadest above the middle, untoothed; stem lvs few, bract-like, opposite. Flheads large, yellow, 4·5-6cm, long-rayed, usually solitary. Meadows and open woods, to 2850 m. T−except B and extreme north. May-Aug. **6a** *A. alpina* is shorter with smaller fls, 3·5-4·5cm, stem lvs *in pairs*. S.

7 AUSTRIAN LEOPARDSBANE *Doronicum austriacum*. Med/tall, patch-forming, hairy per. Basal lvs oval-heart-shaped, slightly toothed, stalked; stem lvs clasping, the *lower stalked as well*. Flheads 5-6cm, in *branched clusters* of 5-12. Shady meadows, woods and streamsides, to 2000 m. July-Aug. P, A, Ap. **7a**, **7b**, p.337.

8 LARGE FLOWERED LEOPARDSBANE *Doronicum grandiflorum* (= *D. scorpioides*). Low/short slightly hairy per. Basal lvs oval, *narrowed into* the long stalk, toothed; upper lvs lance-shaped, clasping stem. Flheads *solitary*, 3·5-6·5cm, ray florets with a pappus. Stony meadows, rocks and screes, 1300-2900 m. July-Aug. P, Cev, A. **8a**, p.338.

9 HEART-LEAVED LEOPARDSBANE *Doronicum columnae*. Short/med patch-forming, hairless or slightly hairy per. Lvs mostly basal, triangular-heart-shaped, stalked, toothed, stem lvs *stalkless, clasping*. Flheads solitary, 2·5-5cm. Stony places, open woods and scrub, to 2300 m. May-Aug. s, e, seA, Ap. **9a**, p.338.

2

3

4

5

6

7

8

9

249

Daisy Family (contd.)

CARLINE THISTLES *Carlina*. Biennials or perennials with spiny lvs. Flheads large with spiny lf-like outer bracts and narrow pointed silvery or yellowish inner ones, surrounding a disc of many tubular florets. Flheads closing in bad weather.

1 STEMLESS CARLINE THISTLE *Carlina acaulis*. Low bien/per, *stemless*. Lvs in a flat rosette, pinnately-cut, spiny, often slightly downy beneath. Flhead solitary, 5-10 cm, with conspicuous wide-spreading *silvery or pinkish* inner bracts; disc whitish or purplish-brown. Stony meadows, rocky slopes and open woods, to 2800 m. July-Sept. P, Cev, A, Ap. **1a** *C.a. simplex* has a *stem* up to 60 cm carrying up to 6 flheads.

2 ACANTHUS-LEAVED CARLINE THISTLE *Carlina acanthifolia*. Low stemless per rather like 1 but with broader lvs, *white-velvety beneath*. Flhead solitary, 12-15 cm, with wide-spreading, *straw-coloured* inner bracts; disc lilac. Meadows and stony places, usually on limestone, to 1800 m. July-Sept. P, Cev, swA–French, Ap. **2a** *C.a. cynara* has narrower lvs and *clear-yellow* inner flbracts. P, Cev, nAp.

3 COMMON CARLINE THISTLE *Carlina vulgaris*. Short/med erect per. Lvs short, narrow oblong or oval, spine-toothed, the lower often woolly, *not* in a basal rosette. Flheads *small*, 2-4 cm, with straw-coloured inner bracts; solitary or in groups of 2-3. Grassy and stony places, open woods, to 1750 m. July-Sept. T.

4 JURINEA *Jurinea mollis*. Med/tall grey-hairy per, branched or unbranched, lower lvs pinnately-lobed with wavy or rolled-under edges; upper lvs often unlobed, linear-lance-shaped. Flheads purple, 25-45 mm, solitary or several together, all florets tubular; flbracts with a recurved purple tip. Dry grassy and stony places, to 2000 m. May-Aug. s, eA, Ap. **4a** *J. humilis* (= *J. bocconi*) is low with stems not exceeding 4 cm; flheads reddish-pink, 20-25 mm, solitary, in middle of a lfy rosette. Dry limestone slopes. June-Aug. P.

SAUSSUREAS. Non-spiny perennials. Flheads solitary or clustered, brush-like, rayless; stamens projecting conspicuously.

5 DWARF SAUSSUREA *Saussurea pygmaea* (= *Carduus* or *Cnicus pygmaeus*). Low lfy, tufted, downy per. Lvs linear to narrow-lance-shaped, pointed, usually tufted. Flheads *solitary*, violet-purple, 25-35 mm. Rocks and screes, usually on limestone, 1600-2550 m. July-Aug. eA.

6 ALPINE SAUSSUREA *Saussurea alpina* (= *Serratula alpina*). Short/med erect tufted per, stems often downy. Lvs oval to broad-lance-shaped, *narrowed* at base, the lower stalked, toothed or untoothed, green above, grey-downy beneath. Flheads closely clustered, purple, 10-15 mm. Meadows, rocky places and screes, 1500-3000 m. July-Sept. B, S, c, eP, A. **6a**, **6b**, p.338.

7 HEART-LEAVED SAUSSUREA *Saussurea discolor* (= *S. lapathifolia* and *Serratula discolor*). Short, erect, downy per. Lvs triangular-lance-shaped, *rounded or heart-shaped* at base, toothed, stalked, green above, white-downy beneath, the upper narrower and clasping the stem. Flheads clustered, bluish-violet, 12-18 mm, fragrant. Rocks and stony places, often on granite, 1400-2800 m. July-Sept. A, Ap.

8 STEMLESS COTTON-THISTLE *Onopordum acaulon*. Low *stemless* bien, greyish or white-woolly. Lvs in a rosette, lance-shaped to elliptical, lobed and *spiny-toothed*, long-stalked. Flheads white, rayless, 40-60 mm, surrounded by numerous narrow spine-tipped bracts, solitary or up to 6 in a cluster. Dry grassland and rocky places, to 1900 m. July-Aug. P, Corbières. **8a** *O. rotundifolium* (= *Berardia subacaulis*) has more rounded, *non-spiny* scarcely toothed lvs, and solitary whitish or lilac flheads surrounded by long cottony flbracts. Stony places, screes, 1500-2500 m. wA, MA.

9 COTTON-THISTLE *Onopordum acanthium*. Tall white or greyish-downy bien; stems stout, *spiny winged*. Lvs oblong-oval to lance-shaped, spiny, *stalkless*. Flheads purple, sometimes white, 35-50 mm, usually solitary. Flbracts numerous ending in yellowish spines. Bare dry and rocky places, to 1500 m. July-Sept. T. **9a** *O.a. gautieri* has flheads *in clusters* of 3-5, each 25-40 mm across. c, eP.

3

2

1

4

6

9

5

7

8

Daisy Family *(contd.)*

THISTLES *Carduus* and *Cirsium*. Perennials with spiny winged stems and spine-edged, alternate, lvs. Flheads rayless, rounded and brush-like, often purplish, surrounded by an involucre of many bracts, often spine-tipped. Fr smooth with a feathery pappus in *Cirsium* and one of non-feathery hairs in *Carduus*.

1 MUSK THISTLE *Carduus nutans*. Variable med/tall per with *white-cottony* stems. Lvs pinnately-lobed, usually deeply so, with long or short spines. Flheads bright red-purple, 20-45mm, solitary or clustered, slightly nodding, on spineless stalks; flbracts purplish. Grassy and waste places, to 2500m. June-Sept. T–except S.

2 APENNEAN THISTLE *Carduus chrysacanthus*. Short/med per with white-cottony stems. Lvs pinnately-lobed, *white-cottony beneath*. Flheads purple, 30-50mm, on spiny stalks; flbracts hairy, long spined, the lower *usually recurved*. Dry grassy and stony places, to 2300m. June-Aug. c,sAp.

3 GREAT MARSH THISTLE *Carduus personata*. Med/tall per, stems narrowly winged, short spined. Lvs lance-shaped or oval, softly spiny, white-cottony beneath, the *upper undivided*. Flheads purple-red, 15-25mm, in *tight clusters;* flbracts narrow, not spine-tipped, the outer recurved. Damp meadows and woods, streambanks, to 2300m. July-Aug. P,Cev,A,n,cAp.

4 ITALIAN THISTLE *Carduus litigiosus*. Med bien, stems broadly winged, white-cottony. Lvs pinnately-lobed, white-cottony *above and beneath*. Flheads purple, 20-30mm, in clusters of 2-5; flbracts, white-hairy on edges, *not* recurved. Dry grassy and stony places to 1400m. May-July. MA,Ap.

5 ALPINE THISTLE *Carduus defloratus*. Med/tall per, stems hairless or slightly white-hairy, wingless and lfless in the upper part. Lvs lance-shaped, more or less *hairless*, pinnately-lobed, spiny-toothed. Flheads purple to rose-purple, 20-30mm, solitary and *slightly nodding;* flbracts hairless, erect but the outer curled into an *S-shape*. Meadows, stony places and open woods, usually on lime, to 3000m. June-Oct. P,A–Pyrenean plant is often called *C. medius*.

6 CARLINE-LEAVED THISTLE *Carduus carlinifolius*. Variable low/tall per, stems hairless or white-hairy, *strongly branched* in the upper part, wingless and spine-less at top. Lvs thick, pinnately-lobed, more or less hairless. Flheads purple, 20-30mm, usually solitary; flbracts more or less spiny, often recurved. Dry grassy and stony places, screes, to 2500m. June-Oct. P,A,Ap.

7 WELTED THISTLE *Carduus acanthoides*. Tall bien to 1·5m, stems slightly white-hairy. Lvs pinnately-lobed, with long weak spines, hairless except on the veins beneath. Flheads red-purple, 20-25mm, solitary or clustered *on spiny stems;* flbracts erect or spreading, narrow, long-pointed. Grassy and waste places, hedgerows, to 3000m. June-Sept. T,(S). **7a** *C. crispus* has short, narrowly-winged flstalks, up to 8cm long; flbracts all erect. To 1900m. T.

8 SOUTH-EASTERN THISTLE *Carduus carduelis*. Med/tall per, stems almost hairless, wingless and spineless at top. Lvs pinnately-lobed, hairless above, but *white-cottony* beneath. Flheads purple, 15-30mm, solitary; flbracts slender, the outer curled into an *S-shape*. Meadows, to 3000m. June-Aug. e,seA.

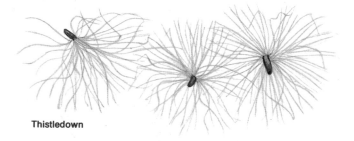

Thistledown

Daisy Family *(contd.)*

1 CORYMBOSE THISTLE *Carduus affinis.* Tall per, stems white cottony, with narrow wings. Lvs pinnately-lobed, long spined, hairy beneath. Flheads purple, 10-25mm, in branched *flat-topped clusters*; flbracts linear, erect, pointed. Grassy places and woods, to 1850m. June-Aug. c,sAp.

2 PYRENEAN THISTLE *Carduus carlinoides* (=*C. pyrenaicus*). Short/med *white-downy* per, stems winged and very spiny up to the fls. Lvs pinnately-lobed, long-spiny. Flheads rose-red or purplish, sometimes white, 18-25mm, in *tight clusters*, flbracts erect, spine-tipped. Grassy and stony places, screes, to 2200m. July-Sept. P.

3 STEMLESS THISTLE *Cirsium acaulon.* Low/short per, *generally stemless* but occasionally with a stem up to 15cm. Lvs in a *flat rosette*, pinnately-lobed. Flheads solitary, red-purple, 30-50mm; flbracts erect, closely overlapping. Dry grassy places, usually on calcareous soils, to 2550m. July-Sept. T.

4 WOOLLY THISTLE *Cirsium eriophorum.* Tall stout bien to 2m, stems unwinged, white-cottony. Lvs pinnately-lobed, *very spiny*, white-cottony beneath. Flheads large red-purple, 40-70mm, usually solitary; flbracts spreading, mostly enveloped in *white-cobwebby-wool*. Grassy and waste places, scrub, on calcareous soils, to 2100m. July-Sept. B,P,A,nAp. **4a.** *C. spathulatum* is shorter with slightly smaller flheads, the flbracts hairless or white hairy, *not woolly*. swA, LA, nAp. **4b** *C. morisianum* has stems branched in the upper part. Lvs pinnately-lobed half clasping the stem, white-hairy beneath usually, with long spines. Flheads purple, 45-65mm; bracts white-hairy beneath usually, with long recurved spines, 10-30mm long. July-Sept. swA, Ap. **4c** *C. richteranum* is shorter than 4b, never more than 60cm, with crowded flheads in *flat topped clusters*, flbract spines 4-8mm long. P.

5 MELANCHOLY THISTLE *Cirsium helenioides* (=*C. heterophyllum*). Med/tall per, spreading by underground runners; stems lfless towards the top, not winged, white-woolly. Lvs lance-shaped, pinnately-lobed or not, toothed, *not spiny*, white-woolly beneath. Flheads red-purple 30-50mm, solitary or in clusters of 2-4; flbracts erect. Damp meadows and woods, scrub, to 2350m. June-Aug. T–except Ap.

6 BROOK THISTLE *Cirsium rivulare.* Med/tall per, stems lfless above the middle, not winged. Lvs elliptical to oblong, usually pinnately-lobed, but often only at the base, weakly spiny, *green above and beneath*. Flheads purple, sometimes white, 25-30mm, solitary or in clusters of 2-5; flbracts erect. Damp grassy places, on acid soils, to 1750m. June-Aug. P, A–except sA. **6a** *C. montanum* is taller with broader lvs and *spreading or recurved* fl bracts. Damp woods and meadows, to 1600m. sA, n, cAp.

7 WALDSTEIN'S THISTLE *Cirsium waldsteinii.* Med/tall per, stems unbranched, lfy to the top. Lvs oval, lobed or deeply toothed, teeth weakly spiny, *white-cottony beneath*. Flheads purple, 25-35mm, in clusters of 3-8; flbracts erect but spreading out at tip. Damp and shady places, on acid soils, to 1650m. June-Aug. e,seA.

3 Stemless Thistle

1

2

3

4

5

6

7

Daisy Family *(contd.)*

1 CARNIC THISTLE *Cirsium carniolicum.* Med/tall slightly branched per; stems with dense long *reddish-brown* hairs. Lvs flat, oval to elliptical, lobed and edged with weak spines. Flheads pale yellow, 17-21mm, in clusters of 2-7 or solitary, short stalked; flbracts spreading, weakly spiny. Grassy places and scrub, over limestone, to 1600m. eA. **1a** *C.c. rufescens* has slightly *larger* fls. w,cP.

2 PALE YELLOW PYRENEAN THISTLE *Cirsium glabrum.* Short/med per; stems short-hairy. Lvs hairless, oblong, *narrowed* at base, pinnately-lobed, with stout spines. Flheads pale yellow, 20-35mm, usually solitary, *surrounded by lvs*; flbracts erect to spreading, short spined. Damp screes and streamsides, to 3000m. July-Sept. P.

3 SPINIEST THISTLE *Cirsium spinosissimum.* Rather like 2 but stems with long hairs. Lvs *clasping the stem*, usually hairy. Flheads pale yellow, 20-25mm, in clusters of 2-10, surrounded by long lvs; flbracts erect, long-spined. Damp meadows, stony places and screes, to 3100m. July-Sept. A, AA, n, cAp.

4 YELLOW MELANCHOLY THISTLE *Cirsium erisithales.* Med/tall per, usually branched, Ifless in the upper part. Lvs dark green, pinnately-lobed, lobes toothed, scarcely spiny, clasping stem at base. Flheads lemon yellow, *nodding* 20-30mm, solitary or 2-5 clustered; flbracts erect, pointed. Meadows, woods and stony slopes, usually on limestone, to 2000m. July-Sept. A, Cev, n, cAp.

KNAPWEEDS or CORNFLOWERS *Centaurea.* Perennials, sometimes annuals, with alternate lvs. Flheads with an involucre of many overlapping bracts, each bract terminated by a scale which may be toothed, spiny, feathery or scale-like; florets tubular surrounded by a ring of conspicuous sterile florets which are deeply lobed and appear 'star-like', as in the Common Cornflower. **(See p.354).**

5 SOUTH-EASTERN KNAPWEED *Centaurea dichroantha.* Short/med slightly branched per. Lvs green, hairless, pinnately-lobed, the lobes linear. Flheads yellow or purple; involucre 10-12mm across, the bracts with a feathery reddish-brown appendage terminating in a fine spine. Grassy and rocky places on calcareous soils, to 1250m. July-Aug. seA–not Austria.

6 ROCK KNAPWEED *Centaurea rupestris.* Low/med per. Lvs *hairy at first*, 1-2-pinnately-lobed, the lobes narrow, pointed. Flheads pale yellow or orange; involucre 12-15mm across, the bracts with a feathery brown spine-tipped appendage. Dry meadows and rocky places, to 1100m. June-July. seA, n, cAp.

7 LEATHERY KNAPWEED *Centaurea grinensis.* Med/very tall, branched, hairy per. Lvs 1-2-pinnately-lobed, the lobes oblong or lance-shaped. Flheads purple, in *flat-topped clusters* usually; involucre 14-18mm across, the bracts with black, brown or yellowish feathery appendages. Dry grassy and rocky places, on calcareous soils, to 1400m. July-Aug. s, seA.

8 KNAPWEED *Centaurea alpestris.* Med generally unbranched per. Lvs pinnately-lobed, the lobes oval, blunt, toothed. Flheads purple; involucre 20mm across, bracts *hidden* by their dark brown feathery, non-spiny appendages. Dry grassy and rocky places, on calcareous soils, to 1600m. July-Aug. P, Jura, A, n, Ap.

9 AUSTRIAN KNAPWEED *Centaurea badensis.* Med/tall hairless, unbranched per. Lvs *shiny-green*, pinnate, the lower with lance-shaped, untoothed Iflets, the upper narrower. Flheads purple; involucre 15-18mm across, bracts brown or black with *white feathery appendages*. Dry scrubby hillslopes on calcareous rocks, to 00m. July-Aug. eA.

5 South-eastern Knapweed, colour forms

1 ALPINE KNAPWEED *Centaurea alpina.* Short/med, erect, hairless per. Lvs pale green, pinnate, the lflets oblong-lance-shaped, toothed at the apex. Flheads pale yellow, solitary. Involucre 18-20mm across, the bracts oval with an oblong or rounded scale. Woodland and scrub on calcareous soils to 1200m. July-Aug. sw,s,seA, not Switzerland or Austria.

2 BLUISH KNAPWEED *Centaurea spinabadia.* Med erect bien. Lvs green or greyish-hairy, the lower 2-pinnately-lobed. Flheads purple, solitary. Involucre 10-12mm across, the bracts with a feathery appendage terminating in a short recurved spine. Rocky slopes, usually calcareous, soils to 1100m. May-Aug. eP, MA.

3 PANICULATE KNAPWEED *Centaurea paniculata.* Med/tall erect, much branched, hairy bien. Lvs green, the lower 1-2-pinnately-lobed. Flheads purple, solitary or in clusters of 2-6. Involucre 3-6mm across, the bracts with a feathery appendage terminating in a short straight spine. Dry grassy and rocky slopes, to 1800m. July-Sept. P, MA, swA - Dauphiné.

4 WHITISH-LEAVED KNAPWEED *Centaurea leucophaea.* Med much branched, hairy bien. Lower lvs pinnately-lobed, at first whitish or greyish, later green. Flheads pinkish-purple or lilac, in flat-topped clusters. Involucre 6-13mm across, the bracts brown, sometimes spotted, or white-hairy, with a pale brown or yellow feathery appendage terminating in a short spine. Dry grassy and rocky slopes, to 1600 m. July-Sept. eP,MA,swA,LA. **4a**, p.338.

5 DOUBTFUL KNAPWEED *Centaurea nigrescens.* Short/tall, erect, unbranched or few branched, hairy per. Lower lvs elliptical to lance-shaped or oval, untoothed, toothed or sometimes slightly lobed, the upper narrower, pinnately-lobed or unlobed. Flheads purple, solitary. Involucre 12-15mm across, the bracts green with a black or pale brown feathery triangular appendage. Meadows or open woods to 2200m. July-Aug. sA,Ap. **5a**, **5b**, p.338.

6 WIG KNAPWEED *Centaurea phrygia.* Med/tall erect, branched or unbranched, hairy per. Lvs green, the lower lance-shaped to oval, toothed or untoothed, the upper ones rounded at base or clasping the stem. Flheads purple, solitary. Involucre 15-20mm across, the bracts with feathery black or deep brown appendages. Meadows, woodland and scrubby places, to 2200m. July-Sept. A.

7 PLUME KNAPWEED *Centaurea uniflora.* Short/med, erect, generally unbranched, hairy per. Lvs green or greyish, the lower oblong-lance-shaped or elliptical, toothed, stalked, the upper clasping the stem. Flheads pale violet, solitary. Involucre 17-22mm across, the bracts with long feathery blackish-brown appendages. Dry meadows, scrub and rocky slopes, to 2600m. July-Aug. A,Ap.

8 RHAETIAN KNAPWEED *Centaurea rhaetica.* Short/med, more or less erect, unbranched or few-branched, hairy per. Lvs green, oblong or linear-lance-shaped, toothed or untoothed, the upper almost clasping the stem. Flheads pink-purple, solitary. Involucre 14-20mm across, the bracts with feathery blackish-brown appendages. Meadows and open woodland, often on calcareous soils, to 2200m. July-Aug. c,s,eA. **8a** *C. procumbens* is shorter, the lower lvs broader, often slightly lobed, rather greyish or whitish-hairy. Dry calcareous rocks, to 1200m. MA.

9 MOUNTAIN CORNFLOWER *Centaurea montana.* Variable, short/tall, more or less erect, hairy per with broadly winged stems. Lvs greyish, *oblong or lance-shaped,* untoothed, sometimes slightly toothed. Flheads with blue outer florets and violet inner ones, solitary. Involucre 10-15mm across, the bracts with a short dark brown comb-like appendage. Meadows and open woods, often on calcareous soils to 2100m. May-July. P, A, Ap. **9a** *C. triumfetti* has narrowly-winged stems and greener linear-lance-shaped lvs. Stony pastures and rocks. w,sA.
9b Cornflower *C. cyanus* is a greyish-downy ann with pinnately-lobed or unlobed lvs. Common cornfield weed to 1800m. T.

Daisy Family *(contd.)*

Species 1-4 have thistle-like flheads; 5-9 have Dandelion-like flheads, opening in fine weather.

1 SAWWORT *Serratula tinctoria.* Slender short/tall, slightly branched, hairless per, thistle-like *but spineless.* Lvs pinnately-lobed, finely toothed; upper lvs often linear, unlobed. Flheads rayless, purple or whitish, 15-20 mm long, short-stalked, *in clusters;* involucre oblong with purplish flbracts. Damp grassy and stony places, open woods and scrub, 1600-2400 m. July-Oct. T−except c,sAp.

2 SINGLE-FLOWERED SAWWORT *Serratula lycopifolia* (=*Caduus lycopifolius*). Med/tall per, stem usually hairless. Lvs rough-hairy, the lower oval, deeply-toothed, long stalked; upper lvs pinnately-lobed, stalkless. Flheads *solitary,* purple, 25-40 mm long; involucre globular. Meadows and scrub, to 1800 m. June-July. swA.

3 CARDOON KNAPWEED *Leuzea centauroides* (=*Rhaponticum centauroides*). Tall stout downy per. Lvs pinnately-lobed, toothed, the upper smaller, stalkless; green above, white-downy beneath. Flheads *large,* rayless, purple, 60 mm, brush-like, solitary; flbracts brown, *narrow lance-shaped,* toothed. Meadows and rocky places, to 2000 m. Aug-Sept. P*.

4 GIANT KNAPWEED *Leuzea rhapontica* (=*Rhaponticum scariosum*). Med downy per. Lvs lance-shaped to oval, toothed, stalked, green above, grey-downy beneath. Flheads large, rayless, rose-purple, 60-70 mm, solitary; *flbracts oval,* pointed, hairy-margined. Damp meadows and rocks, on acid soils, 1400-2600 m. July-Sept. sw,cA*. **4a** *L.r. heleniifolia* (*R.s. lyratum*) is taller with larger flheads, the lvs *with a lobe* on each side at base, white-downy beneath; flbracts blunt. Limestone rocks, 1400-2200 m. A. **4b** *L.r. bicknellii* like 4a, flheads very large, 90-120 mm and *stems lfy to top,* the basal ones pinnately-lobed. MA, LA.

5 PURPLE LETTUCE *Prenanthes purpurea.* Short/tall hairless, blue-green per. Lvs elliptical-oblong or fiddle-shaped, toothed, *clasping the stem.* Flheads violet or purplish, 10-20 mm long, with 3-6 ray florets; in open, long-stalked, *slightly drooping,* clusters. Woods shady places and streamsides, to 2050 m. July-Sept. P,A,n,cAp.

6 ALPINE SOW-THISTLE *Cicerbita alpina* (=*Sonchus alpinus, Lactuca alpina*). Med/tall per with milky juice, usually unbranched, stems *reddish-hairy.* Lvs hairless, the lowest pinnately-lobed with a large triangular end lflet; upper lvs smaller, clasping stem. Flheads pale blue or mauvish, 20-28 mm, in lfy clusters. Grassy and rocky places, open woods, to 2200 m. July-Sept. T−except c,sAp. **6a** *C. plumieri* (=*Sonchus plumieri, Lactuca plumieri*) *is hairless.* P,Cev,w,swA.

7 MOUNTAIN or BLUE LETTUCE *Lactuca perennis.* Med/tall hairless per with milky juice and branched stems. Lvs grey-green, pinnately-lobed, segments toothed or not, the upper lvs clasping the stem, lower short-stalked. Flheads blue or lilac, 30-40 mm, long-stalked, in loose clusters. Open stony and grassy places, on limestone, to 2100 m. May-Aug. T. **7a** *L. quercina* is ann/bien with less cut, lyre-shaped lvs and *yellow flheads.* swA. **7b Edible Lettuce** *L. sativa* is cultivated extensively, to 2500 m.

8 GIANT CATSEAR *Hypochoeris uniflora.* Stout short/med roughly hairy per. Lvs pale green mostly basal, lance-shaped, not lobed, toothed; stem lvs smaller. Flheads *yellow,* 40-50 mm, solitary on stout stems *swollen* at the top; pappus with one row of hairs. Meadows and open woods, on acid soils, to 2600 m. July-Sept. A. **8a Spotted Catsear** *H. maculata* has slightly branched stems with scale-like lvs. Basal lvs slightly lobed, *purple-black spotted.* Grassy places on calcareous soils, to 1800 m. T. **8b** *H. radicata* like 8a but lvs more deeply lobed, *unspotted.* To 1800 m. T.

9 CALYCOCORSUS *Calycocorsus* (=*Willemetia*) *stipitatus.* Low/med per, stems densely *dark-hairy.* Lvs mostly basal, oval or oblong, broadest above the middle, toothed; stem lvs 1-2, linear, or absent. Flheads yellow, 11-14 mm, solitary or in clusters of 2-5; flbracts linear, pointed. Wet grassy places, to 2450 m. July-Sept. eP, A, Ap.

Daisy Family *(contd.)*

All species on this page have Dandelion-like flheads (without disc florets).

1 APOSERIS *Aposeris foetida*. Short slender, slightly hairy per, juice foetid. Lvs all basal, pinnately-lobed, the lobes rhombic with 1-2 teeth on the *lower margin*. Flheads yellow, 25-30mm, with few florets; *pappus absent*. Damp meadows, woods and river banks, often on lime, to 2000m. May-Aug. P,A,nAp.

2 BEARDED VIPERGRASS *Scorzonera aristata*. Short/med, slightly hairy, tuberous-rooted, per with lfless stems. Lvs *linear to narrow lance-shaped*, pointed, usually untoothed. Flhead solitary, golden-yellow, 20-40mm, the florets much longer than the sepal-like flbracts. Meadows and grassy places, on limestone, to 2300m. c,eP,s,scA,n,cAp., July-Aug. **2a, 2b, 2c,** p.338.

3 GOATSBEARD *Tragopogon pratensis*. Med hairless ann/per, slightly branched. Lvs *linear*, grass-like, *clasping stem*. Flheads golden-yellow, 40-60mm, opening fully on sunny mornings; fl bracts often longer than florets; pappus making a large 'clock'. Meadows and grassy places, to 2600m. May-Aug. T. **3a** *T. dubium* is taller, the *stems swollen* just below the flheads. To 2150m. Ap.

4 TOLPIS *Tolpis staticifolia* (= *Hieracium staticifolium*). Short/med, almost hairless per. Basal lvs linear to linear-lance-shaped, usually untoothed; stem lvs few, linear. Flheads yellow, 15-25mm, inner florets often *purplish-brown*; flbracts linear to narrow-elliptical, *downy*. Rocky and stony places, usually on basic rocks, to 2500m. A.

HAWKBITS *Leontodon* are rosette-leaved pers with yellow flheads, solitary, on long stems, rarely branched.

5 PYRENEAN HAWKBIT *Leontodon pyrenaicus*. Low/short, hairy per. Lvs narrow elliptical to oblong, broadest above the middle, *untoothed*, stalked; stem lvs small, bract-like. Flheads 20-25mm, flbracts *dark-hairy*. Meadows, stony places and scrub, often on acid soils, to 3000m. June-Aug. P. **5a, 5b, 5c,** p.338.

6 MOUNTAIN HAWKBIT *Leontodon montanus*. Low hairy per. Lvs linear to oblong, broadest above the middle, lobed and toothed with a *large end segment*. Flhead 20-30mm, flbracts with pale-grey hairs. Rocky and stony meadows and gravels, often on limestone, to 2900m. July-Aug. P,A. **6a, 6b,** p.338.

HAWKWEEDS *Hieracium*. Perennials with milky juice and basal rosettes of lvs. Flheads solitary or grouped, each surrounded by a cluster of irregular rows of overlapping flbracts. Florets usually yellow, sometimes orange or red. A variable and difficult group with many local species.

7 DWARF HAWKWEED *Hieracium humile* agg. Low/short, dark green or blue-green per, *stiffly-hairy*. Basal lvs elliptical, oblong or lance-shaped, deeply-toothed, stalked; stem lvs usually few, small and bract-like. Flheads yellow, 15-25mm, in loose clusters of 4-12. Rocky and stony places, usually on limestone, to 2500m. June-Aug. T.

8 ALPINE HAWKWEED *Hieracium alpinum* agg. Low/short greenish per, shaggily-hairy. Basal lvs elliptical to oblong or spoon-shaped, toothed or not, stalked; stem lvs few, bract-like. Flheads yellow, 25-35mm, *usually solitary*; flbracts blackish. Grassy and rocky places, to 3000m. July-Aug. S,A,n,cAp. **8a, 8b,** p.338.

9 WOOLLY HAWKWEED *Hieracium lanatum*. Short/med per, densely *white-woolly*. Basal lvs oval-elliptical to lance-shaped, untoothed or few toothed, stalked; stem lvs as basal ones but stalkless. Flheads yellow, 20-30mm, long stalked, in clusters of 2-8; flbracts white-hairy. Dry grassy and rocky places, to 2100m. w,swA,cAp. May-July.

10 ORANGE HAWKWEED or FOX AND CUBS *Hieracium aurantiacum* (= *Pilosella aurantiaca*) agg. Short/med, pale or bluish-green creeping per, very hairy. Lvs few, narrow oblong-lance-shaped; stems with dark blackish hairs and 1-4 lvs. Flheads *orange-red*, 15-20mm, in tight clusters of 2-8. Meadows and waste places, often on acid soils, to 2600 m. June-Aug. A. **10a,** p.338.

Lily Family Liliaceae

Lvs undivided, usually with long parallel veins. Rootstock a bulb, rhizome or tuber. Fls with 6 tepals (3 sepals and 3 petals – all looking alike, usually coloured); stamens 6. Fr a three-parted capsule or berry.

1 TOFIELD'S ASPHODEL *Tofieldia calyculata*. Low/short tufted, hairless per; unbranched. Lvs flat, sword-shaped, 5 or more *veins*, pale green; stem lvs smaller. *Bracts* one-lobed, papery. Fls tiny, pale yellowish-green in an *elongated cluster*. Damp meadows, marshes and damp rocks, to 2500 m. June-Sept. S, P, A, Ap. **1a Scottish Asphodel** *T. pusilla* is dwarfer with *3-veined* lvs, fl cluster more rounded; *bracts* three-lobed, green. July-Aug. B, S, A.

2 BLUE APHYLLANTHES *Aphyllanthes monspeliensis*. Short, tufted, 'rush-like' per. Lvs 'sheath-like' encircling the fl stems. Fls solitary or 2-3 together, open saucers, 20-25 mm, tepals bright blue with a darker central stripe. Dry rocky places, to 1400 m. Apr-July. P, w, sA, nAp.

3 MERENDERA *Merendera montana* (= *M. pyrenaica, M. bulbocodium*). Similar to 4 but dwarfer. Lvs narrow, grooved, appearing in the early spring and dying before fls appear. Fls rosy-pink, the strap-shaped tepals falling apart, *not* hooked together at base. Meadows and screes, 1500-2500 m. Aug-Sept. cP.

4 AUTUMN CROCUS *Colchicum autumnale*. Short hairless per. Lvs 2-4 clustered, oblong-lance-shaped, bright shiny green, appearing *in spring* and dying before fls appear. Fls large, 20-45 mm, 'crocus-like' but with *6 stamens*, mauve-pink or white, on long whitish tubes. Damp meadows, to 2000 m. Aug-Oct. T, (S). **4a** *C. alpinum* is a smaller version of 4 with narrower lvs and tepals; fls 20-30 mm long. Meadows, often on acid soils, to 2000 m. Aug-Sept. w, c, sA, Ap.

5 SPRING BULBOCODIUM *Bulbocodium vernum*. Low hairless per. Lvs 2-3, lance-shaped, grooved, *with the fls*. Fls 'crocus-like' but with 6 stamens, rosy-lilac or white, the strap-like tepals *hooked* together at the base. Meadows, to 2000 m. Feb-May. eP*, A, Ap.

6 ST. BRUNO'S LILY *Paradisea liliastrum*. Short/med hairless per. Lvs linear, basal. Fls *trumpet-shaped*, 40-50 mm, white, fragrant, in loose one-sided racemes. Meadows and rocky places, to 2400 m. June-Aug. P, A, Ap.

7 ST. BERNARD'S LILY *Anthericum liliago*. Slender, med, hairless per; unbranched. Lvs all basal, linear, grassy. Fls *starry* white, 30-40 mm, in loose racemes. Dry rocks and pastures over limestone, to 1800 m. May-July. S, A, Ap. **7a** *A. ramosum* □ has *smaller* fls, 20-25 mm, in *branched* clusters. To 2000 m. June-Aug. S, A, Ap.

8 HAIRY GAGEA *Gagea villosa* (= *G. arvensis*). Low, silkily-hairy per. Lvs 2, basal, grassy, grooved, *longer* than stem. Fls starry, yellow, 20-30 mm, in clusters with two large *bracts* below. Meadows and waste ground, to 2200 m. Feb-May. S, c, sA, Ap. **8a** *G. lutea* has a *single hairless* basal lf and smaller fls. Meadows and woods, to 1700 m. Mar-May. B, S, P*, A. **8b** *G. pratensis* like 8a but with a *very long* basal lf and *four* uneven bracts below the fls. Meadows. Mar-Apr. swA–Cottian.

9 PYRENEAN GAGEA *Gagea soleirolii*. Low hairy per. Lvs 2 basal, grassy, *shorter* than stem. Fls starry yellow, 15-20 mm, in loose clusters, each with a *short* bract at base. Meadows and rocky places, to 2500 m. Mar-Apr. c, eP*. **9a** *G. minima* is similar but usually with a *single* delicate basal lf, *hooked* at the top. Fls smaller. Meadows, to 1600 m. May-June. S, A* Ap.

10 YELLOW GAGEA *Gagea fistulosa*. Like 8, 9 but basal lvs *rush-like*, hollow and semi-circular in cross-section. Fls starry yellow, 15-30 mm, with two *very large* pointed bracts below. Meadows, 1200-2800 m. May-July. P, A, Ap.

Lily Family *(contd.)*

ONIONS AND LEEKS *Allium.* Unbranched hairless, bulbous perennials, smelling strongly of onion or garlic. Fls bell-shaped or starry with separate tepals in umbelled clusters, often mixed with bulbils; enclosed initially by 1-2 papery bracts.

1 CROW GARLIC *Allium vineale.* Med stiff per. Lvs cylindrical, grooved. Fls pink or greenish, 5 mm, *mixed* with or entirely replaced by bulbils; stamens protruding the stalks with 2-long projections; bracts *shorter than* flheads. Cultivated and waste places, to 1900 m. June-Aug. B, S, P, w, swA.

2 ROUND-HEADED LEEK *Allium sphaerocephalon.* Med/tall per. Lvs semi-cylindrical, grooved, hollow. Fls pinkish-purple, 5 mm, bell-shaped, numerous in dense heads, stamens protruding, *no bulbils; bracts short,* papery. Grassy and rocky places, to 2650 m. June-Aug. B*, Cev, A, Ap.

3 STRICT ONION *Allium strictum.* Short/med per. Lvs linear, rounded on back, *flat above,* up to middle of stem. Fls pink or purplish, 4-7 mm, in dense rounded heads, tepals narrow with a *purple keel,* stamens protruding; bracts as long as flheads. Grassy and rocky places, to 3000 m. July-Aug. A-specially in the w.

4 MOUNTAIN ONION *Allium montanum.* Short tufted per. Lvs linear, flat above, blunt ended, *mostly basal.* Fls purplish-pink, 5-6 mm, in dense heads, stamens protruding; bracts *shorter than* flheads. Dry places and rocks, to 2300 m. June-Aug. P, A, n, cAp.

5 FRAGRANT ONION *Allium suaveolens.* Short/med per. Lvs linear flattish, up to middle of stem. Fls purplish, in dense heads, stamens protruding; bracts shorter than flheads, *blunt-ended.* Grass, scrub, and rocky places, to 1200 m. Aug-Oct. A.

6 CHIVES *Allium schoenoprasum* (=*A. sibiricum*). Short/med tufted per. Lvs *cylindrical,* fine-pointed, hollow, all from roots. Fls purple or violet-pink, 9-12 mm, bell-shaped, short-stalked, in dense heads, stamens *not protruding;* bracts shorter than flheads. Damp grassy and rocky places, to 2600 m. June-July. T.

7 FIELD GARLIC *Allium oleraceum.* Rather like 1. Med/tall bluish-green per. Lvs semi-cylindrical, grooved, hollow, stems lfy to middle. Fls few, pink, greenish or white, 5-7 mm, mixed with bulbils, stamens *not protruding;* bracts long-pointed, *much longer* than flheads. Grassy, cultivated and waste places, to 2150 m. June-Aug. Probably T. **7a Keeled Garlic** *A. carinatum* has *protruding stamens* and the lvs keeled on the back. T.

8 NARCISSUS-FLOWERED ONION *Allium narcissiflorum.* Short tufted per. Lvs linear, flat, green, all basal. Fls purple-pink, 10-18 mm, bell-shaped, in *rather drooping clusters* of 3-8; bracts shorter than flheads. Limestone rocks and screes, 1500-2000 m. July-Aug. sA.

9 ROCK ONION *Allium saxatile.* Short per. Lvs thread-like. Fls pink or white, 3-5 mm, in dense heads, stamens protruding; bracts pointed, *much longer* than flheads. Rocky places, to 2100 m. Aug-Sept. cAp.

10 ALPINE LEEK *Allium victorialis.* Med/tall per. Lvs *oblong to elliptical,* pleated, stalked. Fls greenish-white to yellowish, 4-6 mm, in rounded heads; bracts as long as flheads. Woods and rocky places, 1400-2600 m. June-Aug. P, Cev, A, Jura.

11 STRAP-LEAVED ONION *Allium ochroleucum* (=*A. suaveolens*). Short/med per. Lvs thin and flat, strap-shaped. Fls yellowish or white, sometimes tinged with pink, in dense heads, stamens protruding; bracts about as long as flheads. Grassy and rocky places, to 2000 m. July-Aug. P, sA, Ap.

12 YELLOW ONION *Allium flavum.* Short/med tufted per. Lvs thread-like, channelled, bluish-green. Fls *clear yellow,* 4-5 mm, drooping in fl but later erect, stamens protruding; bracts pointed, much longer than flheads. Dry grassy and rocky places, to 1500 m. June-Aug. s, swA, Ap.

13 RAMSONS *Allium ursinum.* Short per, often covering large areas. Lvs *broad-elliptical,* bright green, all from roots, smelling strongly of garlic. *Fls white,* starry, 8-10 mm, in flat-topped clusters; bracts shorter than flheads. Woods and shady places, to 1900 m. Apr-June. T.

Lily Family (contd.)

1 WHITE FALSE HELLEBORINE *Veratrum album.* Med/tall, robust, hairy per. Lvs broad, oval, *pleated* lengthwise, downy beneath, sheathing stem. Fls starry white, pale green outside *or* all greenish-yellow, 12-15mm, in branched spikes. Meadows and woodland clearing, to 2700m. June-Aug. T−except B. **Poisonous.**
1a Black False Helleborine *Veratrum nigrum* has purplish-black, 10mm flowers, to 1600m. s,eA,Ap.

2 ASPHODEL *Asphodelus albus.* Med/tall hairless per. Lvs all basal, strap-like, grooved. Fls large, starry, white or pink-flushed, 40-50mm, in dense spikes. Meadows, rocky slopes and thickets, to 1600m. May-July. P,A,Ap.

FRITILLARIES or SNAKESHEAD LILIES. Handsome hairless perennials with thin stems and grassy lvs. Fls large, *nodding lanterns*, often chequered.

3 PYRENEAN SNAKESHEAD *Fritillaria pyrenaica.* Low/med per with *alternate* lvs. Fls, 1 or 2, bell-shaped 25-30mm, dark-purplish-mahogany with pale yellow and purple chequering, rarely all pale greenish-yellow; inner tepals *recurved* at tip. Meadows and rocky slopes, to 2000m. May-July. P.

4 SNAKESHEAD LILY *Fritillaria meleagris.* Short per with alternate *very slender* lvs, grey-green. Fls solitary or in twos, broad lanterns, 30-50mm, pale or dark red purple with darker chequering. Damp meadows, to 1200m. Apr-May. T−except Ap.

5 TYROLEAN FRITILLARY *Fritillaria tubiformis* (=*F. delphinensis*). Low per with alternate *lance-shaped* lvs, *all* on upper part of stem. Fls solitary, broad lantern, 35mm, deep reddish-purple with *faint* chequering. Meadows over limestone, to 2000m. Mar-June. sA−N. Italy*, S. Tyrol*. **5a.** *F.t. moggridgei* has yellow fls, speckled and chequered brown or purple. MA*.

6 SLENDER-LEAVED SNAKESHEAD *Fritillaria tenella.* Short/med per with slender *opposite* or *whorled* lvs. Fls small bells, 20-30mm, dull purple or greenish-yellow flushed purple, often chequered. Rocky meadows, rarely much above 1200m. Apr-May. MA,sA,Ap.

7 THREE-BRACTED SNAKESHEAD *Fritillaria involucrata.* Short/med per with opposite lance-shaped lvs, the uppermost node with *three-leaves* held together above the nodding fl. Fls 1-3, large lanterns, 35-40mm, greenish or yellowish with purple markings. Rocky woods, to 1500m. Apr-May. MA,swA-Provence,LA.*

8 WILD TULIP *Tulipa australis* (=*T. sylvestris australis*). Short/med slender hairless per. Lvs few, linear, grassy with *no* prominent veins or grooves. Fls large yellow, the outer 3 tepals *often* tinged green or reddish-purple and recurved, nodding in bud. Meadows to 2000m. Apr-July. c,eP,wA,AA,Ap,(B). **8a** *T. didieri* □ has crimson-orange *fls*, the tepals edged with yellow. Probably not above 1500m. Apr-May. (Savoy Alps.)

8 Wild Tulips

1
2
3
4
5
6
7
8
8a

Lily Family (contd.)

LILIES. Med/tall perennials with scaly underground bulbs. Fls showy, the tepals often recurved giving a 'turk's-cap' appearance, with prominent anthers and style.

1 MARTAGON LILY *Lilium martagon.* Med/tall hairless per with *red-mottled* stems. Lvs in *whorls* up stem, elliptical, deep green, shiny. Fls in a loose cluster, small turk's-caps, dull pink to purplish with darker spots inside; stamens pinkish. Meadows, woods and scrub, to 2800m. June-Aug. P, A, Ap.(B).

2 ORANGE LILY *Lilium bulbiferum.* Med downy per. Lvs *alternate*, narrow lance-shaped, with black *bulbils* at the base of each. Fls 1-5, very large, *upward pointing*, bright orange with black spots. Meadows, woods and rocky places, to 2400m. May-July. P, w, cA, Ap. **2a.** *L.b. croceum* □ has *no* bulbils. w, cA.

3 CARNIC LILY *Lilium carniolicum.* Med per. Lvs alternate, oblong-lance-shaped. Fls few, often solitary, turk's-cap, bright vermilion or orange-yellow with light spotting; anthers orange. Meadows and scrubland, to 2300m. June-July. seA.

4 RED LILY *Lilium pomponium.* Short/med slender per. Lvs *linear*, crowded, silver-margined. Fls 3-8, small turk's-caps, *glistening scarlet-red* with small black dots inside; anthers red-purple. Rocky places, rarely above 1100m. May-July. MA, LA*.

5 YELLOW TURK'S-CAP LILY *Lilium pyrenaicum.* Med/tall foetid per. Lvs numerous, alternate, narrow-lance-shaped, shiny-green, hairy edged. Fls in a loose cluster, medium turk's-caps, yellow with brown spots, rarely orange; anthers orange. Meadows, clearings and rocky slopes, to 2200m. June-Aug. P—local, (B).

6 DOG'S TOOTH VIOLET *Erythronium dens-canis.* Low/short per. Lvs *two usually*, basal, elliptical or lance-shaped, often mottled. Fls large solitary, bright pink with *recurved tepals* and prominent stamens. Meadows, heaths and woods, to 2000m. Feb-Apr. sP, s, seA.

2a *Lilium bulbiferum croceum*

Lily Family (contd.)

1 LLOYDIA or SNOWDON LILY *Lloydia serotina.* Low slender hairless per. Lvs slender grassy, greyish-green, the stem lvs much shorter than the basal. Fls upright 'cups' 10-20mm, white, *veined* brownish-purple, usually solitary. Rocky places and short turf, often over granite, 1800-3000m. June-Aug. B*, A.

2 ALPINE SQUILL *Scilla bifolia.* Low/short hairless per. Lvs *usually two*, narrow strap-shaped, shiny green. Fls starlike, 10-16mm, bright blue, sometimes rose or white, in loose clusters; *no bracts*. Damp meadows and scrub, to 1500m. March-June. sA, Ap. **2a Spring Squill** *Scilla verna* □ has *3-6* slender lvs and violet blue fls each with a *bluish bract* at the base. Heaths, meadows and woods, to 2000m. Mar-June. B, S, P.

3 PYRENEAN SQUILL *Scilla liliohyacinthus.* Low/med rather coarse, hairless per. Lvs *broad* strap-shaped, in a basal cluster. Fls starry, 18-20mm, blue, in loose racemes on a long stem; bracts greenish, *as long* as fl stalks. Meadows, to 2000m. Apr-May. P.

4 DIPCADI *Dipcadi serotinum.* Slender hairless per. Lvs slender, grassy, grey-green, pointed. Fls narrow bell-shaped, 12-15mm, *yellowish-brown*, outer three tepals *recurved* at tip, in loose racemes. Sandy and rocky places, to 2400m. July-Aug. c, eP*.

5 STAR OF BETHLEHEM *Ornithogalum umbellatum.* Low/short hairless per. Lvs linear, grassy, shiny green with a grooved central *white stripe*. Fls starry, 20-30mm, tepals glistening white, with a *green stripe* on the back of each, in a *umbel-like* cluster. Grassy places and scrub to 1600m. Apr-June. B, A, Ap.

6 BATH ASPARAGUS *Ornithogalum pyrenaicum.* Med/tall hairless per. Lvs linear strap-shaped, grey-green, all basal, usually *withering* at flg time. Fls starry, 10-18mm, greenish-white, in long *spike-like* racemes. Meadows and woods, to 1200m. May-June. B, P, w, sA, Ap.

7 PYRENEAN HYACINTH *Hyacinthus amethystinus.* Low hairless per. Lvs linear, grassy, basal. Fls tubular-bells, 7-10mm, blue, in loose racemes. Meadows, stony places and screes, to 1500m. May-July. cP.

8 GRAPE HYACINTH *Muscari racemosum* (= *M. atlanticum, M. neglectum*). Short hairless per. Lvs thread-like, grassy, grooved, all from roots and usually *sprawling* on the ground. Fls rounded bells, 4-5mm, deep violet-blue or blackish-blue with six white teeth, in dense spikes. Cultivated areas and fields, to 1600m. Mar-May. B*, P, Cev, A, Ap. **8a** *M. botryoides* is similar but with fewer more upright lvs that *broaden* towards the top and pale blue or violet fls. Meadows, fields and woods, to 2000m. Mar-May. s, sw, A.

9 TASSEL HYACINTH *Muscari comosum.* Short hairless per. Lvs long strap-like. Fls bell-shaped, of *two types*, the lower brownish-green, the upper violet-blue, smaller and more upright. Fields and rocky places, to 1300m. Apr-July. P, Cev, w, sA, Ap, (B).

10 MAY LILY *Maianthemum bifolium.* Low/short hairless per, forming patches. Lvs two on each stem, *heart-shaped*, shiny. Fls starry, tiny, 2-4mm, white with *4 tepals*, in small clusters. Fr a small *red berry*. Woods, to 2100m. Apr-July. B*, S, A, Ap.

2a Spring Squill

Lily Family (contd.)

1 STREPTOPUS *Streptopus amplexifolius.* Short/med hairless per, with erect *zig-zagged* leafy stems. Lvs *heart-shaped, clasping* the stem, unstalked. Fls bell-shaped, 9-10mm, greenish-white, solitary, with *jointed* stalks. Fr a red berry. Damp woods and rocks, to 2300m. July-Aug. P, A, Ap.

2 WHORLED SOLOMON'S SEAL *Polygonatum verticillatum.* Med/tall hairless per with erect stems. Lvs in *whorls*, of 3-6, narrow lance-shaped, unstalked. Fls bell-shaped, 6-10mm, white tipped green, in clusters of one to four. Fr a red berry. Woods, meadows and rocky places, to 2400m. May-July. T-B*.

3 SCENTED SOLOMON'S SEAL *Polygonatum odoratum* (=*P. officinale*). Med hairless per with *arching angled* stems. Lvs *alternate*, oval or elliptical, unstalked. Fls narrow bells, 18-22mm, white tipped green, solitary or in twos, fragrant. Fr a bluish-black berry. Woods and rocky places, to 2200m. May-June. T. **3a. Common Solomon's Seal** *P. multiflorum* has *rounded* stems and greenish-white, unscented, fls *waisted* in the middle. T.

4 LILY OF THE VALLEY *Convallaria majalis.* Low patch-forming, hairless per. *Lvs two*, basal, oblong to elliptical, bright shiny green, stalked. Fls *rounded bells*, 6-8mm, white, in loose one-sided spikes, sweetly scented. Fr a small red berry. Woods and scrub, to 2300m. Apr-June. T. Poisonous.

5 HERB PARIS *Paris quadrifolia.* Low/short hairless, patch-forming per. Lvs in a whorl of 4-5 on otherwise lfless stems, oval, short stalked. Fls starry, yellowish-green, with 4-8 tepals and prominent stamens, solitary, erect. Fr.a black berry. Damp woods, often on limestone, to 2000m. May-June. T. Poisonous.

Yam Family Dioscoreaceae

Tuberous rooted perennials with *heart-shaped* lvs and six-petalled fls, male and female parts *separate*, on the same or different plant.

6 BLACK BRYONY *Tamus communis.* Hairless climbing per with twining stems. Lvs alternate, dark green, shiny, stalked. Fls tiny starry, pale green, male in slender erect racemes; female in small drooping clusters on a *different plant*. Fr a shiny *red berry*. Woods, scrub and hedges, to 1200m. May-Aug. T-except S. Poisonous.

7 PYRENEAN YAM *Dioscorea pyrenaica* (=*Borderea pyrenaica*). Low/short hairless per with angular stems branched in the upper half. Lvs clustered or in pairs, long-stalked. Fls tiny greenish, male in loose racemes, female in clusters of one to three, on separate plants. Fr a dry three-winged capsule. Rocky slopes and screes, to 2500m, June-Sept. P*.

Arum Family Araceae

With a single species in the area.

8 LORDS AND LADIES or CUCKOO PINT *Arum maculatum.* Short hairless per. Lvs basal, *arrow-shaped,* dark green, sometimes spotted, long stalked. Fls tiny, in dense whorls; male above female, topped by a purplish *finger-like* spadix, the whole being surrounded by a pale green, purple marked, *hood* or spathe, open above the fls. Fr a cluster of bright orange berries. Woods, scrub and shady banks, to 1250m. Apr-May. T-except S. Poisonous.

8 Lords and Ladies

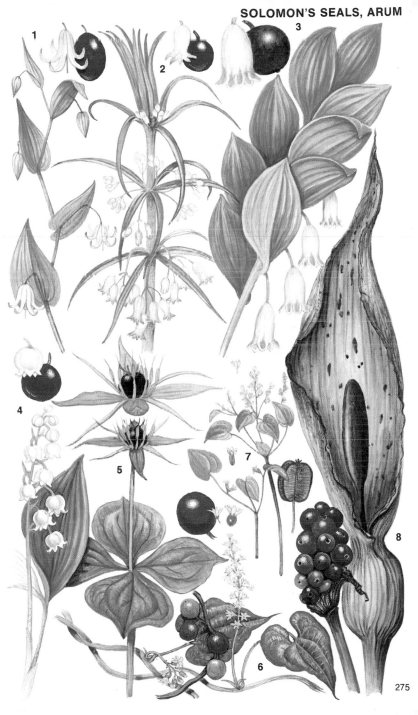

Daffodil Family Amaryllidaceae

Bulbous perennials, hairless, with basal linear lvs and lfless fl stems, with a bract (spathe) toward the top. Fls with six petal-like tepals (3 sepals and 3 petals), the inner three smaller in Snowdrop, with a central trumpet or corona in Daffodil. Stamens six. Fr a capsule developed below the fl.

1 SPRING SNOWFLAKE *Leucoium vernum.* Short per. Lvs strap-shaped, deep green. Fls nodding bells, 15-20mm, with 6 equal tepals, white tipped green, *solitary* or paired. Damp woods and meadows, to 1600m. Feb-Apr. B*, A.

2 SUMMER SNOWFLAKE *Leucoium aestivum.* Short/med tufted per. Lvs long strap-shaped, bright green. Fls nodding bells, 15mm, tepals white tipped green, in *clusters* of three or more. Wet meadows, to 1300m. Apr-June. B*, P, A, Ap.

3 SNOWDROP *Galanthus nivalis.* Low/short per. Lvs usually two, strap-shaped, grey-green. Fls solitary nodding, 20-25mm, white, the *three inner tepals* shorter and green tipped. Damp meadows and woods, to 1600m. Feb-Apr. B, P, sA, Ap.

4 POET'S or PHEASANT'S-EYE NARCISSUS *Narcissus poeticus.* Short/med per. Lvs long strap-shaped, upright, grey-green. Fls solitary, large, 4-6cm, with six white rounded-oval tepals and a *small* cup-like orange and yellow trumpet, sweetly scented. Damp meadows, to 2300m. Apr-May. P, A, (B). **4a** *N.p. radiiflorus* ☐ has narrower lvs and tepals *narrowed* at the base. Ap.

5 RUSH-LEAVED NARCISSUS *Narcissus requienii.* (= *N. juncifolius*) Low/short delicate per. Lvs *thread-like,* grassy 1-2mm wide, deep-green. Fls small, 15-20mm, deep yellow with a short cup-shaped trumpet, sweetly scented, in clusters of two to five. Short turf and stony places, to 1600m. Apr-May. P.

6 ROCK NARCISSUS *Narcissus rupicola.* Similar to 5 but with blue-green lvs 1.5-5mm wide. Fls *larger,* solitary, 15-30mm, fragrant, trumpet flatter, saucer-shaped, *not* cup-shaped. Rocky places, to 1600m. Apr-May. P.

7 WILD DAFFODIL or LENT LILY *Narcissus pseudonarcissus.* Short/med per. Lvs strap-shaped, grey- or bluish-green, blunt ended. Fls solitary, slightly nodding, large, tepals pale yellow, oval; trumpet large 25-35mm long, deeper yellow or golden, with a frilled, flared edge; flstalk above the spathe, 3-12 mm long. Grassy slopes and banks, to 2200m. Mar-June. B, P, S. **7a** *N.p. abscissus* (= *N. bicolor*), ☐ is readily recognised by its medium sized fls with white tepals and a cylindrical orange-yellow trumpet whose edge is straight, *not flared outwards* as in all the other subsp; flstalk 15-35 mm long. P. **7b** *N.p. nobilis* ☐ is taller than 7 with very large fls often *upward pointing,* tepals pale yellow or creamy-white, trumpet golden or orange, *flared;* fl stalk 8-15 mm long. N Spanish Pyrenees. **7c** *N.p. pallidiflorus* smaller than 7 with pale yellow fls, sometimes bicoloured; flstalk 3-10 mm long. P. **7d** *N.p. moschatus* (= subsp. *alpestris*) ☐ similar to 7c but with white or creamy-white, drooping fls, with a long narrow trumpet; flstalk 10-25 mm long. P, especially Spanish.

8 LESSER WILD DAFFODIL *Narcissus minor.* Low per. Lvs narrow strap-shaped, grey or bluish-green. Fls solitary, slightly nodding, *smaller* than 7, tepals pale to deep yellow, trumpet deep yellow, 16-25mm long, slightly flaired; flstalk 5-20mm long. Grassy places and scrub, to 2200m. Mar-June. P.

4 Pheasant's-eye Narcissi

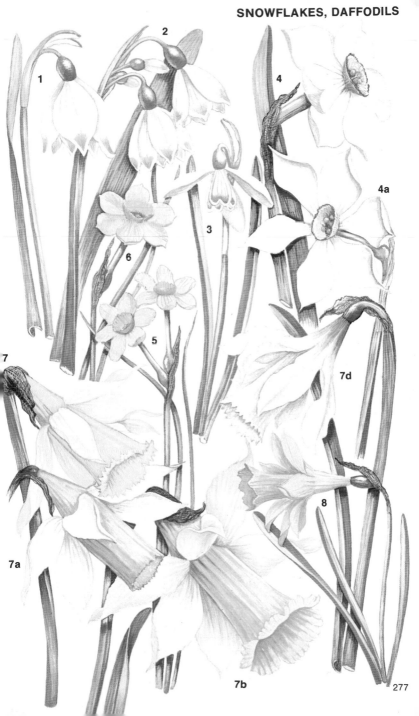

Iris Family Iridaceae

Hairless perennials with bulbs, rhizomes or corms. Lvs flat sword-like, channelled or linear. Fls solitary or several together enclosed by common bracts (spathes) 3-parted. Fr a 3-valved capsule.

CROCUSES. Low corm-rooted pers, stemless. Lvs linear with a white grooved central stripe. Fls goblet-shaped with six similar tepals (3 sepals and 3 petals), *three stamens* and a three-parted orange stigma.

1 WHITE CROCUS *Crocus albiflorus* (= *C. caeruleus, C. vernus albiflorus*). Corm coat with slightly netted fibres. Lvs green, partly developed at flg time. Fls *pure white* but often feathered with violet or violet-purple at base. Style *shorter* than stamens. Wet meadows, often by snow patches, to 2700 m. Feb-June. P, A, nAp, (B).

2 PURPLE CROCUS *Crocus vernus* (= *C. napolitanus, C. purpureus*). Corm coat with slightly netted fibres. Lvs green, partly developed at flg time. Fls twice as large as 1, purple or purple-violet, often darker at base. Styles longer than stamens. Wet meadows, often by snow patches, Feb-June. P, A, nAp.

3 RIVIERA CROCUS *Crocus versicolor*. Similar in size to 2. Corm coat with parallel fibres only. Lvs grey-green, partly developed at flg time. Fls white or pale lilac, the outer three tepals striped with purple. Meadows and stony places, to 1600 m. Feb-Mar. MA, swA- Province & Dauphiné Alps. **3a Leafless Crocus** *C. nudiflorus* has deep purple, *not striped,* fls on very long stalks, flg before the lvs appear. Grassy and scrubby places, rarely much above 1100 m. Sept-Nov. w, cP.

IRISES. Rootstock a rhizome or bulb. Fls in succession with three drooping outer tepals (falls) and three upright inner ones (standards), with three shorter, broad, coloured style arms, each enclosing a single stamen.

4 ENGLISH IRIS *Iris xiphioides*. Med/tall bulbous per. Lvs long linear, channelled, shiny above. Fls very large, violet blue or brilliant blue, sometimes white, the falls with a white and yellow/orange *zone* in the centre, beardless. Meadows, to 2200 m. July-Aug. P.

5 VARIEGATED IRIS *Iris variegata*. Short/med rhizomatous per, stems *branched* in upper half. Lvs sword-shaped, deep green. Fls yellow or yellowish-white, falls *veined* deep violet or brownish-red, *bearded.* Spathe often purple tinged. Rocky and grassy places, seldom above 1200 m. June-July, n, c, eA.

6 GRASSY-LEAVED IRIS *Iris graminea*. Short/med rhizomatous per, with *two-winged* flattened stems. Lvs narrow sword-shaped. Fls with purple standards and falls whitish veined with violet or pinkish-purple. Grassy and scrubby places, to 1100 m. May-June. eP, A, Ap.

7 YELLOW IRIS *Iris pseudacorus*. Med/tall rhizomatous per, with branched stems. Lvs sword-shaped, with a raised midrib. Fls yellow, falls with some green veins or streaks, 1-3 together. Damp and marshy places, often by streams and ponds, to 1200 m. June-Aug. T.

8 STINKING IRIS *Iris foetidissima*. Med tufted per. Lvs sword-shaped, deep green, with a sickly smell when crushed. Fls straw-coloured, flushed and veined with dull violet or grey-purple, 2-3 together, style arms pale yellow. Fr splitting revealing *bright orange* seeds. Open woods, shrub and hedgerows, seldom above 1200 m. May-July. B, P, Cev, wA.

9 GARDEN or FLAG IRIS *Iris germanica*. Med/tall rhizomatous per with branched stems. Lvs broad, sword-shaped, *shorter* than stems. Fls large, fragrant, bluish-violet or white tinged with blue, falls with a conspicuous *yellow beard*, 2-3 together. Dry rocky and grassy places, to 1250 m. May-June. Widely cultivated and probably not truly native in the area. **9a** *I. aphylla* is shorter with some lvs *longer* than stems, fls smaller violet or purple. swA.

Orchid Family Orchidaceae

Tuberous rooted perennials. Lvs oblong to linear, untoothed. Fls generally in dense spikes with the lowermost petal developed into a lip, often 3-lobed, and spurred.

1 LADY'S SLIPPER ORCHID *Cypripedium calceolus*. Shot/med per, lvs oval to lance-shaped, strongly ribbed, pale green. Fls usually solitary, large, 6-9cm, with spreading reddish-maroon sepals and petals and a prominent *pouched yellow* lip. Woods and grassy clearings or slopes, usually on calcareous soils, to 2000m. May-July. B*, S, P*, A, nAp.

2 LATE SPIDER ORCHID *Ophrys holosericea* (= *O. fuciflora*). Short per. Lvs elliptical, pointed. Fls in short spikes, the sepals pink to white, the petals pink, *short*; lip *velvety brown* spider-like, quadrangular, patterned with yellowish green, often with a central blue zone. Grassy places, on calcareous soils, to 1300m. June-July. T–except S, B*. **2a Bee Orchid** *O. apifera* is taller, fls with pink sepals and short *pale green* petals; lip rounded, velvety brown with a variable yellow pattern enclosing a red zone – supposedly like the rear of a bumblebee. Grassy places and thickets, to 1000m. T–except S. **2b Early Spider Orchid** *O. sphegodes* like 2a but sepals and petals *yellowish-green* and the lip with a bluish X- or H-shaped zone. Apr-June. T–except S. **2c**, p. 338.

3 BUG ORCHID *Orchis coriophora*. Short/med per. Lvs narrow lance-shaped. Fls small in a dense spike, unpleasant smelling, sepals and petals brownish-red, joined into a hood; lip reddish-purple, 3-lobed; spur very short, down-pointing. Damp meadows, to 1800m. Apr-July. P, A, MA.

4 LOOSE-FLOWERED RED ORCHID *Orchis laxiflora*. Med per. Lvs narrow lance-shaped, *plain-green*; stem and bracts often purplish. Fls in a loose spike, purple, with two spreading sepals and a hood formed from the upper sepal and the two petals, lip 2-lobed; spur short, horizontal. Damp meadows, to 1150m. Mar-June. T–except B, n, eA.

5 EARLY PURPLE ORCHID *Orchis mascula*. Short/med per. Lvs narrow oblong, pointed, usually with *purplish-black spots*. Fls in loose oblong spikes, purple, sometimes pinkish or white, with a catty odour, the outer sepals *erect-spreading*, the central sepal and petals forming a hood; lip shallowly 3-lobed, spotted; spur long, *upcurved*. Grassy meadows, scrub and woods, to 2650m. Apr-July. T.

6 SOLDIER or MILITARY ORCHID *Orchis militaris*. Short/med per. Lvs large, broad-oblong, shiny, unspotted. Fls in oblong spikes, greyish or lilac-pink with deeper spots and markings, the sepals and petals *all forming* a close helmet; lip with two thin 'arms' and two broad 'legs' with a small central tail in between; spur short, *down-curved*. Grassy and scrubby places, wood and margins, on calcareous soils, to 1800m. April-June. B*, Cev, P*, A, Ap.

7 GREEN-WINGED ORCHID *Orchis morio*. Variable short/med per. Lvs narrow-oblong, unspotted. Fls in loose oblong spikes, purple, pink or white, fragrant, sepals and petals *all* forming a helmet, the sepals or wings conspicuously *veined green*; lip shallowly 3-lobed; spur long, horizontal or upcurved. Meadows and scrubby places, to 1800m. May-June. T.

8 PALE-FLOWERED ORCHID *Orchis pallens*. Short/med per. Lvs broad oblong, shiny, unspotted. Fls in egg-shaped or oblong spikes, *pale yellow*, rarely purplish-red, the outer sepals spreading upwards, the middle sepal and petals forming a hood; lip 3-lobed, *unspotted*; spur horizontal or upcurved, shorter than the ovary. Meadows and woods on calcareous soils, to 2000m. Apr-June. P, A, MA.

9 BOG ORCHID *Orchis palustris*. Med per rather like 4 but more slender. Lvs narrow lance-shaped, unspotted. Fls in a loose spike, magenta or pale purple, rarely white, the outer sepals spreading upwards, lip 3-lobed; spur horizontal or upward pointing. Marshes and damp meadows, to 1150m. Apr-June. T–except B.

10 PROVENCE ORCHID *Orchis provincialis*. Short per. Lvs oblong-lance-shaped, often brown-spotted; bracts pale. Fls in rather dense oblong clusters, pale or dark yellow, the 3-lobed lip *brown* spotted, often orange in the centre; spur long, upward pointing. Grassy and shrubby places, to 1300m. Apr-June. P, w, sA, Ap.

Orchid Family (contd.)

1 LADY ORCHID *Orchis purpurea*. Med/tall per. Lvs broad-oblong, shiny, unspotted, mostly basal. Fls fragrant, in rather dense spikes, the sepals and petals forming a *purplish-green hood*, the lip whitish or pale pink, darker spotted, lobed with 2-narrow 'arms' and 2-broad 'legs'; spur short, downcurved. Woods and scrub on limestone, to 1500 m. May-June. sB*, Cev, P, A, Ap.

2 MONKEY ORCHID *Orchis simia*. Short/med per. Lvs broad-oblong or lance-shaped, shiny, unspotted. Fls in dense blunt spikes, pale rose or violet, flushed and spotted deeper pink, the lip with slender 'arms' and 'legs' *curved forward*, with a short central 'tail-like' lobe; spur short, downcurved. Grassy meadows and open scrub, to 1500 m. May-June. B, Cev, w, sA, Ap.

3 TOOTHED ORCHID *Orchis tridentata*. Short/med per. Lvs oblong-lance-shaped, green, unspotted. Fls in *dense rounded heads*, purple-hooded, the lip whitish, spotted and edged reddish-purple with short *broad* 'arms' and 'legs' and a short central 'tail'; spur as long as the lip. Grassy places, woods and scrub, to 1500 m. Apr-June. P, Cev, sA, Ap.

4 BURNT ORCHID *Orchis ustulata*. Low/short per. Lvs oblong, pointed, unspotted. Fls *tiny, fragrant*, in dense conical-heads, the buds and hoods dark brownish-purple, fading with age, the lip white, dotted with reddish-purple, with very short 'arms' and 'legs'; spur very short. Dry meadows, grassy places over limestone, to 2100 m. May-June. T.

MARSH ORCHIDS *Dactylorhiza*. Like *Orchis* but tubers hand-like, the lvs usually scattered along the stem, bracts lf-like, the sepals spreading; spurs downcurved.

5 ELDER-FLOWERED ORCHID *Dactylorhiza sambucina* (= *Orchis sambucina*). Short per. Lvs elliptical to narrow lance-shaped, shiny pale green, unspotted. Fls in short dense spikes, either *pale yellow* with a purple spotted lip and pale green bracts, or *purple* with reddish bracts, the lip rounded, flat, scarcely 3-lobed; spur thick, about as long as the ovary. Meadows and woodland clearings, to 2100 m. Apr-July. T—except B.

6 EARLY MARSH ORCHID *Dactylorhiza incarnata*. Short/med per. Lvs 4-5, erect, lance-shaped *hooded* at tip, narrowed to the base, usually unspotted. Bracts *longer than* fls. Fls small in cylindrical spikes, variable, purple, pink or whitish, the lip 5-6mm long, slightly 3-lobed with the sides folded backwards; spur half as long as ovary. Damp meadows, marshy places, to 2100 m. May-July. T. **6a, 6b**, p.338.

7 BROAD-LEAVED MARSH ORCHID *Dactylorhiza majalis*. Short/med per. Lvs 4-6, *broad*, oblong to broad-lance-shaped, bluish green, *usually brown spotted*. Lower bracts longer than fls. Fls in dense egg-shaped or oblong spikes, purplish or reddish-purple, the lip 9-10mm long, spotted and lined, rounded, 3-lobed. Damp meadows and marshes, to 2500 m. May-July. T. **7a**, p.338.

8 COMMON SPOTTED ORCHID *Dactylorhiza fuchsii*. Variable short/med per. Lvs 7-12, rounded to elliptical, *usually dark spotted*. Lower bracts *shorter than fls*. Fls in oblong spikes, pink, pale lilac or reddish-purple, the lip 6-9mm long, dotted and lined crimson or purple, *markedly* 3-lobed, the central lobe longer than the outer ones; spur about as long as the ovary. Grassy places and scrub, usually on limestone, to 2200 m. June-Aug. T. **8a**, p.338.

9 BLACK VANILLA ORCHID *Nigritella nigra* (= *N. miniata*). Low/short per. Lvs lance-shaped to linear, pointed. Fls small, fragrant, *blackish-purple* sometimes pinkish, in rounded heads, petals and sepals *all* spreading; lip uppermost, like petals. Meadows, grassy places, to 2800 m. May-Sept. T—except B,S. **9a**, p.338.

10 SERAPIAS *Serapias vomeracea*. Short/med per. Lvs lance-shaped, unspotted. Bracts reddish-violet, pointed, *longer than fls*. Fls large, few to a spike, the sepals and petals forming a pointed violet helmet; lip reddish-brown, pointed, *curved under*. Damp grassy places and woods, to 1100 m. Apr-June. Cev, sA, Ap.

11 MAN ORCHID *Aceras anthropophorum*. Short per. Lvs lance-shaped, keeled, shiny. Fls *greenish-yellow* in long dense spikes; lip with long 'arms' and 'legs', often tinged red-brown, *spurless*. Grassy places and scrub, on limestone, to 1500 m. May-June. T—except S.

Orchid Family (contd.)

1 LIZARD ORCHID *Himantoglossum hircinum*. Short/med pale- or greyish-green per. Lvs oblong-lance-shaped, soon withering. Fls in a long tangled spike, strong-smelling, with a purplish-grey hood and a *very long* purplish, red-spotted, 3-lobed, strap-like lip, the outer two lobes much shorter; spur short. Grassy banks and slopes, woodland margins and scrub, to 1800m. May-July. sB*, P, Cev, w, sA, Ap.

2 PYRAMIDAL ORCHID *Anacamptis pyramidalis*. Med per. Lvs narrow lance-shaped, pale green, unspotted. Fls bright pink, rarely white, in a *flattened pyramidal spike*, foxy smelling, sepals spreading, petals hooded, lip 3-lobed; spur long, curved. Meadows and woodland clearings on limestone, to 1900m. May-Aug. T.

3 MUSK ORCHID *Herminium monorchis*. Low/med per. Lvs oval-oblong, yellow-green. Fls tiny, yellowish-green, in a loose thin spike, *honey scented*, sepals and petals spreading; lip 3-lobed, the middle one longer; spur very short. Dry meadows on limestone, to 1800m. May-Aug. T–except Ap.

4 FALSE ORCHID *Chamorchis alpina*. Low rather insignificant per. Lvs linear, grassy. Fls tiny, green tinged purple, with a yellowish-green slightly 3-lobed lip, sepals and petals spreading, unspurred, few in a loose spike. Damp short grassland, 1600-2700m. July-Aug. S, A.

5 FROG ORCHID *Coeloglossum viride*. Low/short per. Lvs bluish-green, rounded to oblong or lance-shaped. Fls small, yellowish-green tinged red-brown, in a thin spike, sepals and petals hooded; lip oblong 3-lobed, the *central lobe smaller*; spur very short. Meadows and scrub, to 2500m. May-Aug. T.

6 SMALL WHITE ORCHID *Gymnadenia albida* (= *Leucorchis albida*). Low/med per. Lvs oval, the upper narrower. Fls tiny, greenish-white with a yellow-green 3-lobed lip, the central lobe rather larger, in a dense blunt-spike; spur short and thick. Meadows and pastures, to 2500m. June-Sept. T.

7 FRAGRANT ORCHID *Gymnadenia conopsea*. Med per. Lvs narrow lance-shaped, green, unspotted. Fls purplish-pink *fragrant*, in a dense spike, sepals spreading, petals hooded, lip short 3-lobed; *spur long and slender*. Meadows and woodland margins, generally on lime, to 2500m. June-Aug. T. **7a** *G. odoratissima* is a frailer plant with greyish-green lvs and pale pink or white, *vanilla scented*, fls in a short dense spike. To 2700m. May-Aug. S, nP, A, AA.

HELLEBORINES. Perennials with oval or lance-shaped ribbed lvs, clasping the stem. Fls rather drooping in one-sided spikes, each fl with a narrow lfy bract, petals and sepals spreading; lip short lobed or unlobed, unspurred.

8 DARK RED HELLEBORINE *Epipactis atrorubens*. Med red tinged per, stems downy. Lvs large, in two rows up the stem. Fls *dark wine-red* or reddish-violet, fragrant, lip shorter than the sepals. Grassy and rocky places, woods to 2200m. May-Aug. T. **8a Broad-leaved Helleborine** *E. helleborine* has *spirally arranged* lvs and larger greenish-yellow, red-tinged fls, unscented; lip rounded and curled under at tip. Woodland clearings to 1800m. June-Sept. T–except Ap. **8b** *E. leptochila* like 8 but shorter, the fls *green* tinged with white or yellow; *lip pointed*, the same size as the sepals. Beech and coniferous woods, rarely above 1000m. July-Aug. B, nwA, J. **8c** *E. microphylla* like 8a but with much smaller, more rounded lvs and smaller greenish, purple-margined, fls; lip round with a short point. Woods, to 1350m. May-Aug. eP, A, Ap. **8d** *E. muelleri* like 8a, the fls pale green, the inside of the lip cup *reddish*. Forest margins, on lime. wA. **8e, 8f**, p.339.

9 VIOLET BIRDSNEST ORCHID *Limodorum abortivum*. Stout med/tall unbranched per, stem *violet- or bluish-tinged*. Lvs small, *scale-like* along stem, violet. Fls in long spikes, large, violet, with spreading sepals and petals, the lip yellowish and violet, triangular; spur long, pointed. Woodland clearings or shady pastures, to 1200m. May-July. sA, Ap.

Orchid Family (contd.)

1 RED HELLEBORINE *Cephalanthera rubra*. Short/med slightly purple tinged per. Lvs narrow lance-shaped, clasping stem at base, strongly ribbed. Fls *bright carmine-pink*, not opening widely, few in a loose spike; lip whitish with a down-curved pointed tip, unspurred. Open woods, especially Beech and scrub on limestone, to 1800 m. May-July. T-B*. **1a** C. *longifolia* has longer grassy lvs and *white fls*, the lip ridged with orange-yellow; bracts small and inconspicuous. T. **1b White Helleborine** C. *damasonium* like 1a but with broader lvs and *large* lf-like bracts. To 1300 m. May-July. T.

2 SUMMER LADY'S TRESSES *Spiranthes aestivalis*. Med slender yellowish-green per. *All lvs linear*, grassy, the upper shorter. Fls very small, white, unspurred, in a thin *spiralled spike*. Damp meadows and riverbanks, to 1250 m. June-Aug. P,A,Ap*. **2a Autumn Lady's Tresses** S. *spiralis* has a basal rosette of *oval* bluish-green lvs, withering before flg, leaving a few *scale-like* lvs up the stem. Fls white with a greenish lip, sweetly scented. Dry grassy places, rarely much above 1000 m. Aug-Oct. T-except S.

3 LESSER BUTTERFLY ORCHID *Platanthera bifolia*. Short/med per. *Basal lvs 2*, oval-elliptical, shiny-green, stem lvs smaller and scale-like. Fls yellowish or greenish-white, vanilla-scented, in loose spikes, with spreading sepals and petals, the lip long, narrow and undivided; spur long, curved; pollen masses parallel. Meadows, woods and moors, to 2300 m. May-July. T. **3a Greater Butterfly Orchid** P. *chlorantha* is larger with broader lvs; spur swollen towards the tip; pollen masses *diverging*. To 1800 m. T.

4 COMMON TWAYBLADE *Listera ovata*. Med per. Lvs a *single pair* at base of the stem, oval, ribbed, dull green. Fls small, yellow-green, unspurred, in long slender spikes; sepals and petals half-hooded; lip long, deeply forked. Woods, scrub and clearings, to 2100 m. May-July. T. **4a Lesser Twayblade** L. *cordata* is smaller, almost insignificant, the single pair of lvs heart-shaped, shiny. Fls reddish-green, musky-fragrant, the lip with two short upper lobes and a long forked central one. Wet places and woods, often pine, moors, 1300-2300 m. May-Sept. T.

5 BIRDSNEST ORCHID *Neottia nidus-avis*. Short/med saprophytic per; stem *honey-coloured* with scale-like lvs. Fls pale brown with a sickly fragrance, unspurred, in broad spikes; petals and sepals small, half-hooded, the lip large and forked. Beech woods mainly, to 1700 m. May-July. T. Rather like some Broom-rapes (p.222) but these have a 3-lobed lip.

6 CREEPING LADY'S TRESSES *Goodyera repens*. Low/short per with *creeping runners*. Lvs oval, pointed, in a basal rosette; stem lvs smaller, scale-like. Fls tiny, white, fragrant, in a slender *spiralled spike*; sepals spreading, petals hooded, lip short and pointed, unspurred. Woods, especially coniferous, and mossy places on acid soils, to 2200 m. May-Aug. T.

7 CORALROOT ORCHID *Corallorhiza trifida*. Low/short saprophytic per; stem yellowish-green with *long sheathing scales*. Fls yellowish-green or yellowish-white, unspurred, in loose few-fld spikes; sepals and petals spreading, lip *small* 3-lobed. Woods, especially coniferous, meadows, 1400-2700. June-Aug. T.

8 GHOST ORCHID *Epipogium aphyllum*. Low/short saprophytic per; stem *mauvish-yellow*, translucent, with long sheathing, pale brown, scales. Fls large, solitary or 2-3 together; petals and sepals yellowish-white, spreading, the lip pinkish *upturned* and hoodlike; spur short, blunt. Beech, oak and pine woods, to 1900 m. June-Sept. T*.

9 ROUND-HEADED ORCHID *Traunsteinera globosa*. Short/med per. Lvs bluish-green, oblong-elliptic, all on stem. Fls small pale pink, purple-spotted, in dense globular clusters; sepals and petals spreading, the lip short, 3-lobed; spur short. Meadows and woods, to 2600 m. May-Aug. P*,A,Ap.

Nettle Family   Urticaceae

Lvs opposite, armed with stinging hairs. Individual fls small; male and female separate, on the same or on different plants.

1 COMMON NETTLE *Urtica dioica*. Variable med/tall tufted per armed with strongly stinging hairs. Lvs dull green, heart-shaped, toothed and stalked. Fls pale green, in long drooping catkin-like spikes, male and female *on separate plants*. Woods, banks and waste places, often near farms and cowsheds, to 3150 m. June-Sept. T. **1a.** *U. kioviensis* is bright green, the stems not hairy. eA.

2 ANNUAL NETTLE *Urtica urens*. Short/med *annual* armed with weakly stinging hairs. Lvs clear green, oval, heart-shaped or narrowed at the base, short-stalked. Fls greenish, in short erect or spreading spikes, male and female separate but *on the same* plant. Cultivated ground and waste places, to 2700 m. June-Sept. T.

Sandalwood Family   Santalaceae

Semi-parasitic perennials growing on the roots of various herbs and shrubs. Lvs alternate, untoothed. Fls small, whitish, cup or bell-shaped, 4-5 lobed, with sepals, each with a bract and two shorter bracteoles; stamens 4-5, style 1. Fr a small green nut.

3 ALPINE BASTARD TOADFLAX *Thesium alpinum*. Short, usually slightly branched, per. Lvs linear to narrow-oblong, one-veined. Fls *4-lobed*, in one sided spikes, rarely branched; bracts 2-3 times longer than the fls. Dry meadows and stony places, to 2600 m. May-Aug. P,A,Ap. **3a Pyrenean Bastard Toadflax** *T. pyrenaicum* has 5-lobed fls usually in *two-sided*, zig-zagged spikes; fls 3-4 mm long. P,A,nAp. **3b** *T.p. alpestre* like **3a** but fls *larger*, 5-5.5 mm long. A.

4 BAVARIAN BASTARD TOADFLAX *Thesium bavaricum*. Short/med per, stems usually unbranched. Lvs dark green, lance-shaped, *3-5 veined*. Fls in lax-branched spikes; bracts as long or up to twice as long as the fls. Dry meadows and scrub, to 1250 m. June-Aug. A,Ap. **4a** *T. linophyllon* has narrow elliptical, 1-3-veined lvs, often rather yellowish-green. A,Ap.

5 BRANCHED BASTARD TOADFLAX *Thesium divaricatum*. Short, rather robust stemmed, *much branched*, per. Lvs linear, one-veined. Fls 5-lobed, in branched clusters; bracts *shorter than* the fls. Dry meadows and scrub, to 2150 m. June-Aug. P,Cev,sw,s,seA,Ap.

Mistletoe Family   Loranthaceae

6 MISTLETOE *Viscum album*. Woody, regularly branched *parasitic* shrub occurring in rounded masses on various trees. Lvs *opposite*, elliptical or oblong, broadest above the middle, yellowish-green, untoothed. Fls inconspicuous, green, 4-petalled, male and female on separate plants. Fr a juicy *white berry*, sticky. Parasitic on dicotyledonous trees, Apples and Poplars in particular. Feb-Apr. T. **6a** *V.a. abietis* is parasitic on Silver Fir and other Firs. A,Ap. **6b** *V.a. austriacum* is parasitic on Pines and Larch. A,Ap.

Birthwort Family   Aristolochiaceae

7 ASARABACCA *Asarum europaeum*. Low creeping, patch-forming, downy-stemmed per. Lvs kidney or heart-shaped, untoothed, long-stalked, shiny. Fls dull brownish-purple, bell-shaped, 12-15 mm, 3-lobed, solitary and hidden amongst the lvs. Woods on calcareous soils, to 1300 m. A,Ap.(B).

289

Dock Family Polygonaceae *(contd. from p. 30)*

DOCKS or SORRELS *Rumex*. Mostly hairless perennials with alternative lvs and sheaths at the lf bases, forming a tube round the stem (*ochrea*). Fls small, green, in whorls forming long branched spikes. Fls small, green, in whorls forming long branched spikes. The lower accompanied by a small lf; 3 petals and 3 sepals all alike and often reddish, the inner 3 enlarging and hardening to become the valves round a 3-sided nut. Each species has a characteristic shaped fr important in identification.

1 SHEEPS SORREL *Rumex acetosella*. Variable low/short slender per. Lvs *arrow-shaped*, the basal lobes sometimes divided, spreading or forward pointing, often very narrow. Fls greenish, on erect stems, branched from the middle, male and female usually on separate plants. Dry meadows and heaths, on acid soils, to 2400 m. May-Aug. T.

2 RUBBLE DOCK or FRENCH SORREL *Rumex scutatus*. Short/med tufted per, branched from the base. Lvs green or bluish-green, *helmet-shaped*, as long as broad, with spreading basal lobes. Fls few on erect branched stems, stamens and stigmas in the same fl. Fr 4.5-6mm. Rocky places and screes, to 2500m. May-Aug. T, except s, (B).

3 SNOW DOCK *Rumex nivalis*. Low/short tufted per. Lvs small, *often all basal*, the outermost oblong, without basal lobes, the inner often arrow-shaped. Fls reddish, in lax spikes, seldom branched, male and female on separate plants. Fr 3mm. Snow patches in limestone areas, 1600-2750m. July-Sept. e, seA.

4 MOUNTAIN DOCK *Rumex arifolius* (= *R. montanus*). Med/tall coarse per with lfy stems. Basal lvs oval-heart-shaped, twice as long as broad, strongly veined. Fls greenish, in loose *branched clusters*, male and female on separate plants. Fr 2.5-3mm. Meadows and woods, often on acid soils, to 2500m. June-Sept. T probably—except B. **4a** *R. amplexicaulis* has *fringed ochrea* and half-clasping stem lvs and larger fls. Woods and river banks. P, sw, sA, Ap.

5 APENNINE DOCK *Rumex gussonii*. Short tufted, *often* loosely cushion-forming per. Basal lvs narrow arrow-shaped, twice as long as broad, the basal lobes forward pointing; stem lvs few, very narrow. Fls greenish, in short branched clusters. Fr 2.5-3mm; nut dark brown. Limestone crevices and screes, to 2000m. June-Aug. Ap.

6 COMMON SORREL *Rumex acetosa*. Short/tall acid tasting per. Lvs arrow-shaped; 2-4 times as long as broad, the basal lobes *backward pointing*, the upper lvs clasping the stem; ochrea with a fringed margin. Fls in slightly branched clusters, male and female on separate plants. Fr 3-3.5mm; nut *shining black*. Meadows and woods, to 2100m. May-Aug. T—rarer in the south.

7 MONKS RHUBARB *Rumex alpinus*. Tall coarse per, stems often reddish. Basal lvs *heart-shaped*, as long as broad, long-stalked, upper lvs narrower; *ochrea papery*. Fls yellowish-green, in crowded branched clusters, stamens and stigmas in the same fls. Fr 4.5-6mm, valves untoothed. Meadows, often around farms and cattle sheds, to 2500m. June-Aug. P, A, Ap, (B). **7a Northern Dock** *R. longifolius* has broad lance-shaped lvs, 3-4 times longer than broad, wavy-margined. Often by streams. B, S, P. **7b** *R. nepalensis* has oblong-oval lvs, heart-shaped at the base; fr valves with *hooked teeth*. Mountain woodland clearings. Ap.

basal lf

1 fr

2 tr

3 fr

4 fr

5 fr

6 fr

7 fr

Pink Family (contd. from pp. 32-6)

STITCHWORTS *Stellaria*. Annuals or perennials. Fls white, in branched clusters, each with 5 separate sepals and 5 deeply notched or cleft petals (usually cleft over halfway to the base); stamens normally 10; styles 3. Fr capsule with 6 teeth.

1 WOOD STICHWORT *Stellaria nemorum*. Short/med patch-forming per, stems hairy *all round*. Lvs oval, pointed, lower long-stalked, the upper stalkless. Fls 18-24mm, petals cleft almost to the base, twice as long as the sepals. Damp woods, to 2400m. May-July. T.

2 COMMON CHICKWEED *Stellaria media* agg. Variable low/short often sprawling annual, stems hairy all round or with a single line of hairs. Lvs oval, pointed, all but the uppermost stalked. Fls 6-10mm, petals cleft almost to the base, *equalling sepals*, sometimes absent. Cultivated and bare ground, a common weed, to 2500m. In fl most of the year except at higher altitudes. T.

3 GREATER STICHWORT *Stellaria holostea*. Short/med, straggling, rather rough, *square-stemmed* per. Lvs lance-shaped, long-pointed, stalkless. Fls 15-30mm, petals cleft to *halfway*, about twice as long as the sepals. Woods, hedgerows and banks, usually on heavy soils, to 2000m. Apr-June. T. **3a Lesser Stichwort** *S. graminea* has smooth stems and lvs; fls smaller, 5-12mm, petals usually equalling the sepals. May-Aug. T.

4 BOG STICHWORT *Stellaria alsine* (= *S. uliginosa*). Low/short rather sprawling, patch-forming, *hairless* per, stems square. Lvs elliptical to oval lance-shaped, smooth, usually stalkless. Fls small, 4-6mm, the petals shorter than the sepals. Wet places, often by streams and pools, to 2300m. May-Aug. T.

5 LONG-LEAVED STICHWORT *Stellaria longifolia*. Low/short patch-forming hairless per, stems square, *rough*. Lvs narrow lance-shaped, rough-edged. Fls small, 4-8mm, in clusters, the petals equalling the sepals. Damp wooded places, to 2100m. June-July. S, A – except France and Yugoslavia. **5a** *S. crassipes* is lower with oval lvs and fls 8-10mm, solitary. June-July. S. **5b** *S. calycantha* like **5a** but lvs often narrower and *yellowish-green*. Fls tiny, 3-5mm, petals *shorter* than the sepals, or absent. S.

PEARLWORTS *Sagina*. Small insignificant, mossy, tufted or cushion-forming annuals or perennials, with slender stems and lvs. Fls 4-5 parted; sepals separate; petals usually white, *not notched*, sometimes absent. Fr capsule with 4-5 teeth.

6 KNOTTED PEARLWORT *Sagina nodosa*. Low/short tufted per. Lvs linear, diminishing in size up the stem, the upper *with tufts* of lvs at the nodes. Fls 5-10mm, solitary or 2-3 together; petals 5, 2-3 times longer than sepals. Damp, often sandy places, usually on calcareous soils. July-Sept. T.

7 CUSHION PEARLWORT *Sagina caespitosa*. Low per forming small cushions. Lvs linear. Fls 6-8mm, *amongst* the cushion lvs; petals 5, *longer than* the violet-edged sepals. Damp rocky places, gravels, by snow patches, to 1670m. June. S*. **7a** *S. intermedia* has smaller fls, 3-6mm, the petals 4-5, shorter than the sepals. June-Aug. B*, S. **7b** *S. glabra* is laxer than 7, the fls 5-10mm, on short footstalks; sepals *not* violet-edged. Meadows and stony places, on acid soils, 1600-2700m. July-Aug. eP, A, Ap.

8 ALPINE PEARLWORT *Sagina saginoides* (= *S. linnaei*). Low loosely tufted, hairless per. Lvs linear, *bristle-tipped*; stems lvs shorter. Fls 6-7mm, solitary or two together, petals 5, equalling the sepals. Damp places, to 2750m. June-Aug. T.

9 PROCUMBENT PEARLWORT *Sagina procumbens*. Low hairless, mat-forming per, spreading outwards from a *central lfy-rosette*. Lvs linear, bristle-tipped. Fls tiny, greenish-white, 2-4mm, solitary; petals 4 or absent, smaller than the sepals. Damp bare places, to 2600m. May-Sept. T. **9a** *S. apetala* is a variable low/short *annual* without a basal lfy rosette; petals often absent. T–except S.

10 KNAWEL *Scleranthus perennis*. Short more or less erect, almost hairless, rather spiky-looking per. Lvs lance-shaped to linear, pointed, often hairy-edged. Fls small, in clusters, petalless; sepals fused in the lower half, membraneous-edged. Fields and waste places, on acid soils, to 2250m. May-Oct. T–except extreme N. **10a** *S. p. polycnemoides* forms dense *low cushions* with shorter lvs. P.

Pink Family *(contd.)*

MOUSE-EAR CHICKWEEDS *Cerastium*. Tufted perennials. Fls white, in small loose clusters, or solitary, petals 5 shallowly or deeply cleft, but seldom beyond halfway; sepals separate from one another; stamens 5 or 10; styles usually 5 (3 in Starwort Mouse-ear). Fr capsule with as many teeth as styles.

1 STARWORT MOUSE-EAR *Cerastium cerastoides* (= *C. trigynum*). Low loosely matted per, hairless except for a *single line of hairs* down the stem, rooting at the nodes. Lvs pale green, linear to oblong. Fls 9-15mm, solitary or 2-3 together, the petals deeply cleft, styles 3; bracts lflike. Damp rocky and grassy places, 1500-3000m. July-Aug. nB,S,P,A,Ap.

2 SNOW IN SUMMER *Cerastium tomentosum*. Low/short mat forming, *white-woolly* per. Lvs lance-shaped. Fls 15-25mm, in loose clusters, the petals twice as long as the sepals, deeply cleft. Grassy and rocky places, banks, to 2250m. May-July. c,sAp–widely cultivated elsewhere, sometimes naturalised.

3 NARROW-LEAVED MOUSE-EAR *Cerastium lineare*. Low/short, slender, tufted, hairy per. Lvs linear to elliptical, those of non-flg shoots *in loose rosettes*, hairy only along edges. Fls 16-20mm, solitary or 2-3 together, the petals deeply cleft, sepals hairy; bracts lflike. Rocky places, to 1600-2100m. July-Aug. swA.

4 JULIAN MOUSE-EAR *Cerastium julicum*. Low densely tufted hairy per. Lvs narrow oblong to elliptical, hairy along edges at base only, *margin rolled under*. Fls 14-18mm, solitary or 2-3 together, the petals deeply cleft, sepals hairy; upper-most bracts with membranous edges. Rocky and grassy places, 2100-2250 m. July-Aug. seA.

5 ITALIAN MOUSE-EAR *Cerastium scaranii*. Short tufted per, hairy. Lvs oval to elliptical, hairy on *both sides*. Fls 12-20 mm, solitary or in clusters of 2-7, the petals shallowly notched; bracts with membranous edges. Rocky places, to 2000 m. May-July. AA,Ap.

6 FIELD MOUSE-EAR *Cerastium arvense* agg. Variable low/short tufted or loosely matted per, sparsely hairy. Lvs narrow lance-shaped. Fls 12-20mm, in clusters of 3-7, rarely solitary, the petals deeply cleft, twice as long as the sepals; bracts with *thin membranous edges*. Fr *curved*. Dry fields and rocky places, to 3100m. Apr-Sept. T–except S.

7 ALPINE MOUSE-EAR *Cerastium alpinum*. Low mat-forming greyish-green hairy per, hairs *long and soft*. Lvs oblong or elliptic, *broadest* above the middle. Fls 18-25mm, in clusters of 2-5 or solitary, the petals deeply cleft; bracts with membranous edges. Grassy and rocky places, mainly on acid rocks, 1800-2850 m. June-Aug. T–except B. *C.a. lanatum* has *white-woolly* lvs and stems. T. **7b** *C.a. squalidum* like 6a but with *glandular hairs* mixed in the wool. P.

8 ARCTIC MOUSE-EAR *Cerastium arcticum*. Low/short greyish, hairy per, hairs *short* and stiff. Lvs elliptical, broadest above the middle. Fls 20-30mm, solitary or 2-3 together, the petals shallowly cleft; *bracts green* and lflike. Rocky places and screes, to 1700m. June-Aug. B,S. **8a Glacier Mouse-ear** *Cerastium uniflorum* has broader, soft, bright green lvs and smaller fls, often rather creamy-white. On granites and schists, 1900-3400m. July-Aug. A.

9 BROAD-LEAVED MOUSE-EAR *Cerastium latifolium*. Low loosely tufted per, glandular hairy. Lvs *oval to oval-elliptical*, pointed. Fls 15-20mm, solitary or 2-3 together, the petals shallowly cleft, twice as long as the sepals; bracts green and lflike. Rocky places, moraines and screes, on limestone, 1600-3500m. July-Aug. A,nAp. **9a** *C. pyrenaicum* has petals *hairy-edged* at base, scarcely longer than sepals. eP.

Pink Family *(contd.)*

1 BELL-FLOWERED MOUSE-EAR *Cerastium pedunculatum.* Low loosely tufted, slightly hairy per. Lvs lance-shaped, *stiff.* Fls 7-8mm long, *bell-shaped,* usually solitary, the petals deeply cleft. Rocks, moraines and screes, 2000-3600m. July-Sept. c,eA.

2 CARINTHIAN MOUSE-EAR *Cerastium carinthiacum.* Low/short loosely matted, almost hairless, per. Lvs *shiny-green,* oval to lance-shaped, pointed. Fls 10-18 mm, in clusters of 2-7; bracts with wide membranous edges. Rocky places, usually on calcareous or dolomitic rocks, to 2400m. June-Aug. eA. **10a** *C.c. austroalpinum* is *more hairy.* seA.

3 SLOVENIAN MOUSE-EAR *Cerastium subtriflorum.* Low/short tufted hairy bien/per. Lvs elliptical to oval or lance-shaped, *stalkless.* Fls 12-20 mm, in clusters of 3 or more, the sepals with membranous edges; lowest bracts lflike. Rocky places, 1650-2200m. July-Sept. seA–Giulie and Julian Alps.

4 COMMON MOUSE-EAR *Cerastium fontanum.* Variable low/short hairy per with lfy nonflg shoots. Lvs lance-shaped, stalkless. Fls 6-14mm; petals deeply cleft, equalling sepals; lower bracts lflike, the upper membranous edges. Grassy places and bare ground, to 2400m. Apr-Nov. T.

5 GREY MOUSE-EAR *Ceratium brachypetalum.* Variable low/short hairy, often greyish, ann. Lvs oval to elliptical, the lowermost spoon-shaped. Fls 6-10mm, in loose clusters, the petals deeply cleft, shorter or slightly longer than the sepals; flstalks *bent* just below fls; *all bracts* lflike. Dry open places, often on calcareous soils, to 1300m. Apr-Nov. T.

6 STICKY MOUSE-EAR *Cerastium glomeratum.* Low/short stickily-hairy, often yellowish, ann. Lvs oval or elliptical, broadest above the middle. Fls 5-8mm, often not opening fully, in *tight clusters;* petals deeply cleft, about as long as the sepals; *all bracts lflike.* Bare ground, waste places, to 1300m. Apr-Oct. T.

7 LITTLE MOUSE-EAR *Cerastium semidecandrum.* Low/short semi-prostrate or erect, glandular-hairy, ann. Lvs elliptical to oval, the lowermost broadest above the middle. Fls 5-9mm, in small clusters, the petals slightly notched, *shorter than* the sepals; sepals and bracts with membranous edges. Dry, often stony, places, to 1650m. Mar-May. T.

8 DWARF MOUSE-EAR *Cerastium pumilum.* Variable low, more or less erect, glandular-hairy ann, often reddish tinged. Lvs elliptical, broadest above the middle, the upper ones oval. Fls 6-9mm, in loose clusters, the petals often *purplish tinged,* about as long as the sepals; bracts and sepals with *membranous edges.* Dry grassy and bare places, on calcareous soils, to 1300m. Apr-June. T, except S.

9 SEA MOUSE-EAR *Cerastium diffusum* (= *C. tetrandrum*). Variable low/short glandular-hairy ann. Lvs oval to elliptical, the lowermost spoon-shaped. Fls 6-14mm, in loose clusters, the petals shallowly notched, *shorter than* the sepals; bracts usually all lflike. Dry grassy and stony places, to 1300m. Mar-July. T.

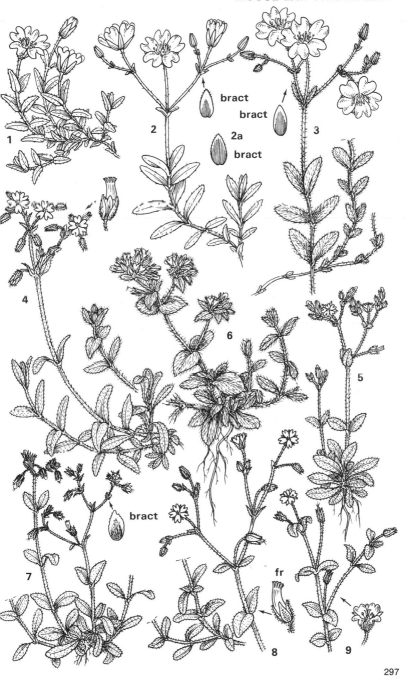

bract

bract

2a

bract

bract

fr

Pink Family (contd.)

SANDWORTS *Minuartia*. Like *Arenaria* but the fr capsule has the same number of teeth as styles. Lvs often bristle-tipped or awl-like and forming moss-like tufts.

1 SICKLE-LEAVED SANDWORT *Minuartia recurva* (= *Alsine recurva*). Low densely tufted per, slightly glandular-hairy. Lvs *sickle-shaped* 4-10 mm, 3-veined. Fls white 7-9 mm, solitary or in clusters of 2-8. Grassy and stony places, usually on acid soils, to 1900 m. P, sw, c, seA, Ap.

2 APENNEAN SANDWORT *Minuartia graminifolia*. Low/short dense cushion per, glandular-hairy. Lvs narrow lance-shaped, stiff, 10 mm long or more. Fls white, 10-16 mm, in clusters of 2-7 on 4-14 cm stems. Rocky places, 1400-2000 m. July-Aug. a, cAp. **2a** subsp. *clandestina* is *hairless* with shorter lvs. c, sAp. **2b** *M. cerastiifolia* like 2, but lvs *less than* 10 mm long, distinctly keeled beneath and fls 6-10 mm. cP.

3 ROCK SANDWORT *Minuartia rupestris* (= *Alsine rupestris*). Low loosely tufted or creeping per, glandular-hairy, stems *often rooting* at the nodes. Lvs small lance-shaped, 2-4 mm, *hairy-edged*. Fls white 7-9 mm, usually solitary on 1-3 cm stems. Rocks and screes, 1900-2500 m. A. **3a** *M. lanceolata*. □ has larger lvs with 5-7 cm fl stems carrying 2-3 fls. swA–Cottian.

4 BERGAMASQUE SANDWORT *Minuartia grignensis*. Low *hairless* per. Lvs crowded towards base of plant, linear, 5-15 mm, 3-veined. Fls small white, 5-7 mm, in branched clusters of up to 12; petals slightly longer than the *3-veined sepals*. Limestone dolomitic cliffs, 1300-1900 m. sA–Bergamasque.

5 *Minuartia cherlerioides* (= *M. aretioides*). Very low dense cushion-forming per. Lvs tiny, 1.5-3 mm, oblong-elliptical, blunt, *hairless*, 3-veined. Fls white, 4-7 mm, solitary on *very short* stalks; petals slightly shorter than the 3-veined sepals. Limestone rocks and screes, 2000-2500 m. e, seA. **5a** subsp. *rionii* has *hairy-edged* lvs. Acid rocks and screes. c, sA.

6 AUSTRIAN SANDWORT *Minuartia austriaca* (= *Alsine austriaca*). Low/short loose, cushion-forming per, hairless or slightly glandular-hairy. Lvs *narrow lance-shaped to linear*, 10-20 mm, hairless. Fls large white, 12-20 mm, solitary or 2-3 together; petals *slightly-notched* about twice as long as sepals. Limestone and calcareous rocks and screes, 1900-2100 m. sc, seA. **6a** *M. villarsii* has slightly broader lvs and smaller, 8-15 mm, fls. P, swA.

7 VERNAL SANDWORT *Minuartia verna* (= *Alsine verna*). Very variable low/short, loose cushion-forming, glandular-hairy per. Lvs linear, pointed, keeled below, to 20 mm long. Fls white, 6-8 mm, with *purplish anthers*, in clusters of 2-7, sometimes solitary; petals slightly longer than sepals. Grassy and stony places, screes, to 3000 m. May-Sept. T–except S. **7a** subsp. *collina* has fls in clusters of 6 or more, *anthers yellow*, the petals slightly shorter than the sepals. c, eA, Ap.

8 *Minuartia capillacea* (= *Alsine liniiflora*). Short, loose cushion-forming, glandular-hairy per. Lvs linear, *stiffly-pointed*, 10-20 mm, hairy-edged, 1-3 veined. Fls white, 14-22 mm, solitary or in clusters of 2-6, petals about twice as long as sepals. Calcareous rocks and screes, to 2000 m. A–except Austrian, Ap. **8a** *M. laricifolia* □ is more delicate with smaller, 5-12 mm, lvs and fls; sepals and flstalks *downy*. Acid rocks and screes, 1300-2000 m. July-Aug. P, sA, Ap. **8b** *M.l. kitaibelii* has hairless sepals and flstalks. Calcareous rocks and screes. eA.

9 NORTHERN SANDWORT *Minuartia biflora*. Low slender, slightly hairy, per, forming loose tufts. Lvs linear, 4-10 mm, l-veined. Fls white, rarely pale lilac, 7-12 mm, solitary or 2-3 together; petals slightly longer than sepals. Damp open places, often by snow patches, 2000-2800 m. June-Aug. sc, eA.

10 MOSSY CYPHEL *Minuartia sedoides* (= *Alsine sedoides*). Low hairless per forming dense *flattish yellowish-green* cushions. Lvs small, 3-6 mm, narrow lance-shaped, *closely overlapping*, 3-veined. Fls greenish, 4-8 mm, solitary, almost stalkless; petals absent. Grassy and stony places, rock ridges, moraines, 1800-3800 m. nB–Scotland, P, A.

Pink Family (contd.)

SANDWORTS *Arenaria and Moehringia*. Fls in small loose clusters or solitary, white, rarely pale purplish-pink; petals not notched, stamens 10, styles 3-5. Fr a capsule splitting with twice as many teeth as styles.

1 PINK SANDWORT *Arenaria purpurascens*. Low mat-forming per, stems hairy in the upper half. Lvs oblong or lance-shaped, pointed, l-veined, hairy at base. Fls *pale purplish-pink*, sometimes white, 7-14mm, in clusters of 2-4; sepals hairless. Damp rocky places, 1800-2800m. July-Aug. P.

2 IMBRICATE SANDWORT *Arenaria tetraquetra*. Low almost hairless per forming dense flat cushions. Lvs oval, blunt, *closely overlapping*. Fls white, 5-9mm, solitary, with 4 or 5 petals and sepals. Dry places, 1800-2700m. July-Aug. c,eP. **2a** *A. aggregata* has narrower, *pointed,* lvs and fls in dense clusters of 3 or more. P,MA. **2b** *A.a. erinacea* (= *A. erinacea*) usually has solitary fls. se France–Mt. Ventoux.

3 LARGE-FLOWERED SANDWORT *Arenaria grandiflora*. Low/short hairy per, forming loose cushions. Lvs *narrow lance-shaped*, pointed. Fls white, 8-15mm, solitary or 2-3 together, stalks downy; sepals hairy. Dry rocky and stony places, to 2000m. May-Aug. P,c,s,seA,Ap.

4 CARNIC SANDWORT *Arenaria huteri*. Low hairy tufted per. Lvs oblong, *broadest above* the middle. Fls white, 15-20mm, solitary or 2-3 together; sepals hairy. Dolomitic rock crevices, to 2000m. June-Aug. seA-Carnic.

5 TWO-FLOWERED SANDWORT *Arenaria biflora*. Low sprawling, usually hairless per; stems *rooting at the lower nodes*. Lvs *oval* or almost rounded, l-veined. Fls white, 5-7mm, two together or solitary. Damp places, often over granite rocks or by snow patches, 1700-3200m. June-Aug. c,eP,A,Ap.

6 NORWEGIAN SANDWORT *Arenaria norvegica*. Low loosely tufted ann/per, almost hairless. Lvs oblong, broadest above the middle, usually dark green. Fls white, 5-6mm, stalked, solitary or 2-3 together; anthers *white*. Open stony places and screes on calcareous soils, to 1450m. June-July. nB,S. **6a** *A. humifusa* is more mat-forming with short fl-stalks and *pale purple anthers*. S*.

7 CILIATE-LEAVED SANDWORT *Arenaria ciliata*. Variable low per with *rough-edged stems*. Lvs elliptical to spoon-shaped, hairy on edges, at least in the lower half. Fls white, 6-10mm, in clusters of 2-7; anthers white. Meadows and rocky places on calcareous soils, 1400-3200m. July-Aug. P,A,nAP.

8 CADI SANDWORT *Arenaria ligericina*. Low/short tufted, *glandular-hairy* per. Lvs narrow-elliptical or oblong, pale green. Fls white, 6-9mm, in loose clusters of 3-10, limestone rocks, to 2000m. July-Aug. eP–Sierra del Cadi.

9 SOUTH-EASTERN SANDWORT *Moehringia diversifolia*. Low/short almost hairless per, with thin stems. Lvs oval or spoon-shaped, the lower stalked. Fls small, white, 3-4mm, on *downy stalks*, solitary or in clusters of 2-5. Acid rocks and screes, to 1800m. May-July. seA.

10 *Moehringia dielsiana*. Low hairless, fragile per. Lvs narrowly-oblong or spoon-shaped, pointed, blue-green. Fls white, 10-12 mm, usually solitary, long-stalked. Cliff crevices, 1300-1400 m. June-July. sA–Bergamasque. **10a** *M. papulosa* often has pendant stems and small 4-5 parted fls. MA,c&sAAP. **10b** *M. tommasinii* always has 4-parted fls. Limestone rocks. seA.

11 NARROW-LEAVED SANDWORT *Moehringia bavarica*. Low/short sprawling or erect, hairless, *bluish-green*, per. Lvs linear, appearing veinless. Fls white, 5-8mm, solitary or in small clusters, 5-petalled. Limestone rocks to 1600m. May-July. seA–from Monte Baldo eastwards. **11a** *M.b. insubrica* has shorter lvs and smaller fls. sA–Brescian.

12 MOSSY SANDWORT *Moehringia muscosa*. Variable low hairless per with sprawling weak stems. Lvs *linear*, pointed, *1-3 veined*. Fls white, 4-8mm, in clusters of 3-6, usually 4-petalled. Shaded damp rocks and mossy places, to 2350m. May-Sept. P,A,Ap. **12a** *M. glaucovirens* has smaller, *5-petalled* fls, usually solitary or 2-3 together. Shady limestone rocks. sA.

13 CREEPING SANDWORT *Moehringia ciliata*. Low *creeping*, mat-forming per; stems hairless. Leaves linear, with a few hairs at base. Fls small, white, 4-5mm, solitary or 2-3 together, 5-petalled. Limestone screes, to 3000m. June-Aug. eP,A.

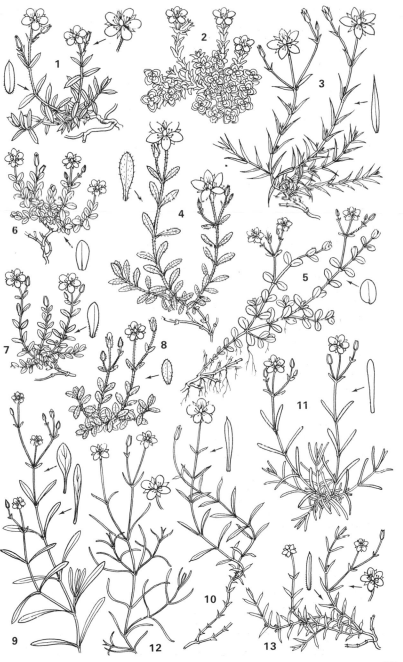

Cress Family – Rockcresses *(contd. from p. 64)*

ROCKCRESSES *Arabis*. Annuals or perennials with undivided lvs, often in basal rosettes and with few stem lvs. Fls white or pinkish, the inner 2 sepals often pouched at the base. Fr a long, parallel-sided, rather flattened pod.

1 HAIRY ROCKCRESS *Arabis hirsuta*. Low/med hairy bien/per; stems rather stiff, branched above. Basal lvs in a rosette, oval, only slightly toothed, *tapered* to the base; stem lvs smaller, oval to lance-shaped, *clasping* the stem. Fls white, 3-4.5mm, numerous. Pod slender, erect, 15-35mm. Wet or dry calcareous rocks, to 2050m. May-Aug. T. **1a** *A. allionii* is shorter with *hairless* stems; fls 5-6mm. Wet rocks. P, A.

2 CORYMBOSE ROCKCRESS *Arabis corymbiflora*. Low/short hairy bien/per; stems stiff, mostly unbranched. Basal lvs in a rosette, oval, broadest above the middle, toothed; stem lvs oval to elliptical, *rounded* at base. Fls white, 3-5mm, in *compact* clusters. Pod slender 12-22mm, erect. Rocky slopes and gravels, to 2700m. May-July. P, A, Ap. **2a** *A. serpyllifolia* □ has the lower lvs *long-stalked*, the stems and lvs with starry hairs. P, wA, Ap.

3 COMPACT ROCKCRESS *Arabis vochinensis*. Low tufted per with branched hairs. Basal lvs in compact rosettes, oblong, broadest above the middle, blunt, *untoothed*; stem lvs oblong, narrowed at the base. Fls white, 6-8mm, in lax clusters. Pod 8-15mm long. Rocky places, to 2200m. seA.

4 ANNUAL ROCKCRESS *Arabis recta*. Short hairy ann. Lvs oval to oblong, untoothed, the lowermost stalked and withered at flg time; stem lvs clasping the stem, *arrow-shaped* at base. Fls tiny, white, 3-4mm, petals *erect*. Pod slender, 10-35mm, less than 1mm wide. Rocky places, to 1500m. Apr-June. T–except B, S. **4a** *A. nova* is taller, often bien, fls 4-5mm, pod 25-70mm, *over* 1 mm wide. P, A, Jura. **4b Towercress** *A. turrita* □ Med softly hairy biennial/per often with reddish stem and *very long pods*, all *twisted* to one side and curving downwards.

5 SCOPOLI'S ROCKCRESS *Arabis scopoliana*. Low tufted per, *hairless except* on lf edges. Lvs oval or oblong, broadest above the middle, *narrowed* into the stalk, untoothed, mostly in basal rosettes. Fls white, 7-11mm, in compact clusters. Pod short, 6-10mm. Rock crevices and screes, to 2200m. May-July. seA, –not Austria.

6 BRISTOL ROCKCRESS *Arabis stricta* (= *A. scabra*). Low/short, rather horny-stemmed, tufted per. Lvs mostly in basal rosettes, oblong, toothed and stalked, thick, *glossy dark green* above. Fls cream, 4-8mm. Pod slender, 35-50mm long. Dry grassy and rocky places, to 1400m. Mar-May. swB*, P, wA–French and Jura.

7 DWARF ROCKCRESS *Arabis pumila*. Low/short tufted, hairy per. Lvs mostly in basal rosettes, oval, broadest above the middle, narrowed into the stalk, with several teeth or untoothed. Fls white, 6-7mm, in small clusters. Pod slender, 20-40mm. Damp rocky places, gravels, to 2850m. June-July. A, Ap.

8 SOYER'S ROCKCRESS *Arabis soyeri*. Short/med *almost hairless* tufted per. Lvs mostly in basal rosettes, *glossy* dark green above, oval, broadest above the middle, stalked and finely toothed; stem lvs clasping. Fls white, 5-7mm, in small clusters. Pod slender, 25-50mm. Damp places on calcareous soils, 1800-2800m. June-Aug. P. **8a** *A.s. jacquinii* (= *A. jacquinii, A. bellidifolia*) has stem lvs *not* clasping. A.

9 ALPINE ROCKCRESS *Arabis alpina*. Low/med somewhat creeping, mat-forming, starry-haired, grey-green per. Basal lvs oblong, *deeply toothed*, stalked, the stem lvs clasping, unstalked. Fls white, 6-10mm, in loose clusters, with spreading petals. Pod 20-35 mm. Damp rocks and gravels, streamsides, often on calcareous soils, to 3000 m. Apr-Aug. T.

10 CEVENNE ROCKCRESS *Arabis cebennensis*. Med/tall hairy per, stems branched in the upper half. Lvs oval, coarsely toothed, stalked. Fls pale or deep violet, sometimes white, 7-10mm. Pod long, 30-45mm. Rocky places, to 1550m. May-July. Cev. **10a** *A. pedemontana* is short, smaller in all its parts and almost hairless; fls always white, 6-7mm. swA–Piedmont.

1

2

2a

fl

3

4b

4

6

5

8

basal lf

9

7

10

Cress Family *(contd.)*

ALYSSUMS *Alyssum*. Perennials with star-shaped hairs and untoothed lvs. Fls yellow in compact clusters, sometimes elongating in fruit. Fr oblong or rounded, somewhat inflated, with a persisting style (see also p.64).

1 DIFFUSE ALYSSUM *Alyssum diffusum*. Low loosely tufted, grey-green per with straggling non-flg lfy stems. Lvs oval to elliptical, often broadest above the middle, the upper lvs narrower. Fls mid-yellow, 5-7mm, in clusters greatly elongating in fr, petals *slightly notched*. Pod rounded, 4-6mm. Rocky and stony places, to 2400m. May-July. P, swA, Ap. **1a** *A. cuneifolium* is more closely tufted with *grey or whitish* stems and lvs. Fl clusters short and compact. Pod more elliptical, grey-hairy. c, sAp*.

2 ITALIAN ALYSSUM *Alyssum argenteum*. Short/med erect hairy per *with* long non-flg lfy shoots. Lvs elliptical, broadest above the middle, greenish above, grey beneath, the basal lvs *much smaller*. Fls yellow, 3-5mm, in loose clusters. Pod oblong, 3.5-6mm, hairy. Rocky places, to 1250m. June-July. sA–Italian.

3 ALPINE ALYSSUM *Alyssum alpestre*. Low spreading *whitish* or sometimes grey-hairy per, with numerous non-flg lfy rosettes. Lvs elliptical, broadest above the middle or spoon-shaped. Fls yellow, 3-4mm, in loose clusters. Pod elliptical, 2.5-4.5mm, *rather* assymetrical. Grassy and rocky places, 1500-3100m. June-Aug. w, cA. **3a** *A. serpyllifolium* has *larger* lvs and smaller, 2-3 mm, fls. Pod broader. P.

4 SHEPHERD'S PURSE *Capsella bursa-pastoris*. Very variable low/med ann/bien, hairy or hairless. Lvs mostly in a basal rosette, lance-shaped, pinnately-lobed, toothed or not; upper lvs clasping the stem. Fls white, 2-3mm, in lax racemes, petals *twice* as long as the green sepals. Pod small, *triangular heart-shaped*, 6-9mm long. Fields and waste places, a frequent weed, to 2000m. Flg most of the year except at higher altitudes. T. **4a** *C. rubella* has *reddish tinged* petals B*, P, A, Ap.

5 HOLLY *Ilex aquifolium* (Holly Family – Aquifoliaceae), p. 339.

Spindle-Tree Family   Celastraceae

Trees or shrubs with opposite lvs and 4-5 parted fls. Fr a rather fleshy, brightly coloured, capsule; seeds with a conspicuously coloured pulpy aril.

6 SPINDLE-TREE *Euonymus europaeus*. Deciduous shrub or small tree to 6m; twigs green, *square*, bark smooth, greyish. Lvs elliptical to oval, finely toothed, stalked, red tinged in the autumn. Fls greenish-white, 8-10mm, *4-petalled*, in lax clusters at base of lower lvs. Fr capsule *bright coral pink*, 4-lobed, seeds with a bright orange aril. Woods, scrub and hedges on calcareous soils, to 1600 m. May-June. T.

7 ALPINE SPINDLE-TREE *Euonymus latifolius*. Deciduous shrub or small tree, to 5m; twigs greyish-brown, *square*. Lvs larger than 5, oval, finely toothed, short stalked. Fls greenish-brown, 8-10mm, usually 5-petalled, in lax, long-stalked, clusters. Fr capsule pink, 5-lobed. Woods and scrub, to 1600m. May-June. eP, A. Poisonous. **7a** *E. verrucosus*☐ is smaller with *round stems* and 4-petalled fls. A, except w.

Maple Family   Aceraceae

Deciduous trees and shrubs with long-stalked, opposite, palmate leaves. Fls small, greenish-yellow, in clusters; sepals and petals 4-5 or petals absent, stamens usually 8; male and female fls separate but on the same plant. Fr a pair of samaras (keys), each with a single wing.

1 NORWAY MAPLE *Acer platanoides*. Large spreading tree to 30 m; bark pale grey, smooth or shallowly ridged. Lvs large, with 5 *long-pointed* lobes, each further lobed, bright green but with bright yellow and red autumn colours. Fls yellowish-green, 8 mm, in erect branched clusters, appearing before the lvs. Fr with widely diverging wings. Woods, but widely planted, to 1600 m. April. T–except B. **1a** *A. lobelii* □ is a rather erect tree with dark grey bark. Lvs with lobes *not* further lobed, pale shiny green above. Fls smaller, 5 mm, pale green. c,sAp.

2 FIELD MAPLE *Acer campestre*. Small tree or shrub to 25 m; bark pale grey, fissured, *twigs downy*. Lvs small, divided into 5 *blunt* lobes, each lobe itself shallowly lobed, reddish when young. Fls pale green, 6 mm, in erect branched clusters with the lvs. Fr with horizontal wings, usually downy. Open woods and hedgerows, to 1500 m. Apr-May. T.

3 SYCAMORE *Acer pseudoplatanus*. Large spreading tree to 30 m; bark smooth grey, flaking when old, twigs hairless. Lvs with 5 blunt, *toothed*, lobes, soon hairless beneath. Fls pale green, 6 mm, in *drooping* clusters, with the young lvs. Fr with wings at right angles. Woods, hedges and streamsides, to 2000 m. Apr-May. T(B,S). **3a Italian Maple** *A. opalus* □ is a smaller tree with rather leathery, less toothed, smaller lvs, usually hairy beneath; bark smooth. Fls pale yellow, with the young lvs. To 1900 m. Apr. P,swA,Ap.

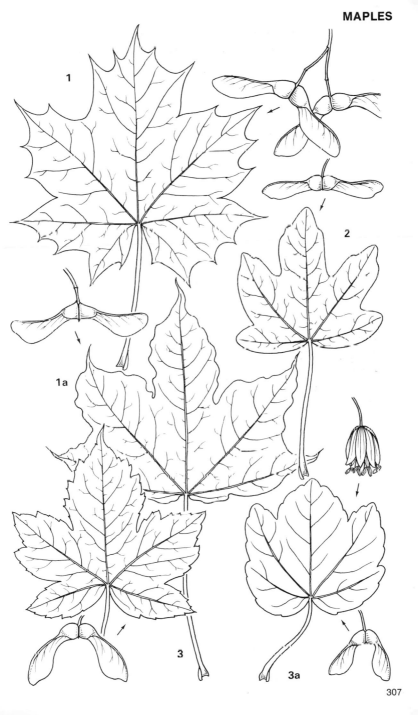

Box Family   Buxaceae

1 BOX *Buxus sempervirens.* Evergreen shrub or small tree to 5m. Lvs small, *opposite*, elliptical, shiny and leathery, untoothed. Fls tiny, whitish-green, 2mm, petaless, in dense clusters at base of upper lvs, lower fls in each cluster male, the upper female. Fr a small three-horned capsule. Dry hillslopes and open woods, to 1600m. Mar-Apr. Widely cultivated. B*,sP,sA,Ap.

Buckthorn Family   Rhamnaceae

Deciduous shrubs or small trees, sometimes spiny, with opposite or alternate lvs, stipules present, soon falling, petals sometimes absent. Fls usually 4 parted, calyx bell-shaped, petals small; style divided. Fr a black berry, bitter tasting, with 2-4 seeds. 2-4 *Rhamnus* have 4-parted fls and buds with scales.

2 BUCKTHORN *Rhamnus catharticus.* Deciduous shrub or small tree, to 6m, spiny; old branchlets ending in a single spine. Lvs opposite, broad oval-elliptic, 4-7 cm. long, finely toothed, stalked; lateral veins 2-4 pairs. Fls 3-4 mm, greenish, in dense clusters arising from the old branches; male and female on separate plants. Berry red, then black. Woods, scrub and hedges, usually on calcareous soils, to 1500 m. May-June. T. **2a Rock Buckthorn** *R. saxatilis* □ is a much branched very spiny shrub, often prostrate but sometimes to 2 m tall. Lvs *smaller*, 1-3 cm long, elliptic to lance-shaped, with a very short stalk. Old branchlets with *several* spines. P,A,Ap.

3 DWARF BUCKTHORN *Rhamnus pumilus.* Prostrate deciduous shrub much branched with contorted old branches, *not spiny*. Lvs *alternate*, oval to elliptical, finely toothed, or untoothed with 4-9 pairs of veins. Fls yellowish-green, in clusters at base of young branches. Berry bluish-black. Rocky places, cliffs and screes, on limestone, 1100-3050m. May-Aug. P, A, Ap.

4 ALPINE BUCKTHORN *Rhamnus alpinus.* Erect deciduous shrub to 3.5m, branches *seldom contorted*, not spiny. Lvs alternate, oval, blunt, often almost heart-shaped at the base, with 9-15 pairs of veins. Fls greenish, 3-4 mm, in clusters at base of young branches. Berry bluish-black. Open woods, rocky places and streamsides, to 2150m. May-June. P, A, Ap.

5 ALDER BUCKTHORN *Frangula alnus.* Erect deciduous shrub or small tree to 5m, *not* spiny. Lvs alternate, oval, broadest above the middle, shiny green, *untoothed*. Fls greenish, 3mm, in small clusters on young stems. Berry red, but becoming black. Damp woods, heaths and hedges, to 1800m. May-June. T.

6 ROCK BUCKTHORN *Frangula rupestris.* Small shrub to 80cm tall. Lvs elliptical to rounded, finely toothed or untoothed, hairy beneath, stalked. Fls small, greenish, 5-petalled, *in small clusters*, flstalks hairy. Fr a red berry, becoming black. Rocky places, to 1200m. May-June. seA.

♀fl

♀fl

♂fl

2

2a

3

4

fl

5

6

Water Starwort Family Callictrichaceae

Aquatic or terrestrial herbs with opposite, untoothed lvs. Fls tiny, green, male and female separate, solitary or sometimes one of each together, at lf bases, stamens one, stigmas two. Fr 4-lobed. Very variable and often difficult to distinguish.

1 WATER STARWORT *Callictriche stagnalis*. Aquatic or terrestrial ann/per. Lvs pale green; submerged ones narrowly-elliptical; floating lvs about 6, broadly-elliptical to almost rounded, smaller if terrestrial. Fls solitary or one male and one female together; *styles recurved*. Fr pale brown, almost rounded, 1.7 mm, deeply grooved between the lobes, each lobe *broadly winged*. Still or slow-moving water, mud-flats, to 2000 m. Apr-Sept. T. **1a** *C. obtusangula* has 12-20, often fleshy, rhombic floating lvs, rather yellowish-green. Fr brown, elliptical, 1.5 mm, shallowly grooved between each lobe, *lobes rounded*. T—except S. **1b** *C. lophocarpa* (= *C. polymorpha*) like 3a but with smaller more rounded frs, 0.8-1.2 mm. T—except P.

2 ABORTIVE STARWORT *Callictriche platycarpa*. Aquatic or terrestrial ann/per. Lvs dark green; submerged lvs linear, floating lvs elliptical, smaller in terrestial form. Fls solitary, styles erect or spreading. Fr brown, rounded, 1·5 mm, often only *2-3 lobed*, lobes narrowly-winged. Still or slow-moving, often calcareous, waters and mudflats, to 2000 m. Apr-Sept. T. **2a** *C. palustris* (= *C. verna*, *C. vernalis*) has a male and female fl at each lf axis. Fr *blackish*, oblong, 1 mm, winged at the top only. To 2600 m. T. **2b** *C. hamulata* is more robust than 4 with styles of female fls *recurved onto sides* of fr; fr short-stalked, 1.5 mm, rounded, lobes narrowly winged. T—except P, Ap.

Olive Family Oleaceae

3 ASH *Fraxinus excelsior*. Spreading deciduous tree to 40 m; bark grey, smooth at first but becoming fissured; buds black. Lvs *opposite, pinnate*; lflets toothed, *unstalked*. Fls without sepals or petals, anthers dark purple, becoming greenish, in tufts, appearing before the lvs. Fr a one-winged samara. Woods and hedges, to 1600 m. Apr-May. T.

4 MANNA ASH *Fraxinus ornus*. Small spreading deciduous tree to 10-20 m; bark grey, smooth; buds greyish or brownish. Lvs opposite, pinnate; lflets toothed, *short-stalked*. Fls *creamy-white*, in branched clusters, appearing with the lvs, petals linear. Mixed woods, thickets and rocky places, to 1500 m. May-June. P,sA,Ap.

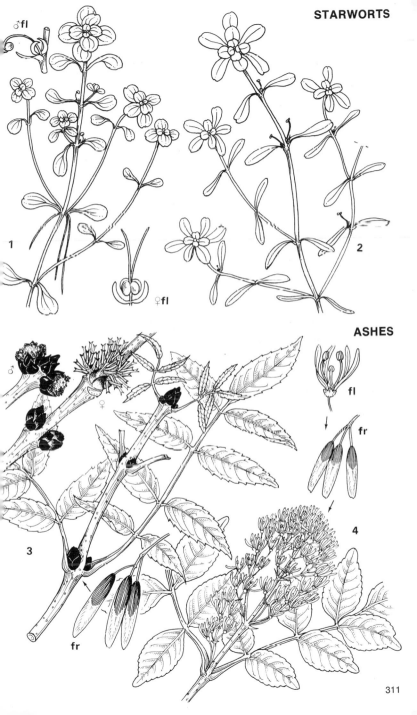

STARWORTS

♂fl

♀fl

1

2

ASHES

fl

fr

♂

♀

3

fr

4

311

Bedstraw Family Rubiaceae (contd. from p. 224)

BEDSTRAWS *Galium*. Annuals or perennials, often with rather weak square stems and lvs in whorls, sometimes prickly. Fls usually white or yellow, in branched terminal or lateral clusters, 3-4 parted. Fr 2-parted, each with one seed, sometimes covered with hooked bristles. A difficult group with many closely related species.

1 NORTHERN BEDSTRAW *Galium boreale*. Variable short/med rather stiff erect per, sometimes slightly hairy; stems square. Lvs lance-shaped, *usually in fours*, dark green, 3-veined, rough on edges. Fls white, 3-4mm, in dense terminal clusters. Fr with hooked bristles. Grassy, shrubby and rocky places, to 2200m. June-Aug. T.

2 WOODRUFF *Galium odoratum* (= *Asperula odorata*). Short *carpeting* per; stem erect, square, hairless, except at the lfjoints. Lvs elliptical, in whorls of 6-9, rough edged. Fls white, 4-7mm, in loose heads. Fr with hooked bristles. Open deciduous woods, to 1600m. Apr-June. T.

3 CONIFEROUS BEDSTRAW *Galium triflorum*. Short/med per, stems rather weak, square, slightly hairy Lvs narrow-elliptical, in whorls of 6-8. Fls tiny, white, 1.5-3.5mm, in loose terminal or lateral clusters. Fr with hooked bristles. Rocky coniferous woods, to 1200m. June-Aug. S, cA*.

4 MARSH BEDSTRAW *Galium palustre*. Variable, rather straggling per, stems square, rough-edged. Lvs elliptical, broadest above the middle, in whorls of 4-6. Fls white, 2-3mm, 4-parted, in loose, stalked, clusters, anthers red. Fr *black, smooth*. Wet places, to 2100m. June-Aug. T. **4a Fen Bedstraw** G. *uliginosum* has narrow lance-shaped lvs ending in a *sharp point*, in whorls of 6-8; anthers yellow. Marshes and wet places. T. **4b** G. *trifidum* has lvs in whorls of 4 and 3-parted fls. S, eP, eA.

5 HEDGE BEDSTRAW *Galium mollugo*. Variable med/tall, often sprawling per, sometimes hairy; stems square, smooth. Lvs oblong, l-veined, in whorls of 6-8. Fls white, 2-3mm, in loose, branched, clusters. Fr black. Grassy places and hedgebanks, to 2100m. June-Sept. T. **5a** G. *album* has larger fls, 3-4mm. June-Sept. T.

6 *Galium lucidum* agg. Short/med per, stems smooth, usually hairless. Lvs narrow lance-shaped. Fls *pale yellow to greenish*, 3-5mm, in branched clusters. Calcareous rocks and screes, to 1600m. May-Aug. P, A, Ap.

7 WOOD BEDSTRAW *Galium sylvaticum*. Tall hairless per, stems *rounded*. Lvs elliptical to lance-shaped, bluish-green, rough edged. Fls white, 2-3mm, often nodding before flg, in lax clusters. Fr blue-green, smooth. Woods and scrub, to 1600m. June-Sept. P, A, Ap. **7a** G. *aristatum* is shorter and usually hairless but with *square stems* and bright green lvs. Open woods. **7b** G. *laevigatum* has stems rooting at the lower nodes. s, seA.

fl

2

fl

fr

fr

fl

3

fr

fl

fr

fr

5

fr

4

fr

fr

6

fr

7

Bedstraw Family *(contd.)*

1 CUSHION BEDSTRAW *Galium saxosum*. Low/short *cushion-forming* per, stems square, hairless. Lvs lance-shaped to linear, 5-6 in a whorl, with an apical point. Fls white, 1.5-2mm, in rounded clusters. Fr smooth. Calcareous screes, to 2500m. swA. **1a** *G. cometerhizon* – siliceous screes. c, eP. **1b** *G. palaeoitalicum* – rocky and grassy places. AA, sAp. **1c** *G. pyrenaicum* has very narrow, closely overlapping lvs. P. **1d** *G. caespitosum* like **1c** but stems *rounded* and lvs in whorls of 8-10 and fls yellowish-white, 2-4mm. P. **1e** *G. austriacum* like **1d** but fls smaller and whiter. Grassy places and coniferous woods. eA.

2 REDDISH BEDSTRAW *Galium rubrum*. Short/med per, stems rather weak, rounded, hairy at the base. Lvs elliptical, broadest above the middle, in whorls of 7-9. Fls *dark purple*, 1.5-2mm, in large oblong clusters. Woods, to 2000m. June-Aug. sA, nAp. **2a** *G. obliquum* has yellow or greenish fls usually and narrower lvs. Dry places. Ap.

3 *Galium anisophyllum*. Low/short cushion-forming per, almost hairless. Lvs elliptical, broadest above the middle, in whorls of 7-9, rough. Fls *yellowish-white*, 2-4mm, in rounded clusters. Grassy and rocky places, to 2900m. June-Aug. Cev, A.

Lime Tree Family   Tiliaceae

Deciduous trees with heart-shaped, alternate, toothed lvs. Fls in drooping clusters half-joined to a large oblong bract; 5-parted with many stamens. Fr a globular nut.

4 SMALL-LEAVED LIME *Tilia cordata*. Large spreading tree to 30m, bark smooth, dark brown, with large bosses, young twigs more or less hairless. Lvs heart-shaped, pointed, 3-8 cm, greyish, hairless except for reddish hair tufts at vein angles beneath. Fls *fragrant*, in small clusters of 4-10. Woods, but widely planted, to 1500m. June-July. T.

5 LARGE-LEAVED LIME *Tilia platyphyllos*. Large spreading tree to 30m, bark smooth, dark brown, *without* bosses, young twigs usually downy. Lvs larger than 7, 6-12cm, *greyish-white downy beneath.* Fls yellowish-white, fragrant, in small clusters of 2-5. Woods, but widely planted, to 1800m. June-July. T except S. **5a** *T.* × *vulgaris* a hybrid with intermediate characters exists between 7 and 8 where they grow together.

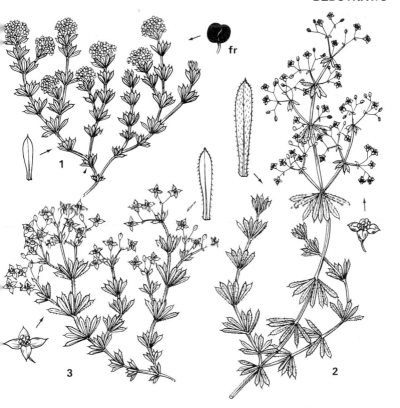

fr

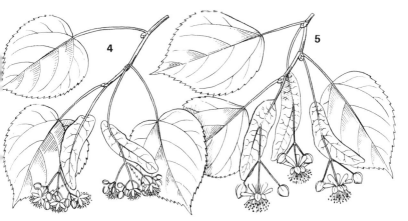

315

Nightshade Family   Solanaceae

Herbs with alternate lvs, no stipules, often poisonous. Fls 5-parted, petals joined into a short or long tube; stamens joined to petal tube. Fr a fleshy berry or a capsule.

1 DEADLY NIGHTSHADE *Atropa belladonna.* Med/tall hairy or hairless per, well branched. Lvs large, oval, pointed, untoothed, short-stalked. Fls brownish-violet or greenish, *bell-shaped*, 25-30 mm, solitary usually. Fr a large *shiny-black berry*, very poisonous. Damp and shady places on lime, to 1700 m. June-Sept. T – except S.

2 SCOPOLIA *Scopolia carniolica.* Short/med hairless per, usually slightly branched. Lvs elliptical or oval, pointed, untoothed, stalked, the *lower scale-like.* Fls dark brownish-violet outside, yellowish-green inside, drooping, funnel-shaped, 15-25 mm, solitary. Fr a rounded capsule. Waste & rocky places, banks, to 1200 m. May-Aug. e, seA, Ap.

3 HENBANE *Hyposcyamus niger.* Med/tall *stickily-hairy*, unpleasant smelling, ann/bien. Lvs oval to oblong, untoothed, sometimes slightly lobed, the lower stalked, the upper clasping the stem. Fls pale yellow with a *network of purple veins*, broad bells, 20-30 mm, in dense lfy, one-sided spikes. Fr a capsule. Bare and disturbed ground, frequently as a weed, to 1900 m. May-Sept. T.

4 BITTERSWEET *Solanum dulcamara.* Scrambling hairless to downy per, to 2 m. Lvs oval, pointed, often *heart-shaped or two lobed* at the base, untoothed, stalked. Fls dark purple, rarely white, 10-15 mm, petals turned-back, with prominent yellow anthers in a column; in loose branched clusters. Fr a shiny red berry. Damp woods and hedgerows, to 1700 m. May-Sept. T.

5 BLACK NIGHTSHADE *Solanum nigrum.* Variable low/med, branched, hairless or downy ann. Lvs oval-rhombic to lance-shaped, blunt-toothed or not, stalked. Fls white, 10-14 mm, petals turned back, anthers yellow in a column; in small loose clusters opposite lvs or in between. Fr a *dull black or green* berry – poisonous. Bare and waste places, cultivated ground, to 1750 m. July-Oct. T.

Besides these, **Box Thorn**, *Lycium barbarum* (= *L. halimifolium*) is frequently cultivated as a hedging plant and the **Potato**, *Solanum tuberosum*, **Tomato**, *Solanum lycopersicum* and **Tobacco**, *Nicotiana rustica* are common crop plants, the latter two mainly in the south of the area.

1

fl

5

fr

3

fr

2

fr

4

Figwort Family *(contd. from pp. 204-18)*

FIGWORTS *Scrophularia*. Perennials with square stems. Lvs usually opposite, toothed and stalked. Fls small, in slender branched terminal clusters, usually two-lipped, with five blunt-lobes.

1 YELLOW FIGWORT *Scrophularia vernalis*. Med/tall softly-hairy bien/per. Lvs oval to heart-shaped, stalked; bracts like lvs. Fls *yellowish-green, not* two-lipped, 6-8mm. Woods, scrub and damp waste places, to 1800m. Apr-June. P*,A*,(B).

2 PYRENEAN FIGWORT *Scrophularia pyrenaica*. Short/tall, hairy per. Lvs *rounded*, often heart-shaped at base; bracts mostly like the lvs. Fls yellowish with a reddish-brown upper lip, 8-11mm. Rocky places, usually shaded, to 1900m. June-July. P*.

3 ITALIAN FIGWORT *Scrophularia scopolii*. Med/tall, more or less hairy per. Lvs oval to lance-shaped, often heart-shaped at the base; bracts narrow lance-shaped, *mostly not like the lvs*. Fls greenish with a purple-brown upper lip, 7-12mm. Woods and damp places, to 2300m. June-Aug. s,e,seA,Ap. **3a** *S. alpestris* has larger lvs. P.

4 COMMON FIGWORT *Scrophularia nodosa*. Med/tall *hairless* per; stems narrowly winged. Lvs oval to lance-shaped, sometimes heart-shaped at the base; bracts narrow lance-shaped, sometimes heart-shaped at the base. Fls green with a purple-brown upper lip, 7-10mm. Damp and shady places, to 1850m. June-Sept. T. **4a Water Figwort** *S. auriculata* often has *two lobes* at the base of the lf; sepal teeth white-edged. Damp places, particularly riversides. B,P,w,swA,Ap. **4b Green Figwort** *S. umbrosa* like 4a but stems with broader wings; fls *olive-brown*. B,A,Ap.

5 ALPINE FIGWORT *Scrophularia canina*. Short/med per. Lvs often purple tinged, *2-pinnately-lobed*, only the lowest opposite, lobes linear, pointed; bracts linear. Fls dark *purple-violet* with white markings, the upper lip only one third the length of the fl tube; flstalks glandular, not hairy. Stony places, screes and waste ground, often on limestone, to 2200m. June-Aug. P,Cev,A,Ap. **5a** *S.c. hoppii* (= *S. hoppii*) has *glandular-hairy* flstalks; fls with the upper lip at least half as long as the fl tube. Jura,sA,Ap.

6 MUDWORT *Limosella aquatica*. Low hairless ann with *creeping runners*. Lvs in basal tufts, spoon-shaped, untoothed, long-stalked. Fls tiny, white, often tinged with pinkish-purple, 4-5 petalled, solitary on slender stalks, downturned in fr. Wet muddy places, to 1800m. June-Sept. T.

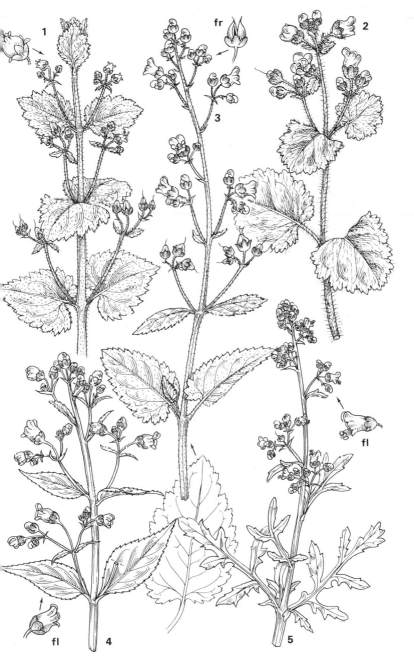

Figwort Family – Speedwells *Veronica (contd. from pp. 208-10)*

1 NETTLE-LEAVED SPEEDWELL *Veronica urticifolia.* Short/med erect, slightly hairy per. Lvs triangular-oval, sharply toothed, *unstalked*. Fls lilac, 7 mm, in *opposite stalked spikes* at the upper lvs. Woods and shady places, usually on calcareous soils, to 2000 m. May-July. P,A,n,cAp.

2 COMMON or HEATH SPEEDWELL *Veronica officinalis.* Low creeping hairy per. Lvs oval to elliptical, toothed, short-stalked. Fls dull lilac-blue, 8 mm, in opposite stalked spikes. Woods and heaths, to 2150 m. May-Aug. T. **2a** *V. allionii* is more or less hairless; fls *deep blue* in solitary spikes usually. Dry meadows, 1800-2700 m. July-Aug. swA.

3 WOOD SPEEDWELL *Veronica montana.* Low/short, rather sprawling per, stems hairy *all round*. Lvs broad-oval, toothed, stalked, pale green. Fls pale lilac-blue, 8-10 mm, in *alternate spikes*. Woods, to 1400 m. Apr-July. T–except S. **3a Marsh Speedwell** *V. scutellata* is usually *hairless*, more slender with narrow-oblong, slightly toothed lvs, *stalkless*; fls smaller, 5-6 mm. Marshes and wet places, to 1800 m. June-Aug. T.

4 WATER SPEEDWELL *Veronica anagallis-aquatica.* Short/med fleshy hairless per; rootstock creeping and rooting. Lvs pale green, oval or lance-shaped, toothed, the upper *half-clasping* the stem. Fls blue with violet veins, 5-10 mm, in opposite stalked spikes. Streambanks and other wet places, to 1450 m. June-Aug. T. **4a** *V. catenata* has dark green, narrow lance-shaped lvs and smaller pinker fls, 3-5 mm. T–except S.

5 WALL SPEEDWELL *Veronica arvensis.* Low/short erect or somewhat sprawling, hairy ann. Lvs triangular-oval, toothed, the lower short-stalked. Fls tiny, blue, 2-3 mm, in dense lfy spikes *at shoot tips*. Dry places, walls and cultivated ground, to 2100 m, Mar-Oct. T. **5a Spring Speedwell** *V. verna* is shorter, the upper lvs *pinnately-lobed*. T.

6 FIELD SPEEDWELL *Veronica agrestis.* Low sprawling hairy ann. Lvs *mostly alternate*, except the lowest, oval, toothed, short-stalked. Fls whitish with a pink or blue upper petal, 3-6 mm, *solitary*, long-stalked. Fr 2-lobed. Cultivated land, often on acid soils, to 1800 m. Mar-Nov. T. **6a** *V. polita* has *blue fls* and more rounded lvs, sepal teeth oval, overlapping at base. To 2100 m. T. **6b** *V. opaca* like 6a but sepal teeth oblong lance-shaped, not overlapping. T–except B,P.

7 IVY-LEAVED SPEEDWELL *Veronica hederifolia.* Low sprawling downy ann. Lvs mostly alternate, rounded in outlines, but *with 3-7 lobes*. Fls blue or lilac, 4-9 mm, *solitary*, long-stalked. Fr 4-lobed. Cultivated areas mainly, to 1800 m. Apr-Aug. T.

8 LONG-LEAVED SPEEDWELL *Veronica longifolia.* Variable med/tall erect per, hairy or hairless. Lvs *opposite or in whorls* of 3-4, lance-shaped to almost linear, pointed, toothed, short-stalked. Fls pale blue or lilac, 6-8 mm, in dense spikes, often with short branches. Woods, riverbanks and other damp places, to 1250 m. June-July. S,A.

1

2

fr

fr

3

fr

fr

4

5

fr

6

fr

7

8

fr

fl

Plantain Family   Plantaginaceae

PLANTAINS *Plantago*. Perennials with strongly veined or ribbed lvs in basal rosettes, sometimes flat on the ground. Fls tiny, usually 4-parted, in dense, long-stalked spikes, often elongating in fruit; stamens conspicuous, long-stalked. Fr a small capsule.

1 ALPINE PLANTAIN *Plantago alpina*. Low hairless or slightly hairy tufted per, with several or many rosettes. Lvs *linear*, soft, 3-veined, the innermost triangular-shaped. Fls whitish-green in narrow spikes, 10-30mm; bracts oval, shorter than the fls. Meadows, sometimes screes, to 3000m. July-Aug. P, A, Ap.

2 DARK PLANTAIN *Plantago atrata* (= *P. fuscescens, P. montana*). Variable low tufted per with several or many rosettes, usually slightly hairy. Lvs *green*, narrow lance-shaped, long-pointed. Fls brownish-green, in oblong spikes, 15-30mm; anthers yellowish or violet; bracts *rounded* with membranous edges. Meadows and stony places, usually on calcareous soils, 1500-2500m. May-Aug. P, A, Ap. **2a** *P. monosperma* has shorter silvery-hairy lvs and shorter spikes; anthers white. 1500-2800m. July-Aug. c, eP.

3 FLESHY PLANTAIN *Plantago maritima serpentina* (= *P. serpentina*). Low/short tufted, *usually hairless* per, with several or many rosettes. Lvs linear, thick and slightly fleshy, 2mm or more wide, sometimes with a few teeth. Fls whitish-green in *long spikes*, 30-70mm, anthers yellowish; bracts oval, long-pointed. Meadows, rocky places, screes and gravels, often on poor calcareous soils, to 2400m. June-Aug. A, Ap. **3a** *P. holosteum* (= *P. carinnata, P. recurvata*) has *very narrow* rigid lvs, less than 2mm wide and shorter spikes. Usually on acid soils, to 2600 m. May-Sept. eP, MA, sA, Ap.

4 RIBWORT PLANTAIN *Plantago lanceolata*. Low/med tufted, hairy or almost hairless per, with several rosettes. Lvs *lance-shaped*, slightly toothed or untoothed, 3-5-veined. Fls brown, in blackish oblong spikes, 5-50mm long, on *5-ridged* stalks; anthers pale yellow. Meadows, grassy, rocky and waste places, to 2300m. Apr-Oct. T. **4a** *P. argentea* □ has a single or a few lf-rosettes; lvs un-toothed, often *silvery-hairy* beneath; spikes 5-20mm long, the stalks with 6 or more ridges, anthers *white*. Dry places on calcareous soils. c, eP, sw, s, seA, Ap.

5 GREATER PLANTAIN *Plantago major*. Low/short downy or hairless per, often with a solitary rosette only. Lvs *broad-oval to elliptical*, stalked, toothed or not, 3-9-veined. Fls pale greenish-yellow, in long pointed spikes equal in length to their *unridged* stalks; anthers pale purplish, then yellowish-brown. Grassy and waste places, paths, banks, to 2800m. June-Oct. T.

6 HOARY PLANTAIN *Plantago media*. Low/short *downy-greyish* per with a soli-tary or few rosettes. Lvs elliptical or oval-elliptical, stalked, toothed or not, 7-9-veined. Fls whitish, fragrant, in blunt spikes, 20-60mm long, shorter than the long *ridged stalks*; anthers lilac, sometimes whitish. Grassy, stony and waste places, on calcareous soils, to 2450m. May-Aug. T.

2

1

3

bract

bract

bract

act

4

4a

6

5

Daisy Family (contd. from p. 240-64)

ARTEMISIAS – usually aromatic perennials with alternate, pinnately-divided lvs. Flheads small, rayless, in branched spikes.

1 SEA WORMWOOD Artemisia maritima vallesiaca. Short/med stout white-downy per with many basal lfy rosettes; strongly aromatic. Lvs 1-3-pinnately-lobed; uppermost lvs not lobed. Flheads egg-shaped, yellow or orange-yellow, 2.5-5mm long. Grassy and stony places, on limestone, to 1300m. Aug-Oct. wA.

2 DIGITATE-LEAVED WORMWOOD Artemisia petrosa (= A. eriantha). Low/short tufted, silvery-white-downy per with basal lfy rosettes. Lvs 2-trifoliate or digitate, stalked, the upper stalkless. Flheads yellow, 3.5-4.5mm, in slightly branched spikes, florets hairy; flbracts oval membranous with a brown edge, hairy. Rocks and screes, 2000-3150m. July-Sept. P, sA, MA, Ap. **2a Genipi** A. genipi (= A. nivalis) is less hairy with hairless florets. 2000-3800m. P, A.

3 Artemisia insipida. Short/med tufted per, not aromatic. Lvs 1-2-pinnately-lobed, lobes linear, green above, silvery-hairy beneath. Flheads nodding, yellow, 3mm, in branched spikes, florets hairy; flbracts with a membranous margin, hairy. Grassy banks and rocky places, to 1400m. July-Sept. swA–French.

4 Artemisia chamaemelifolia. Med hairless per, aromatic. Lvs 2-3-pinnately-lobed, the upper stalkless, half-clasping the stem. Flheads yellow, 2.5-3mm, in crowded branched spikes; flbracts linear or oblong, usually hairless. Rocky places, to 2400m. July-Aug. P, s, swA–French*.

5 NARROW-LEAVED WORMWOOD Artemisia nitida. Short/med white-woolly tufted per. Lower lvs digitate, upper pinnately-lobed, lobes all stalked, linear. Flheads yellowish, 4-5mm, in one-sided spikes; florets hairy. Calcareous rocks, 1300-2000m. July-Sept. seA, AA.

6 YELLOW GENIPI Artemisia mutellina. Low/short cushion-forming, silvery-hairy per. Lvs digitately-lobed, lobes linear. Flheads yellow, 3.5-4.5mm, in slender spikes, florets downy; flbracts silvery-hairy. Rocks, screes and moraines, usually on acid rocks, 1300-3700m. July-Sept. eP, A. **6a Glacier Wormwood** A. glacialis has fls in small rounded clusters; florets hairless, 1900-3200m. eP, s, wA*.

7 DARK ALPINE WORMWOOD Artemisia atrata. Short/med slightly hairy per, not aromatic. Lvs green, 2-3-pinnately-lobed, fern-like, dotted with glands. Flheads greenish-yellow, 3.5-4mm, in slender branched or unbranched spikes, florets hairy at the top. Dry grassy and stony places, on acid soils, 1800-2400m. July-Aug. sw, s, seA*. **7a Norwegian Wormwood** A. norvegica has fewer than 10 flheads to a spike, each long-stalked and slightly nodding, 8-9mm. To 1900m. B–w Scotland, sS.

8 PYRENEAN WORMWOOD Artemisia herba-alba (= A. aragonensis). Short/med, grey-downy aromatic per, stem branched from the base. Lvs small, pinnately-lobed, short stalked or stalkless. Flheads yellowish, 2.5-3mm, in branched spikes, florets hairless. Rocky places, to 2000m. July-Aug. P*.

9 FIELD WORMWOOD Artemisia campestris. Variable short/tall, slightly aromatic, tufted per, stems hairless, brownish-red. Lvs 2-3-pinnately-lobed, silvery-hairy when young, but becoming hairless, the lowest stalked. Flheads yellowish or reddish, 1.5-4mm, in wide-branched spikes. Waste places, often sandy, to 2000m. Aug-Sept. T–B*. **9a** A.c. alpina is shorter with narrow fl clusters, 1000-2000m. A. **9b Arctic Wormwood** A.c. borealis (= A. borealis) is shorter than 9a, seldom reaching 25cm tall and with larger flheads, 5-6mm. Dry rocky ridges and screes, 1500-2800m. n, c, eA*.

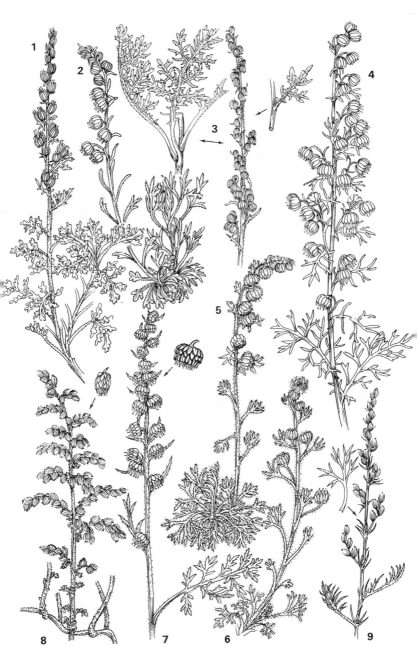

Daisy Family *(contd.)*

GROUNDSELS or RAGWORTS *Senecio.* Perennials, sometimes annuals with alternate lvs. Flheads usually in flat-topped clusters, rarely solitary, yellow or orange-yellow, usually with ray florets. A large and difficult group with numerous species and forms.

1 SOUTHERN RAGWORT *Senecio ovirense.* Variable short/tall grey-downy or greenish per. Basal lvs erect, oval to lance-shaped, *coarsely-toothed*, with broad-winged stalks, withered at flg time; stem lvs short-stalked or unstalked. Flheads *large*, yellow or golden-yellow, 30-40mm, in flat-topped clusters of 3-15, each with 18-21 rays. Damp or shady places, to 2500m. May-July. P,s,e,seA,n,cAp. **1a** *S. helenites* has lvs untoothed or only slightly so and *smaller* flheads, 20-25 mm. P,w,cA. **1b** *S. rivularis* like 7 but with *heart-shaped* lvs, the upper half-clasping the stem. s,e,seA. **1c** *S. balbisianus* is stouter than 7 but with broad oval or oblong lvs, the uppermost clasping the stem. MA,AA,Lig.

2 ROCK RAGWORT *Senecio squalidus* (= *S. rupestris*). Variable short/med downy or almost hairless ann/bien/per with *branched stems*. Lower lvs *pinnately-lobed*, with a broad-winged stalk; upper lvs clasping the stem. Flheads bright yellow, 15-25mm, in flat-topped clusters, each with 10-13 rays; flbracts black-tipped. Open stony or sandy ground, waste places, to 2300m. c,eA,Ap.

3 PINNATE-LEAVED RAGWORT *Senecio abrotanifolius.* Short slightly hairy per. Lvs 2-3-*pinnately-lobed*, stalked, lobes linear; uppermost lvs less lobed. Flheads orange-yellow with *brown stripes*, 25-40mm, in clusters of 2-5, each with 10-12 rays. Rocky slopes, to 2700mm. July-Sept. c,eA. **3a** *S. adonidifolius* has small yellow flheads, 10-16mm, in dense clusters, each with 3-6 rays. Acid rocks. P,Cev.

4 TOURNEFORT'S RAGWORT *Senecio tournefortii.* Short/med hairless or slightly hairy per. Lvs *crowded* in middle of stem, oblong to narrow lance-shaped, pointed, the lower short-stalked. Flheads yellow, 25-40mm, in flat-topped clusters, each *with* 10-16 rays. Grassy and rocky places, 1300-2000m. July-Aug. P.

5 CHAMOIS RAGWORT *Senecio doronicum.* Variable short/med somewhat hairy per. Lvs elliptical to oval, toothed, stalked, *white-downy beneath*; upper lvs narrower, half-clasping the stem. Flheads *large*, deep yellow or orange-yellow, 30-60mm, solitary or 2-4 together, each with 10-22 rays. Grassy and rocky places, often on limestone, to 3100m. July-Aug. P,A,Ap. **5a** *S. lanatus* has *always* solitary, pale yellow flheads. n,cAp.

6 FIELD FLEAWORT *Senecio integrifolius.* Variable short/tall per, greyish- or white-downy at first. Lvs mostly in *flat basal rosettes*, oval to almost rounded, stalked, often untoothed; upper lvs smaller, unstalked. Flheads yellow or golden-yellow, 15-25mm, in flat-topped clusters of 3-15, each with 12-15 rays. Dry grassy places, to 2200m. May-June. T−except sAp. **6a** *S. i. capitatus* (= *S. capitatus*) is med, the whole plant grey-white-downy *even when mature*. Flheads yellow or red, with or without rays; 1800-2500m. A. **6b** *S.i. aurantiacus* (= *S. aurantiacus*) like 6 but flheads *orange or brownish-red*, sometimes without rays. Grassy places and open woods. eA.

7 ALPINE RAGWORT *Senecio alpinus* (= *S. cordatus*). Med/tall hairless per. Lvs *heart-shaped*, often lobed at the base, coarsely-toothed, stalked, greyish-hairy beneath; upper lvs smaller, oval. Flheads yellow or orange-yellow, 25-40mm, in flat-topped clusters of 5-20, each with 12-21 rays. Meadows, open woods and damp places, to 2150 m. July-Sept. A,n,cAp. **7a** *S. subalpinus* has lvs green beneath, *the upper* pinnately-lobed. eA.

1

2

fr

3

basal
lf

basal
lf

7

6

4

5

Daisy Family (contd.)

1 GREY ALPINE GROUNDSEL *Senecio incanus.* Low/short *grey- or silvery-white* downy per. Lvs broadly oval in outline, long-stalked, pinnately-lobed with linear segments; upper lvs less deeply lobed or unlobed. Flheads deep yellow, 10-13 mm, in *dense clusters* of 4-10 usually, each with 3-6 rays. Pastures and rocky places on acid soils, 1700-3500 m. July-Sept. w, scA, Ap. **1a** *S.i. persoonii* has *lance-shaped* segments to the lower lvs. MA, s, seA. **1b** *S.i. carniolicus* has narrower basal lvs than 1, *broadest above* the middle. eA. **1c** *S. leucophyllus* □ is more robust than 1 with larger flheads in clusters of 10 or more, *with* 10-16 rays to each flhead. Screes, 1500-2700 m. eP.

2 ONE-FLOWERED ALPINE GROUNDSEL *Senecio uniflorus.* Low *silvery-white* downy per. Basal lvs oblong in outline, deeply toothed, long-stalked. Flheads orange-yellow, 20-25 mm, *solitary*, with 10-16 rays. Pastures and rocky places on acid soils, 1900-3600 m. July-Sept. sw, sA. Hybrids between 1 and 2 occur where the two species grow together and these have 2-4 flheads in a cluster.

3 WOOD RAGWORT *Senecio nemorensis.* Med/tall, slightly hairy, densely lfy per. Lvs oval to lance-shaped, toothed, stalked, the upper smaller and *clasping the stem.* Flheads yellow, 15-25 mm, in loose flat-topped clusters, each with 5-6 rays; flbracts often *black-tipped.* Damp meadows and woods, to 2200 m. July-Sept. P, A. **3a** *S. n. fuchsii* (= *S. fuchsii*) is often purple-tinged, the upper lvs short-stalked, *not* clasping the stem. A, Ap. **3b** *S. cacaliaster* is *usually rayless*, but sometimes with 1-3 whitish-yellow rays to each flhead. To 1900 m. s, seA.

The **Common Groundsel**, *Senecio vulgaris* and the **Sticky Groundsel**, *S. viscosus* are common annual weeds that may reach 2300 m, throughout the area.

DANDELIONS *Taraxacum.* Perennials with loose rosettes of rather fleshy lvs; juice milky. Flheads on hollow lfless stems, usually yellow, with ray florets only; flbracts in 2 rows, the outer shorter. Fr a typical 'clock'. A difficult and variable group with many microspecies – the more important montane species are described here.

4 BRENNER DANDELION *Taraxacum pacheri.* Low slightly hairy per. Lvs thin, bright green, 3-5 cm long, spoon-shaped with shallow blunt lobes. Flheads 20-30 mm, the rays yellow, *striped brown or grey*, stems hairy just below flhead; flbracts oval to lance-shaped, black or dark green. Dry pastures and heaths, grassy places, 2000-2900 m. July-Sept. A−except se.

5 SHRÖTER'S DANDELION *Taraxacum shroterianum.* Short hairless per. Lvs narrow spoon-shaped, 5-15 cm, unlobed or shallowly lobed, untoothed, with *long reddish stalks.* Flheads 28-32 mm, the rays *plain yellow*; flbracts oval, pointed greyish-green, often red tinged. Grassy meadows and marshes, to 2750 m. July-Sept. w, cA.

6 GLACIER DANDELION *Taraxacum glaciale.* Low hairless per. Lvs oblong or spoon-shaped, 3-7 cm, unlobed to pinnately-lobed. Flheads *small* 10-15 mm, rays yellow, striped grey or red; flbracts *linear, black.* Meadows and rocky debris, to 2600 m. Ap.

7 *Taraxacum phymatocarpum.* Low hairless per. Lvs bright green, narrow spoon-shaped, 2-7 cm, unlobed or with shallow triangular lobes. Flheads 15-25 mm, rays white or yellow, striped grey, violet or purple; flbracts oval, greyish-green or blackish. S-Norway, cA.

Daisy Family (contd.)

1 *Taraxacum ceratophorum*. Short hairy per. Lvs large dark green, broad lance-shaped, 8-20cm, with large triangular lobes; stalk winged. Flheads 35-50mm on *hairy stems*, rays narrow yellow, striped red, brown or purple, flbracts oval to lance-shaped, green. Grassy and rocky meadows, banks, to 2650 m. June-August. S.A–except s,se.

2 MARSH DANDELION *Taraxacum palustre*. Low/short hairless or slightly hairy per. Lvs *linear to narrow lance-shaped*, 5-20cm, finely toothed or with narrow lobes; stalk long and slender, purplish often. Flheads 25-50mm on hairless purple stems, the rays pale yellow, sometimes striped grey or purple; flbracts oval, purple or violet flushed, with a membranous margin. Wet places, stream banks, to 2200m. Apr-Aug. T-except S,P.

3 DARK DANDELION *Taraxacum nigricans*. Short slightly hairy per. Lvs bright to pale green, spoon-shaped, 8-15cm, with variable often toothed lobes; stalk long, winged. Flheads 25-35mm, on slightly hairy stems, the rays dark to orange-yellow, striped grey or purple; flbracts lance-shaped, *dark violet-purple*. Grassy and rocky meadows, to 2200m. June-Sept. A. **3a Brownish Dandelion** *T. cucullatum*. Like 3 but lvs larger, dark bright green. Flheads 30-45mm, on hairy stalks, *rays yellow-brown* fading white at edges, sometimes purple striped; flbracts oval-lance-shaped, dark green, often flushed purple. Rocky and grassy meadows, to 2200m. June-Sept. cA.

4 ALPINE DANDELION *Taraxacum apenninum* (= *T. alpinum*). Very low, slightly hairy to white-downy per. Lvs in *flat rosettes*, soft, mid-green, lance-shaped, 3-10cm, lobed, or unlobed; stalks long, winged. Flheads 15-20mm, on delicate stems, rays yellow with a grey or brown stripe; flbracts narrow lance-shaped, green. Meadows, rocky debris and by snow patches, 1500-3350m. July-Sept. P,A,Ap.

5 BROAD-LEAVED DANDELION *Taraxacum fontanum*. Low/short hairy per. Lvs bright green, oval, unlobed or sometimes lobed and toothed; lfstalks widely winged. Flheads 25-35mm, on hairy stalks, rays long and narrow, orange-yellow, striped purplish-brown; flbracts lance-shaped, green with a paler edge. Grassy and rocky meadows, to 2500m. July-Sept. P,A.

6 CUT-LEAVED DANDELION *Taraxacum dissectum*. Low/short per. Lvs numerous, mid to dark green, spoon-shaped, *much cut* into narrow, toothed, lobes, hairy beneath often; lf stalks short, narrow. Flheads 20-30mm, on slender *hairless* stalks, the rays pale yellow, striped grey or red. Rocky and grassy meadows, to 3100m. June-Aug. P,A.

7 HOPPE'S DANDELION *Taraxacum hoppeanum*. Short hairy per. Lvs dark olive-green, lance-shaped with long narrow pointed lobes, toothed towards top; stalk winged. Flheads large, 35-45mm, on hairy stalks *red flushed* in the upper part, rays bright yellow striped purple; flbracts oval-lance-shaped, dark green. Dry grassy and rocky places, to 2300m. P,Cev,A,Ap.

Daisy Family (contd.)

HAWKSBEARDS *Crepis*. Perennials with lvs in basal rosettes and alternate up the stem. Flheads solitary or clustered, dandelion-like, the outer series of flbracts noticeably *shorter* than the inner and often spreading. Pappus white, sometimes yellowish.

1 MARSH HAWKSBEARD *Crepis paludosa*. Med/tall per, stems lfy, slightly hairy, branched above. Lvs *hairless*, lance-shaped, deeply and sharply toothed, the lower stalked; upper clasping the stem with *pointed* basal lobes. Flheads yellow or dull orange-yellow, 15-25mm, in lax flat-topped clusters; flbracts linear, with sticky blackish hairs. Damp grassy places, streamsides, to 2150m. July-Sept. T–except c,sAp.

2 NORTHERN HAWKSBEARD *Crepis mollis*. Slender med/tall per, yellow-hairy or hairless. Lvs elliptical to oblong, toothed or not, stalk winged, the upper lvs *clasping* the stem with rounded basal lobes. Flheads yellow, 20-30mm, few in a loose flat-topped cluster; flbracts linear with blackish or yellowish hairs. Woods and streamsides, to 1400m. July-Aug. B,P,A. **2a** *C. lampsanoides* has *lyre-shaped* lower lvs; all lvs toothed. P.

3 MOUNTAIN HAWKSBEARD *Crepis pontana* (= *C. montana*). Short/med per, stem stout hairy. Lvs oblong, broadest above the middle, toothed, the lower with short winged-stalks, the upper *clasping the stem* and untoothed, all hairless except on the edges. Flheads yellow, 40-55mm, solitary; flbracts lance-shaped, green or yellowish hairy on both sides. Meadows, open woods and stony slopes, usually on limestone, 1100-2500m. June-Aug. A. **3a** *C. conyzaefolia* (= *C. grandiflora*) has finely *glandular-hairy* lvs, often more deeply toothed and with up to 9 flheads on a stem *branched* above the middle. Meadows, to 3000m. P,A. **3b** *C. pyrenaica* is very variable, low/med, with a single or up to 5 flheads; flbracts hairless *on the inside*. To 2200m. P,A.

4 ALPINE HAWKSBEARD *Crepis alpestris*. Med hairy per. Lower lvs oblong, broadest above the middle, deeply toothed to *pinnately-lobed* with winged stalks; the upper lvs more or less clasping the stem. Flheads yellow, 20-30mm, *usually solitary*; flbracts lance-shaped, grey or yellowish-hairy. Meadows and stony places, usually on limestone, to 2650m. June-Sept. A,nAp.

5 PYRENEAN HAWKSBEARD *Crepis albida*. Variable low/short downy per. Lvs *mostly basal*, lance-shaped to oblong, toothed to pinnately-lobed. Flheads yellow, 25-45mm, *solitary* or 2 together; flbracts white or yellowish-hairy. Grassy and stony places, usually on limestone, to 2000m. June-Aug. P,MA.

6 PINK HAWKSBEARD *Crepis praemorsa* (= *C. incarnata*). Variable short/tall, hairy or almost hairless per. Lvs usually all in a *basal rosette*, lance-shaped or oblong-oval, broadest above the middle, narrowing into a short stalk. Flheads small, yellow, pink or white, 15-18mm, in loose, oblong clusters, the stem *branched* near the top. Dry meadows and stony places, usually on limestone, to 1800m. May-June. w,s,e,seA.

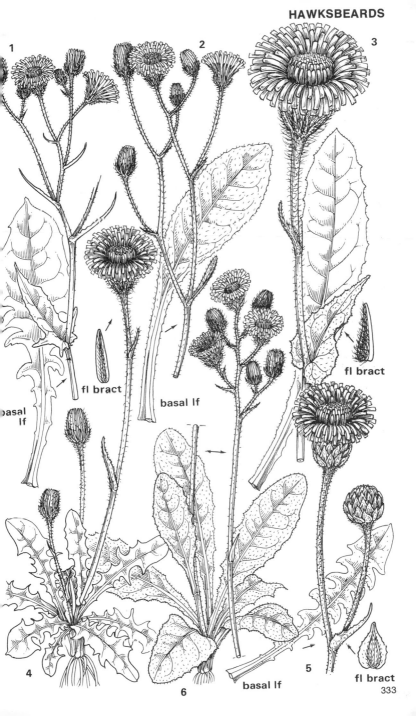

1

2

3

fl bract

basal lf

fl bract

basal lf

4

5

6

basal lf

fl bract

Daisy Family (contd.)

1 PYGMY HAWKSBEARD *Crepis pygmaea.* Low/short downy per. Lvs often reddish, *not* in a basal rosette, elliptical to rounded, toothed, with a *wavy winged-stalk.* Flheads pale to golden yellow, 20-25mm, the outer rays often reddish-purple beneath, several on long stalks from the lfjoints; flbracts lance-shaped, densely hairy. Damp calcareous rocks and screes, 1600-3000m. July-Aug. P, MA, w, swA, cAp.

2 TRIGLAV HAWKSBEARD *Crepis terglouensis* agg. (including *C. jacquinii*). Low/short slightly hairy per. Lvs *mostly basal,* oblong to linear, pinnately-lobed or unlobed, often hairless. Flheads yellow, 20-45mm, usually solitary; flbracts narrow lance-shaped, *dark-hairy.* Stony places, screes, on limestone, 1500-1800m. July-Aug. A–except France. Often confused with *Leontodon montanus.* **2a** *C. rhaetica* □ is smaller with untoothed or slightly toothed lvs, usually hairy; stalks winged. Flheads solitary; flbracts oblong with long *yellowish-hairs.* Limestone rocks and screes, 1950-3000m. cA.

3 GOLDEN HAWKSBEARD *Crepis aurea.* Low/short, slightly hairy per. Lvs *all basal,* oblong, broadest above the middle, toothed or pinnately-lobed. Flheads yellow or orange, 20-30mm, usually solitary on *long slender* stalks; flbracts lance-shaped, with dark green hairs. Meadows and stony places, often on acid soils, to 2800m. June-Sept. A. **3a** *C.a. glabrescens* usually has hairless flbracts. Ap.

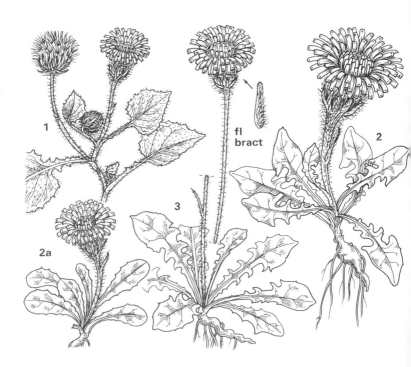

fl bract

Appendix 1 Additional species and aggregates

To allow adequate space for description of the illustrated species on certain plates, some of the subsidiary species have been removed to this Appendix. As with the subsidiary species in the main text, however, descriptions refer back to the main numbered species. This, and where appropriate also the family and genus description, must therefore also be taken into account.

Page 22
2a *Picea abies alpestris* is smaller, cones more rounded, 6-8cm. A.
8a Savin *Juniperus sabina* is a low spreading shrub. Lvs tiny. Cone 4-6mm, bluish-black when ripe. Rocky and stony places, meadows and woods, to 2300m. A, Ap.

Page 30
11b *Petrocoptis crassifolia* is like 11a but calyx larger, 9-13mm. cP, Bielsa region.

Page 32
1a *Silene furcata* has sticky, branched stems; sepal-tube 10-12mm. S.
2a *S. italica* is branched, with erect fls; petals less recurved. P, A, Ap.
5a Eared Catchfly *S. auriculata* is taller with lance-shaped lvs densely hairy along margins; fls smaller, white petals with two small lobes at base. To 2000m. AA.
6a *S. vulgaris glareosa* is smaller and more sprawling; lvs lance-shaped, 2-5mm broad. Fls with scales in throat. P, A, Jura. **6b** *S.v. prostrata* like 6a but lvs oval, 3-10mm broad and fls without throat-scales. A, Ap. **6c Red Campion** *S. dioica* has larger, bright pink fls; sepal-tube hairy, not inflated. Woods, specially on limestone, to 2400m. T.
13a *S. veselskyi* is less robust and densely hairy. seA, at low altitudes. **13b** *S. pusilla* (= *S. quadrifida*) is a frail version of *S. alpestris*; fls sometimes pale pink or lilac, 6-8mm; sepal-tube usually hairless. To 2900m. P, A, Ap.

Page 34
7a *Petrorhagia prolifera* is larger with unbranched, dense fl clusters, fls pink or purplish. Usually on lime. May-Sept. T-except B.
10a *Dianthus barbatus compactus* has the lower lvs stalked and purplish-brown epicalyx scales. Ap.

Page 36
7a *Dianthus pyrenaicus catalaunicus* has blue-green lvs and deep pink fls, 12-16mm. eP.

Page 60
5a *Braya linearis* is taller, the bottom lvs linear. Gravels on lime. S.
6a Woad *Isatis tinctoria* is taller, bien, with smaller fls, 4 mm, and dark brown fr. Waste ground, grassy and rocky places, to 2000m. T(B, S).

Page 74
5a *Sempervivum pittonii* is smaller with grey-green rosettes, not scented of resin. Fls greenish-yellow. eA-Steiermark.
6a *S. arachnoideum tomentosum* has rosettes 15-25 mm across. P, wA.
7a *S. montanum burnatii* has larger rosettes, 30-80mm across. swA. **7b** *S. m. stiriacum* has rosettes 20-45mm across, the lvs sharply pointed and reddish-tipped. eA.
9a Catalonian Houseleek *Sempervivum andreanum*. Low per with 20-25mm rosettes. Lvs oval, broadest above middle, hairy-margined, otherwise hairless. Fls pale pink, 14-18mm in small clusters; anthers yellow. Limestone rocks, to 2500m. eP-Sierra del Cadi.

6c *Sedum alpestre* is a dwarfer plant, lvs broadest above the middle; fls 6-7mm. Rocks and moraines, usually on acid soils, to 3500m. P, A, Ap.
11a *S. magellense* is *hairless*, with smaller lvs and racemes of fls, *not branched*. n, cAp.

2a *Saxifraga stellaris alpigena* is more loosely tufted, with smaller fls, 7-10mm. To 3000m. P, A, nAp. **2b** *S. foliolosa* is like a small version of 2 but rosettes usually solitary. Fls usually *replaced by small bulbils*. S.
3a Arctic Saxifrage *S. nivalis* (☐ on p. 81), a low per with lfless hairy stems; lvs thick in basal rosettes, coarsely toothed, purplish beneath. Fls white or pink, 6mm, in *rounded heads*. Rock crevices, usually basaltic, to 2100m. July-Aug. B, S. **3b Slender Saxifrage** *S. tenuis* is smaller and slenderer. Fls distinctly stalked, in loose clusters. To 1700m. S.
5a *S. hirsuta* has *kidney-shaped or rounded lvs*, hairy above and beneath; lf-stalks long-narrow, rounded. Fls with numerous deep red spots. Shady places and streamsides, 1500-2500m. May-July. B-sw Ireland, P. **5b** *S.h. paucicrenata* has oblong-elliptical lvs with short stalks. Limestone rocks and screes. wP. Hybrids occur between 5 and 5a where they grow together and are known as *S × geum*.
7a Mossy Saxifrage *S. bryoides* has shorter shoots, forming dense mats, lvs incurved; fls *solitary*. 2000-4000m. P. Cev–Auvergne, s, eA.
10a *S. adscendens parnassica* is smaller than 10. c, sAp. **10b** *S. petraea* has the lower lvs cut into numerous narrow lobes; fls 14-18mm with notched petals. scA.

2a *Saxifraga glabella* has narrow, spoon-shaped lvs, *entirely hairless*; fls white, 5mm. cAp.
3a *S. cuneata* is smaller, rarely over 15cm tall; lvs hairless except sometimes on edges, rather sticky. Shady rocks, screes and walls, limestone. wP.
5a *S. corbariensis* has lvs with 3 main lobes, each cut into further segments. Limestone rocks and screes, usually shaded, to 1500m. eP–Spanish; Corbières. **5b** *S. hypnoides* (☐ on p. 83) has widely-spaced lvs, entire or 3-lobed, on slender lateral shoots; fls pure white, dropping in bud. B, S–Vosges.
8a *S. pedemontana prostii* (= *S. prostii*) has lvs more rounded in outline and with *narrow stalks*.
9a *S. seguieri* has lvs *always* unlobed and hairy all over; fls smaller, 5-6mm, dull yellow. July-Aug. c, eA. **9b** *S. depressa* is taller than 9, with *more elongated* lfy shoots and fls 8-10mm in clusters of 3-8. Shady rock ledges and damp screes, mostly on porphyritic rocks, 2000-2850m. sA–Dolomites.
10a *S. aphylla* is similar in habit to 10 but lvs *mostly 3-lobed*; fls pale greenish-yellow, petals *very narrow*, slightly longer than sepals. 2000-3200m. July-Sept. c, eA.
11a *S. muscoides* has shorter shoots with narrow lance-shaped lvs, hairy, but scarcely sticky. Fls white or pale lemon-yellow with *broad petals*, not notched. Rocks and screes, rarely on limestone, 2200-4200m. June-Aug. c, s, eA. **11b** *S. facchinii* is like 11a but with very small, 4mm, dull yellow fls tinged with red, *usually solitary* and among the lvs on short stalks. Limestone rocks and screes, 2200-3250m. July. sA–Dolomites.
12a *S. hariotii* has 3-lobed lvs, shiny green, lobes grooved, pointed, almost hairless. Limestone rocks and stony meadows, 1600-2300m. wP.
13a *S. cebennensis* (☐ on p. 83) is a much larger plant than 13, with more compact, pale green cushions and lvs with long hairs. Shaded limestone rocks. Cev.

4a *Saxifraga rivularis* has bulbils only at the *basal lvs*; lvs 3-5 lobed. July-Aug. B, S.
6a *S. retusa augustana* has long fl-stems, 2-5cm, carrying 2-5 fls, each 10mm across. Limestone rocks, swA.
8b *S. vandellii* has narrow lance-shaped lvs, 8-11mm; fls 16mm across. sA–Italian. **8c** *S. marginata* is like 8a but lvs 3-12mm long, oblong to spoon-shaped; fls larger, white or pale pink. c, sAp.

11a *S. squarrosa* has *greener*, harder, cushions; lvs *smaller*, narrow-oblong, recurved only at the tip. To 2700m. July-Aug. seA.

Page 108

9a *Cytisus sessilifolius* is hairless, all the lvs trifoliate; fls in clusters of 3-12. Woods and scrub to 2300m. eP,MA,sA,Ap.

Page 136

1a *Linum austriacum* has *down-turned* fl-stalks and some lvs generally 3-veined. A,Ap. **1b Flax** *L. usitatissimum* is annual; *all lvs 3-veined.* Widely naturalised and cultivated for linen.

11a *Polygala carueliana* has all lvs *alternate*, linear or spoon-shaped. Fls slightly larger with *greenish*, purple-tinged, wings. AA.

12a *P. amarella* has stem-lvs blunt and broadest near the apex, fls 2-4.5mm. B,S,A.

Page 150

7a *Astrantia major carinthiaca* has bracts *twice as wide* as the umbel. P,s,swA.

9a *A. carniolica* is similar to 9 but *none* of the lf-lobes are separated to the centre, and bracts do not equal the umbel. To 1600m. seA.

Page 170

3a *Androsace carnea laggeri* (☐ on p. 171) has *smaller* lf-rosettes than 3; lvs 4-6mm, hairless. eP. **3b** *A.c. rosea* (☐ on p. 171) has *larger* lf-rosettes than 3; lvs 10-25mm, hairless. eP, wA–French. **3c** *A.c. brigantiaca* (☐ on p. 171) has slightly *toothed* lvs up to 30 mm long and white fls in longer-stemmed clusters than 3. swA.

10b *A. brevis* is similar to 10 but lvs smaller and spoon-shaped, 3-5mm, with *branched hairs* only. 1700-2600m. sA–around Lake Como.

Page 174

7a *Armeria ruscinonensis* is similar to 7 but with *blunt-tipped* lvs, shorter stems and pink or white flheads. Acid cliffs near coast, to 1300m. oP.

11a Greater Dodder *Cuscuta europea* has larger fls than 11, sometimes with yellowish-green stems. Petals and sepals blunt. Parasitic on nettles and other herbaceous plants. T.

Page 204

9a *Cymbalaria pallida* has pale lilac-blue fls with a longer spur, 6-9mm. Rocks and screes to 2100m. June-Aug. eAp.

Page 224

3a *Asperula hirta* (☐ on p. 225) has smaller, *hairy margined* lvs. July-Aug. w, cP.

4a *A. cynanchica* is taller, with lvs always in fours, *all narrow* lance-shaped. Fls whitish or pale purple. To 2100m. June-Sept. T–except S.

8a *Lonicera coerulea pallasii* has hairy lvs and narrower fls. A*.

Page 228

6a *Jasione laevis* (= *J. perennis*) is taller, with numerous, rather *creeping*, non-flg shoots and *deeply-toothed* bracts. Dry meadows, to 1900m. P.

8a *Phyteuma pyrenaicum* has its basal lvs withered at flg time and the lvs *slaty-blue.* P. **8b Dark Rampion** *P. ovatum* (= *P. halleri*) is like 8 but bracts oval and fls *blackish-violet*. To 2400m. July-Aug. P,A,Ap.

10a *P. michellii* is shorter, with narrower linear lvs which are *hairy-margined* at the base. July-Aug. sA. **10b** *P. zahlbruckneri* is like 10 but lvs *rounded or heart-shaped* at the base. Fls deep blue or blue-black. e,seA.

Page 248

2a *Homogyne discolor* has lvs *white-woolly* beneath and larger stem lvs. e,seA. **2b** *H. sylvestris* is like 2 but lvs shallowly-lobed and lower stem lvs *stalked*; stems *often branched* with 2 or more flheads. s,eA.

7a Leopardsbane *Doronicum pardalianches* (☐ on p. 249) is more hairy, the lower stem lvs stalked but *not clasping*; flheads 3-5cm, usually in clusters of 2-8. P,sA,Jura. **7b** *D. cataractarum* is more robust than 7a but the basal lvs are *hairless or almost so*; flheads 4-7cm. eA.

8a *D. plantagineum* is taller, the ray-florets *without a pappus* at base. Woods. P, Cev. **8b** *D. clusii* (= *D. stiriacum, D. glaciale*) is like 8 but stems long-hairy and basal lvs *short-stalked*, not more than 2 cm long. P, A.
9a *D. carpetanum* has the lower stem-lvs *stalked and clasping*; flheads solitary or 2-3 together. P.

Page 250

6a *Saussurea alpina macrophylla* has rounded lvs, *not* narrowed at the base. eA. **6b** *S.a. depressa* is a dwarf form of 6, lvs grey-hairy above and beneath. wA.

Page 258

4a *C. maculosa* is similar to 4a but with 1-fld branches, purple heads and greenish hairy or hairless lvs. wA,Cev,MA.
5a *C. carniolica* has the lower lvs elliptical or rounded-oval and the upper oval or elliptical, often partly clasping the stem. Flheads pink, either solitary or clustered. **5b** *C. transalpina* is shorter and more branched than 5, with solitary or 2-4-clustered pinkish-orange or pink flheads. Involucre 12-18 mm across. sA.

Page 262

2a Austrian Vipergrass *Scorzonera austriaca* has stems with *3-6 scale-like lvs* and pale yellow fls; tuber fibrous at the top. Dry meadows and rocky places, to 2800 m. May-July. eP, A, Ap. **2b** *S. humilis* is like 2a but tuber *not* fibrous at the top; lower stem lvs like the basal ones, upper scaly. To 1700 m. T. **2c** *S. purpurea* (including *S. rosea*) is like 2 but fls pale lilac or purplish-pink. Damp and shady places, to 2000 m. seA, Ap.
5a Swiss Hawkbit *Leontodon pyrenaicus helveticus* is larger in all its parts, stems up to 50 cm; fibracts *whitish*-hairy. w, cA. **5b Alpine Hawkbit** *L. hispidus alpinus* has *toothed* lvs; stem only 5-15 cm tall. wA. **5c** *L. croceus* is like 5b, but lvs linear to narrowly oblong, finely toothed. eA.
6a *L. montaniformis* is smaller, fibracts with dark hairs. c, eA, cAp. **6b** *L. autumnalis* is larger than 6, with *branched stems* and deep yellow fls, the florets red-striped outside. To 2600 m. July-Oct. T.
8a *Hieraceum arolae* agg. has larger stem lvs, similar to the basal ones, and smaller flheads in groups of 2-10. c, ecA. **8b Mouse-ear Hawkweed** *H. pilosella* is *white-hairy*; flheads solitary lemon-yellow, 20-30 mm, on long lfless stems. Grassy places, to 3000 m. May-Oct. T.
10a *H. lactucella* agg. is *sparsely hairy* with narrower bluish-green lvs. Flheads yellow, often red-striped or red-tipped, in clusters of 2-5. Grassy places, usually on limestone, to 2600 m. T.

Page 280

2c Fly Orchid *Ophrys insectifera* (= *O. muscifera*) is more slender with narrow elliptical lvs. Fls smaller than 2 with spreading green sepals, the petals brownish, *thread-like*; lip *fly-shaped*, oblong with two short wings and a forked central lobe, purplish-brown with central a shiny-bluish zone. Grassy places, woods and scrub, on calcareous soils, to 1850 m. May-June. T.

Page 282

6a Flecked Marsh Orchid *Dactylorhiza cruenta* is shorter, the lvs purple-spotted, not hooded at the tips; fls reddish or dark purple, the lip untoothed. B, S, w, ceA. **6b** *D. traunsteineri* is like 6 but lvs *not* hooded; fls larger, purple, the lip 7-9 mm long, strongly marked purple, scarcely folded. B, S, A.
7a *D. majalis alpestris* is shorter, the lower lvs broadest near the top. P, A.
8a Heath Spotted Orchid *D. maculata* has longer fls, the lip 7-11 mm, shallowly 3-lobed, the central lobe small and 'tooth-like'. Heaths and bogs, on acid soils. T–except Ap.
9a Rosy Vanilla Orchid *Nigritella rubra* has rose-pink or flesh-coloured fls, with a *broad* lip and broader sepals and petals than 9. 1600-2300 m. May-July. e, seA.

Page 284

8e Marsh Helleborine *Epipactis palustris* has larger fls than 8, the sepals purple or brown-purple, petals crimson and white; lip white, the cup streaked crimson, frilly-margined, longer than the sepals; fr downy. Marshes and damp places, to 1800 m. July-Aug. T. **8f Violet Helleborine** *E. purpurata* is like 8 but the lvs *grey-purple*, spiralled up the stem. Fls pale greenish-white, purplish green outside, the lip shorter than the sepals, white mottled with violet, tip curled under. Beech woods mainly, to 1400 m. Aug-Sept. sB, P, A.

Page 304

5 Holly *Ilex aquifolium*. Small evergreen tree or shrub to 10 m; bark smooth grey. Lvs alternate, oval, *spiny edged*, deep green, shiny. Fls white, 5-6 mm, 4-petalled, in dense clusters at base of lvs; male and female on seperate trees. Fr a scarlet berry. Woods, shrub and hedges, to 2000 m. Apr-May. T. Poisonous.

Aggregates

Where the abbreviation "agg." follows a scientific name, several very similar species or microspecies are included, as follows:

Fumaria officinalis (p.56) incl. many local forms and more widespread subspecies, mostly confined to lowland and submontane habitats.
Rubus fruticosus (p.90) incl. many species of bramble, far too numerous to list.
Rosa vosagiaca (p.94) incl. *R. caesia, R. rhaetica, R. subcanina* and *R. subcollina.*
Rosa tomentosa (p.94) incl. *R. scabriuscula* and *R. sherardii.*
Rosa villosa agg. (p.94) incl. *R. mollis.*
Linum perenne (p.136) incl. *L. leonii*
Viola canina (p.144) incl. *V. lactea* and *V. persicifolia.*
Myosotis sylvatica (p.188) incl. *M. decumbens* and *M. latifolia.*
Myosotis scorpioides (p.188) incl. *M. nemorosa* and *M. rehsteineri.*
Euphrasia salisburgensis (p.212) incl. *E. illyrica* and *E. portae.*
Asperula pyrenaica (p.224) incl. *A. neglecta,* (c,sAp); *A. neilreichii* (ne A); *A. pyrenaica* (P) and *A. rupicola* (Sw A).
Campanula rotundifolia (p.234) – many forms have been described.
Hieracium humile (p.262) incl. *H. cottetii, H. kerneri* and *H. valoddae.*
Hieracium alpinum (p.262) incl. *H. nigrescens* and *H. pietroszense.*
Hieracium aurantiacum (p.262) hybridises with many other species of Hieracium; hybrids often have orange in the fls.
Stellaria media (p.292) incl. *S. neglecta* and *S. pallida.*
Cerastium arvense (p.294) Many subspecies are recognised, mostly occurring at low altitudes.

Appendix 2   **Key to Monocotyledons** (see p. 15)

FLOWERS WITH A CONSPICUOUS LIP,
OFTEN WITH A SPUR

Orchids 280-6

FLOWERS WITHOUT A SPUR OR LIP
3 stamens (Iris family)

Crocuses 278

Irises 278

6 stamens
OVARY BELOW PETALS (Daffodil family)

Snowflakes 276   Snowdrops 276

Poet's Narcissus
etc 278

Wild Daffodils or
Lent Lily 276

OVARY ABOVE PETALS – inside flower

Gageas, etc. 264; Onions,
Leeks 266; False
Helleborines, Asphodel 268;
Star of Bethlehem, etc. 272;
Herb Paris 274

Lloydia 272

LEAVES HEART-SHAPED

Lords and Ladies
274

May Lily 272   4 petals

Pyrenean Yam
274

male and female
fls on separate
plants

Wild Tulip 268

Orange Lily 270,
Tulip didieri 268

Spring Bulbocodium
and Merendera 264

St Bruno's Lily 264

Streptopus 274

Dog's Tooth Violet 270

Lilies 270

Fritillaries 268

Lily of the Valley 274
Tassel Hyacinth
and Dipcadi 272

Grape Hyacinths 272

Solomon's Seals 274

Pyrenean Hyacinth 272

341

Appendix 3 Key to Trees and Large Shrubs

This chart is based on leaf-shape to all the trees and the shrubs more than 1 m tall included in the book. Trees are indicated by a (t) and shrubs by a (s), the former recognised readily by their single stout woody trunk unless the specimens have been coppiced in the past.

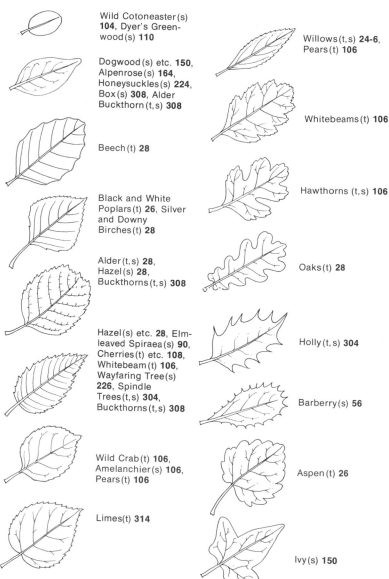

Wild Cotoneaster(s) **104**, Dyer's Green-wood(s) **110**

Dogwood(s) etc. **150**, Alpenrose(s) **164**, Honeysuckles(s) **224**, Box(s) **308**, Alder Buckthorn(t,s) **308**

Beech(t) **28**

Black and White Poplars(t) **26**, Silver and Downy Birches(t) **28**

Alder(t,s) **28**, Hazel(s) **28**, Buckthorns(t,s) **308**

Hazel(s) etc. **28**, Elm-leaved Spiraea(s) **90**, Cherries(t) etc. **108**, Whitebeam(t) **106**, Wayfaring Tree(s) **226**, Spindle Trees(t,s) **304**, Buckthorns(t,s) **308**

Wild Crab(t) **106**, Amelanchier(s) **106**, Pears(t) **106**

Limes(t) **314**

Willows(t,s) **24-6**, Pears(t) **106**

Whitebeams(t) **106**

Hawthorns(t,s) **106**

Oaks(t) **28**

Holly(t,s) **304**

Barberry(s) **56**

Aspen(t) **26**

Ivy(s) **150**

Field Maple (t, s) and
Italian Maple (t) **306**,
Currants (s) **78**,
Gooseberry (s) **78**

Norway Maple (t) and
Sycamore (t) **306**

Laburnum (t) **108**,
Brooms (s) **108-110**

Brambles (s) **90**

Bladder Senna (s)
112, False Senna (s)
126

Mountain Ash (t, s)
106, Roses or
Briars (s) **92-4**,
Elders (s) **226**, Ash (t)
and Manna Ash (t, s)
310

Pines (t) **22**

Firs (t), Spruces (t),
Larch (t), Junipers (s)
and Yew (t, s) **22, 335**

French Alpine
Juniper (t) and
Savin (s) **22, 335**

Gorse **112**

Appendix 4 Crucifer fruits

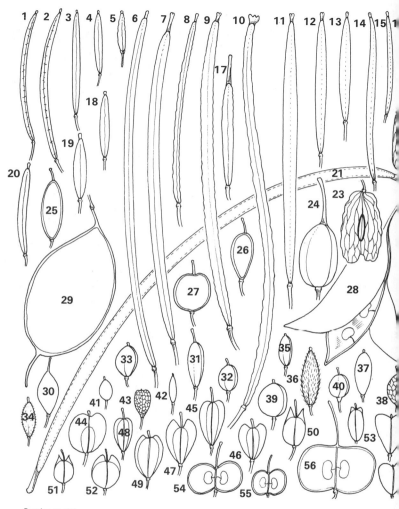

Scale: × 1⅓

1 *Sisymbrium irio* **60**
2 *S. austriaca* **60**
3 *Murbeckiella pinnatifida* **60**
4 *Hugueninia tanacetifolia* **60**
5 *Braya alpina* **60**
6 *Erysimum sylvestre* **60**
7 *E. decumbens* **60**
8 *E. hieracifolium* **60**
9 *Hesperis inodora* **62**
10 *Matthiola fruticulosa* **62**
11 *Cardamine pentaphyllos* **62**
12 *C. kitaibelii* **62**
13 *C. trifolia* **62**

14 *Cardamine amara* **62**
15 *C. plumieri* **64**
16 *C. bellidifolia* **64**
17 *Brassica repanda* **70**
18 *Arabis vochinensis* **302**
19 *A. caerulea* **64**
20 *A. pumila* **302**
21 *A. turrita* **302**
22 *Isatis tinctoria* **60**
23 *I. allionii* **60**
24 *Alyssoides utriculata* **64**
25 *Alyssum wulfenianum* **64**
26 *A. ovirense* **64**

27 *Alyssum montanum* **64**
28 *Lunaria rediviva* **64**
29 *L. annua* **64**
30 *Ptilotrichum pyrenaicum* **64**
31 *Draba aizoides* **66**
32 *D. hoppeana* **66**
33 *D. sauteri* **66**
34 *D. ladina* **66**
35 *D. carinthiaca* **66**
36 *D. tomentosa* **66**
37 *Erophila verna* **68**
38 *Petrocallis pyrenaica* **68**
39 *Cochlearia pyrenaica* **68**

344

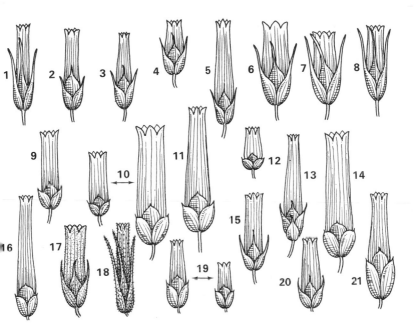

Scale: × 2

1 *D. seguieri* **34**
2 *D. collinus* **34**
3 *D. furcatus* **34**
4 *D. barbatus* **34**
5 *D. monspessulanus* **34**
6 *D. alpinus* **36**
7 *D. glacialis* **36**
8 *D. pavonius* **36**
9 *D. gratianopolitanus* **36**
10 *D. sylvestris* **36**
11 *D. superbus* **34**

12 *D. subacaulis* **36**
13 *D. pyrenaicus* **36**
14 *D. plumarius* **36**
15 *D. pungens* **36**
16 *D. serotinus* **36**
17 *D. deltoides* **36**
18 *D. armeria* **36**
19 *D. pontederae* **36**
20 *D. carthusianorum* **36**
21 *D. giganteus* **36**

40 *Kernera saxatilis* **68**
41 *Rhizobotrya alpina* **68**
42 *Hutchinsia alpina* **68**
43 *Hymenolobus procumbens* **68**
44 *Aethionema saxatilis* **68**
45 *Thlaspi goesingense* **70**
46 *T. montanum* **70**

47 *Thlaspi praecox* **70**
48 *T. alpinum* **70**
49 *T. rotundifolium* **70**
50 *Iberis aurosica* **70**
51 *I. stricta* **70**
52 *I. spathulata*
53 *Lepidium villarsii* **72**

54 *Biscutella laevigata* **72**
55 *B. brevifolia* **72**
56 *B. cichoriifolia* **72**
57 *Capsella bursa-pastoris* **304**
58 *C. rubella* **304**

Appendix 6 Rock Jasmine *(Androsace)* leaves

Scale: **1-7** × 2½; **8-20** × 4. All on p.170 except where otherwise indicated.

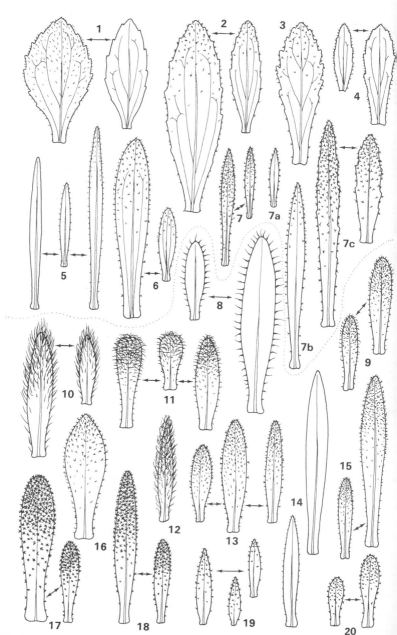

346

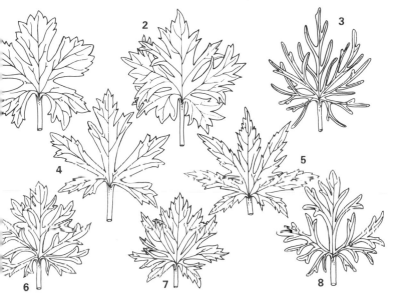

Scale: ×⅓

1 *A. vulparia*
2 *A. lamarkii*
3 *A. anthora*
4 *A. variegatum*

5 *A. paniculatum*
6 *A. napellus*
7 *A. tauricum*
8 *A. compactum*

1 *A. maxima* **172**
2 *A. septentrionale* **172**
3 *A. chaixii* **172**
4 *A. elongata* **172**
5 *A. lactea*
6 *A. obtusifolia*
7 *A. carnea;* **7a** *A.c. laggeri;*
7b *A.o. rosea;* **7c** *A.c. brigan-*
    tiaca
8 *A. chamaejasme*
9 *A. pyrenaica*
10 *A. villosa*

11 *A. helvetica*
12 *A. vandelii*
13 *A. cylindrica*
14 *A. mathildae*
15 *A. pubescens*
16 *A. ciliata*
17 *A. alpina*
18 *A. hausmannii*
19 *A. wulfeniana*
20 *A. brevis* **337**

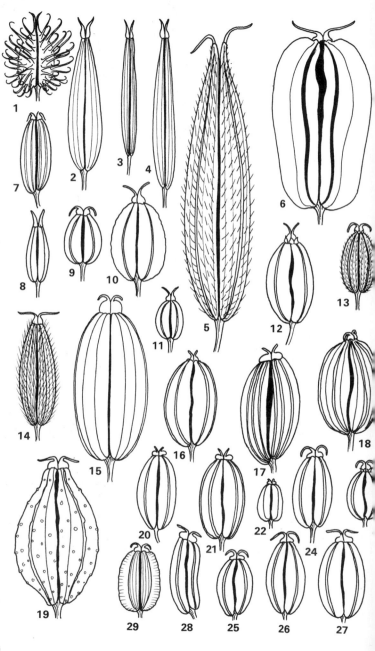

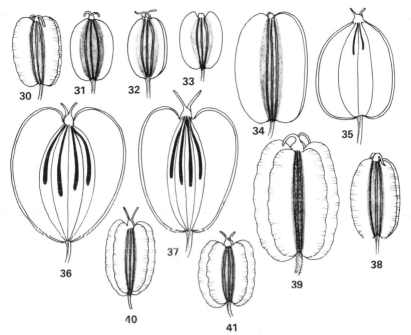

Scale: × 4

1 *Sanicula europaea* **150**
2 *Chaerophyllum hirsutum* **152**
3 *C. elegans* **152**
4 *C. villarsii* **152**
5 *Myrrhis odorata* **152**
6 *Molospermum peloponnesiacum* **152**
7 *Bunium alpinum* **152**
8 *Conopodium pyrenaicum* **152**
9 *Pimpinella major* **154**
10 *P. siifolia* **154**
11 *P. saxifraga* **154**
12 *Dethawia tenuifolia* **154**
13 *Seseli nanum* **154**
14 *Athamanta cretensis* **154**
15 *Grafia golaka* **154**
16 *Trochiscanthes nodiflora* **154**
17 *Meum athamanticum* **154**
18 *Xatardia scabra* **154**
19 *Pleurospermum austriacum* **154**
20 *Bupleurum longifolium* **156**

21 *B. angulosum* **156**
22 *B. ranunculoides* **156**
23 *Endressia pyrenaica* **156**
24 *Ligusticum mutellina* **158**
25 *L. mutellinoides* **158**
26 *L. lucidum* **158**
27 *L. ferulaceum* **158**
28 *Carum carvi* **156**
29 *Selinum pyrenaeum* **156**
30 *Angelica sylvestris* **158**
31 *Levisticum officinale* **158**
32 *Peucedanum venetum* **158**
33 *P. ostruthium* **158**
34 *P. verticillatum* **158**
35 *Heracleum minimum* **158**
36 *H. austriacum* **158**
37 *H. sphondylium* **158**
38 *Angelica archangelica* **158**
39 *Laserpitium latifolium* **158**
40 *L. halleri* **158**
41 *L. peucedanoides* **158**

Appendix 9 **Saxifrage leaves**

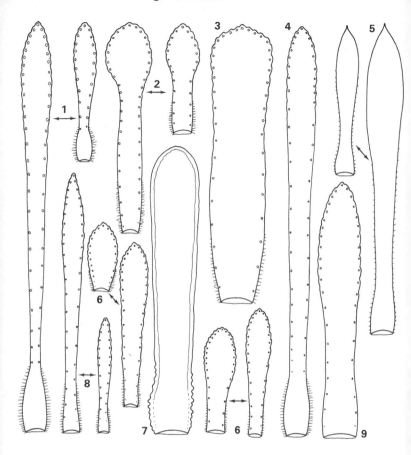

Scale: × 1⅓

1 *S. callosa* **86**
2 *S. cochlearis* **86**
3 *S. cotyledon* **86**
4 *S. longifolia* **88**
5 *S. florulenta* **88**

6 *S. paniculata* **88**
7 *S. mutata* **88**
8 *S. crustata* **86**
9 *S. hostii* **86**

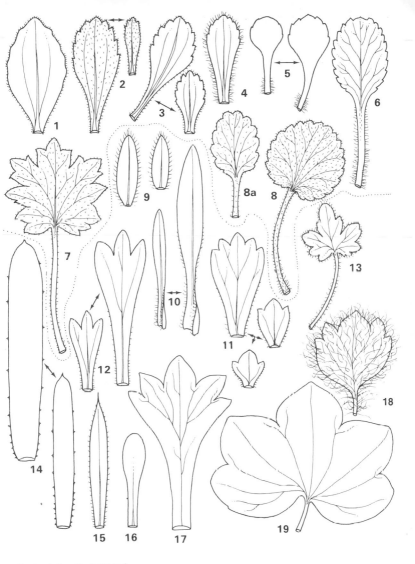

Scale: **1-8** × ⅔; **9-18** × 3

1 *S. hieracifolia* **80**
2 *S. stellaris* **80**
3 *S. nivalis* **336**
4 *S. clusii* **80**
5 *S. cuneifolia* **80**
6 *S. umbrosa* **80**
7 *S. rotundifolia* **80**
8 *S. hirsuta* **336**
8a *S.h. paucicrenata* **336**
9 *S. aspera* **80**

10 *S. hirculus* **80**
11 *S. adscendens* **80**
12 *S. tridactylites* **80**
13 *S. petraea* **336**
14 *S. aizoides* **82**
15 *S. tenella* **82**
16 *S. glabella* **336**
17 *S. praetermissa* **82**
18 *S. arachnoidea* **80**
19 *S. paradoxa* **80**

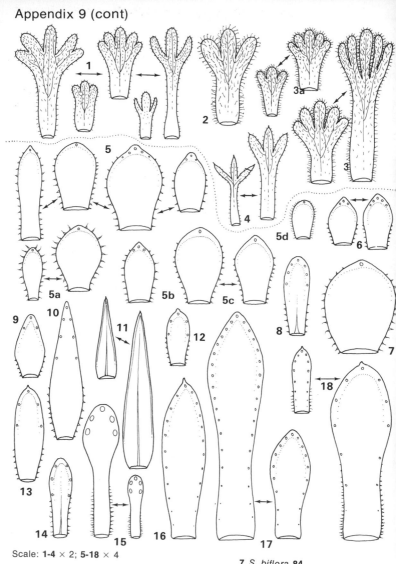

Scale: 1-4 × 2; 5-18 × 4

1 *S. exarata* 82
2 *S. cebennensis* 83, 336
3 *S. pubescens* 84
3a *S. p. iratiana* 84
4 *S. cespitosa* 84
5 *S. oppositifolia* 84
5a *S.o. blepharophylla* 84
5b *S.o. latina* 84
5c *S.o. speciosa* 84
5d *S.o. rudolphiana* 84
6 *S. retusa* 84

7 *S. biflora* 84
8 *S. diapensioides* 84
9 *S. tombeanensis* 84
10 *S. vandelii* 337
11 *S. burserana* 84
12 *S. squarrosa* 337
13 *S. aretioides* 84
14 *S. caesia* 84
15 *S. valdensis* 86
16 *S. porophylla* 86
17 *S. media* 86
18 *S. marginata* 336

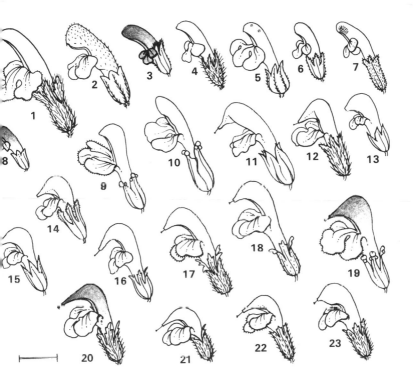

Scale: life-size

1 *P. acaulis* **214**
2 *P. foliosa* **214**
3 *P. recutita* **214**
4 *P. rosea* **214**
5 *P. verticillata* **214**
6 *P. hirsuta*
7 *P. oederi* **214**
8 *P. flammea* **214**
9 *P. palustris* **214**
10 *P. sylvatica* **214**
11 *P. comosa* **214**
12 *P. tuberosa* **214**

13 *P. lapponica* **214**
14 *P. elongata* **214**
15 *P. ascendens* **216**
16 *P. pyrenaica* **216**
17 *P. gyroflex* **216**
18 *P. portenschlagii* **216**
19 *P. rostratocapitata* **216**
20 *P. cenisia* **216**
21 *P. rostratospicata* **216**
22 *P. asplenifolia* **216**
23 *P. elegans* **216**

Cornflower *(Centaurea)* **flower-bracts**

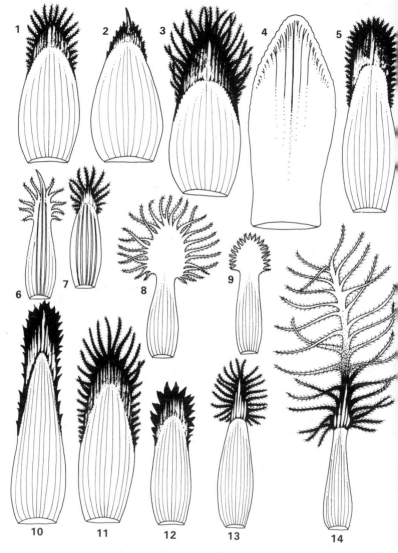

Scale: all × 4

1 *C. dichroantha* **256**
2 *C. rupestris* **256**
3 *C. alpestris* **256**
4 *C. alpina* **258**
5 *C. baldensis* **256**
6 *C. paniculata* **258**
7 *C. maculosa* **338**

8 *C. transalpina* **338**
9 *C. nigrescens* **258**
10 *C. montana* **258**
11 *C. triumfettii* **258**
12 *C. cyanus* **258**
13 *C. procumbens* **258**
14 *C. uniflora* **258**

Glossary

Alternate

Most of the following terms are used frequently throughout the text. Use of the book will be simplified for those unfamiliar with botanical terms such as anther, bract, calyx, corolla, petal, sepal, stamen, stigma, stipule and style, if their meanings are memorised.

Bracts

Achene: a single seeded dry fruit, often in clusters as in the Buttercup.

Acid soil or rocks are those having an acid reaction and are hence non-alkaline. Rocks such as sandstone and granite may give rise to acid soils. Peaty soils are often acid.

Aggregate: groups of closely related species or microspecies that are often very difficult to distinguish except by very close scrutiny.

Alpine meadows and pastures are those occurring above the tree line.

Alpines: those mountain plants that occur above the tree line – as defined in the context of this book.

Alternate: leaves placed alternately along the stem, *not* opposite or whorled

Anther: the male organ of the flower which contains pollen.

Annuals: plants which complete their whole life cycle, from seed to flowering and fruiting, within the same year. High mountain and scree plants are rarely annual.

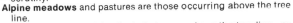

Bulbs

Berry: a fleshy fruit, usually rounded and containing several hard seeds.

Biennials: plants which germinate in the first year, often surviving the winter as leafy rosettes and flower, seed and die in the second year.

Bog: a habitat on wet peat, an environment for various acid loving plants (calcifuge).

Bracts: small leaf-like or scale-like organs that subtend the flowers – not always present. *See also* flower bracts.

Bulbs: underground storage organs, consisting of fleshy swollen leaf-bases that closely overlap one another.

Bulbils: small bulb-like organs, found above ground in the axils of leaves or in some instances replacing flowers; bulbils break off to form new plants.

Bulbils

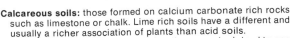

Calyx

Calcareous soils: those formed on calcium carbonate rich rocks such as limestone or chalk. Lime rich soils have a different and usually a richer association of plants than acid soils.

Calyx: refers to the sepals as a whole – these may be joined to one another in whole or part to form a tube (the sepal tube), or separate.

Catkin: tight spikes of usually tiny flowers, erect or drooping, often either male or female, the male catkin producing copious pollen.

Cluster: a loose group of flowers.

Composite: a collective name used to describe members of the Daisy Family, *Compositae*.

Catkin

Composite

Cone: the woody fruit-like structure characteristic of the Pine and related plants. The cone consists of overlapping woody scales enclosing the seeds.

Cone

Corm: 'bulb-like' underground storage organs, consisting of a swollen stem base.

Corolla: refers to the petals as a whole and usually when they are joined wholly or partly into a tube.

Corm

Crucifer: a collective name used to describe members of the Cress Family, *Cruciferae*.

Cushion: plants forming dense rounded tufts, often composed of numerous tight leaf rosettes. A tight cushion form is characteristic of many high alpines, scree and cliff dwellers in particular.

Corolla

Crucifer

Deciduous: trees or shrubs that shed all their leaves in the autumn.

Digitate: Leaves consisting of several similar 'finger-like' leaflets radiating from a central point and joined only at the very base.

Disc florets: of Composites refers to the tubular flowers which make up the central 'eye' of the daisy flowerhead. The flowerhead may sometimes consist solely of disc florets – *see also* Ray florets.

Cushion

Digitate

Epicalyx is a secondary calyx outside the primary; the sepals of the epicalyx may be the same size or smaller than the inner 'true' sepals. Common in the Rose Family in particular – Cinquefoils, Strawberries etc.

Disc florets

Epicalyx

Falls: the outer drooping petals of the Iris flower.

Female flowers: flowers containing only female parts – ovary, style(s) and stigma(s). Female flowers may occur on the same plant as the male flowers e.g. Hazel, or on separate plants e.g. Willow (p.24).

Falls

Florets: the individual, often rather small, flowers making up a tight flowerhead.

Flower bracts: as defined in this book refer to the 'bract-like' structures, often numerous and overlapping, which surround the base of flowerheads, as in Composites or Daisies. Flower bracts may be small and soft, or large, or spiny as in many Thistles.

Flower bracts

Flowerhead: refers to closely, often tightly grouped, flower clusters, usually terminating the stem(s).

Flower parts refer to the sepals, petals, stamens, and styles which make up the individual flower, although these may not all be present in a given species. There may be additional organs such as nectaries, epicalyx-sepals and so on. The parts present and their numbers are often characteristic of a particular species and may give a good key to identification.

Follicle: a dry 'pod-like' fruit or carpel, solitary or in clusters, opening along one side and containing several to many seeds e.g. *Delphinium* (p.42).

Follicle

Fringe: refers to the cut margin as in the flowers of the Snowbell (p.172).

Fruit: the seed bearing organ of plants, which may be dry or fleshy, single to many seeded, winged or unwinged and may or may not split to release the seeds.

Fringe

Glandular hairs: short or long hairs with a gland at the tip, seen as a swelling or blob, and often giving the plant a sticky feel.

Glandular hairs

Heath: a habitat on acid soils dominated by heathers and related shrubs.

Helmet: the upper petal, or petal-like structure of a flower, which forms a 'cap-like' helmet, as in the Monkshoods (p.40).

Helmet

Hip: brightly coloured false fruit consisting of a fleshy 'urn-like' structure enclosing the seeds and characteristic of Roses.

Hood: refers to the upper petals and/or sepals of certain Orchid flowers which form a small 'hood-like' cap.

Hip

Introduced plants: those plants brought in by man and not native to a particular area. Such plants may become widely naturalised.

Keel

Keel: the lowest 'petal' of a pea flower (p.108) which consists of two petals joined together along their lower margins.

Labiate

Labiate: a collective name used to describe members of the Mint Family, *Labiatae*.

Lance-shaped or **Lanceolate:** spear-shaped, oval but broadest below the middle, and pointed.

Lax: refers to flowers forming a loose, spaced cluster.

Legumes: a collective name used to describe members of the Pea Family, *Leguminosae*.

Linear: refers to long narrow leaves with parallel sides, or almost so.

Lobed: leaves which have lobes and not separate leaflets. The lobes may be toothed or not, deep or shallow, alternate or opposite (*see* Pinnately-lobed).

Lanceolate

Linear

Lobed

Male flowers: flowers containing only male parts – stamens – and may occur on the same or on different plants from the female flowers (*see also* Female flowers p.000).

Marsh: a wet habitat, but not on peat.

Midrib: the central vein of a leaf, often thickened and raised.

Moraine: the rocks and debris deposited along the sides and at the snout of glaciers and consisting of damp compacted material, acid or alkaline, and with little soil. Moraines often have characteristic plants, though seldom rich in species.

Moor: an upland area, usually covered in heather or related plants.

Midrib

Needle

Nectary: small glandular organ in flowers which secrete a sugary liquid or nectar, attractive to various insects.

Needle: slender stiff, often leathery leaves, characteristic of Pines, Firs and related trees.

Net-veined: leaf veins forming an interlacing network, not parallel.

Node: the points on stems where leaves and lateral branches arise.

Node

357

Opposite: leaves arising opposite each other on the stem, thus appearing in pairs.

Opposite

Palmate: leaves lobed like a hand as in the Sycamore.

Palmate

Parasites are plants that live on other plants, have no green pigment (chlorophyll) and are thus entirely dependent on the host plant for their nutrients e.g. Broomrapes. Partial or semi-parasites are only partially dependent on the host plant and have some green pigment, e.g. Cow-wheats.

Pinnate

Peat: deposits of only partially decayed plant material, characteristic of acid moors and heaths, and producing distinct plant communities, often with heathers and related plants dominating.

Perennial plants living for more than two years, sometimes woody, and flowering each year. Most alpine and subalpine plants fall into this category.

Pinnately-lobed

Petals: the innermost floral-leaves of a flower, usually showy and brightly coloured, separate from one another, or joined into a tube and collectively referred to as the corolla. Petals are absent in some species e.g. Salad Burnet.

Pod

Pinnate: a leaf consisting of several opposite pairs of distinct leaflets spaced along a central axis or rachis and with or without an end leaflet. 2-pinnate leaves are further divided into pinnate structures – see figures.

Pinnately-lobed: leaves which have several pairs of shallow or deep lobes, but *not* distinct leaflets.

Raceme

Pod: a fruit, long and cylindrical or somewhat flattened, not fleshy and usually splitting into two equal halves when ripe. A characteristic of the Legumes or Pea Family, *Leguminosae*.

Ray floret

Raceme: a flower spike in which the individual flowers are stalked.

Ray floret: the conspicuous outer flowers of a daisyhead, with a flat 'strap-like' corolla, tubular at the base and often sterile. Sometimes, as in the Dandelion, all the florets in the flowerhead are rayed in which case at least the central ones have anthers and style. In some instances ray florets are absent e.g. Thistles.

Rays

Rays in the context of this book refer to the primary branches of an umbel – the spokes in the Carrot Family, *Umbelliferae* (see p.150). Secondary rays are the spokes of subordinate umbels.

Rhizome: a fleshy-swollen, horizontal, underground stem bearing leaf scars.

Rhombic

Rhombic: roughly diamond-shaped.

Rosette: a cluster of leaves, usually at ground level and radiating from one point. Cushion plants are usually composed of numerous small, tightly packed, rosettes.

Rosette

Runners: slender horizontal above ground stems as in the Wild Strawberry and usually rooting at the nodes to form new plants.

Runners

Scale leaves: small appendages, green or sometimes brown or colourless, replacing leaves where they occur on the stem e.g. Broomrapes (p.222).

Scale leaves

Scree: rock detritus which accumulates at the base of cliffs or steep rocks. Screes are often rather dry, but have characteristic plants especially when they have become stabilised.

Sepals form the outer (lower) ring of floral leaves collectively called the calyx. They are usually small and green, but are sometimes large and conspicuously coloured like the petals, as in the Anemones (p.42).

Sepal tube

Sepal tube refers to the tube formed by sepals which are joined together.

Shrub: deciduous or evergreen woody plants branched from the base.

Species: the basic unit of classification. Species consist of a group of individuals, distinct, but interbreeding freely with one another.

Spike

Spike: a dense elongated group of flowers, often pointed, the individual flowers unstalked. In the context of this book the term is used loosely to cover spike-like racemes in which the flowers are short stalked.

Spine: a sharply pointed branchlet or leaf tip.

Spine

Spoon: when referring to leaf or petal shape indicates the general outline whether or not the surface is convex or concave.

Spreading: standing out horizontally or at a wide angle from the stem, or central point when referring to the petals or sepals of a flower.

Spur

Spur: a hollow tubular or sac-like extension to a petal, or sepal, and often containing nectar.

Stamens: the male organs of a flower consisting of a stalk (the filament) and an anther, which contains the pollen. The number of stamens in a flower is usually characteristic of given species.

Stamens

Standard refers to the upper petal in the pea flower (p.108) or the upper three erect petals in an Iris (p.278).

Starry hairs: hairs which are branched in a star shape, often characteristic of certain plants, and usually only clearly seen with the aid of a hand lens (×10).

Standard

Stigma: the tip(s) of a style which receives the pollen.

Stipules are leaf-like organs, often rather small, located at the base of a leafstalk where it joins with the stem.

Starry hairs

Style: the elongated 'stalk-like' organ linking the ovary to the stigma. A single flower may have one or many styles depending on the species; occasionally absent.

Subshrub: a shrub with a woody base and a herbaceous upper part which dies back each year.

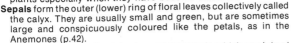

Stipules

Subspecies: the unit of classification immediately below that of species. Subspecies usually differ from one another in one or two characters, such as flower colour or leaf size but inter-breed freely. However, they are normally confined to different geographical locations.

Tendrils: fine twisting filaments forming part of a leaf or stem which enables the plant to climb.

Thorn: sharply pointed straight or curved woody appendage found on the stems or leafstalks of various plants, e.g. Roses.

Toothed: teeth like serrations, along the margins of leaves usually.

Tree: a tall, woody plant with a single basal stem or trunk usually.

Trifoliate: leaves with three distinct leaflets as in the Clovers or Trefoils (p.124).

Two-lipped: corollas or calyces in which the petals or sepals are grouped together into an upper and a lower lobe or lip. Two-lipped flowers are characteristic of the Labiates, *Labiatae*, and related families.

Tendrils

Thorn

Toothed & untoothed

Umbel: a flower cluster in which all the stalks (rays) arise from a central point, like the spokes of a wheel, to give a flat-topped cluster. A characteristic feature of the Carrot Family, *Umbelliferae*.

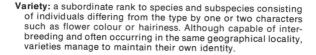

Trifoliate

Variety: a subordinate rank to species and subspecies consisting of individuals differing from the type by one or two characters such as flower colour or hairiness. Although capable of inter-breeding and often occurring in the same geographical locality, varieties manage to maintain their own identity.

Two-lipped

Waste places: places much disturbed by man but not cultivated. Frequently inhabited by weedy species, often annuals.

Whorl: a group of flowers or leaves arising from a central point on a stem.

Whorled spike: a spike composed of whorls of flowers rather than alternate or spirally arranged flowers.

Winged: with a flange or flanges running down the stem as in various Vetches (p.118).

Wings: the lateral two petals of a Pea flower, usually lying on either side of the keel.

Umbel

Winged

Mountain Flowers in Britain

We have already seen (p.7) that as one moves northwards in Europe the zones of true alpine and subalpine vegetation commence at much lower altitudes. At the same time the number of species of plants diminishes rapidly due both to the harsher conditions, but also to the effects of the Quaternary Ice Ages which virtually wiped out the natural flora of the areas covered by glaciation. In Britain and Scandinavia this effect was

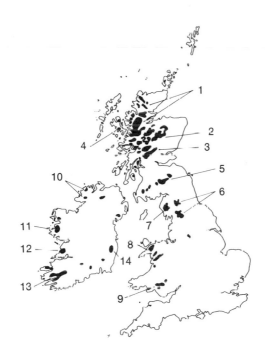

Mountain areas of Great Britain and Ireland where alpine flowers are most abundant.

1 Western Highlands
2 Cairngorm Mountains
3 Central Highlands
4 Mountains of Skye
5 Southern Highlands
6 Pennines
7 Lake District Mountains
8 Snowdonia
9 Black Mountains and Brecon Beacons
10 Donegal Mountains
11 Connacht Mountains
12 Burren
13 Kerry Mountains
14 Wicklow Mountains

most severe, but strangely, isolated pockets of alpine flora survived in both areas, though poorly represented in species numbers. Incomplete glaciation in the Alps, Pyrenees and the other mountains of Central Europe allowed the survival of many species which were quickly able to 'move back' through those mountains after the final retreat of the ice.

The mountain flora of Great Britain is best described as depauperate; it consists of about 130 species in all, of which many are local or rare and none are endemic to the area, though the following can be considered endemic to the Northern European mountain and arctic regions:

Cerastium arcticum	*Koenigia islandica*
Cornus suecica	*Minuartia rubella*
Deschampsia alpina	*Rubus chamaemorus*
Diapensia lapponica	*Sagina intermedia*
Draba norvegica	*Salix lanata*
Erigeron borealis	*Saxifraga cespitosa* & *S. rivularis*

Surprisingly both *Diapensia lapponica* and *Koenigia islandica* were only found comparatively recently in the British Isles.

Besides the northern association there is also a southern link and the following species are found both in Britain and in the Alps and Pyrenees but are excluded from the Scandinavian mountains:

Alchemilla conjuncta	*Oxytropis halleri*
Gentiana verna	*Thaspi alpestre*
Minuartia sedoides	*Viola lutea*
Myosotis alpestris	

Other species, most notably *Dryas octopetala, Pinguicula grandiflora* and *Saxifraga oppositifolia*, occur throughout the mountain areas included in this book.

It is difficult to define exactly the alpine element in the northern and western European countries as the plants involved may descend right down to sea level – the spectacular gentian *(Gentiana verna)* of the Burren in Eire is the classic example. At the same time *Primula farinosa* which is truly alpine in the Alps and Pyrenees never reaches alpine elevations in Britain. The closely related *Primula scotica* is another anomaly for it is restricted to the grassy cliffs in the north-west Scottish Highlands, yet it can be considered a derivative of an alpine species, *P. farinosa*.

Those wishing to study the subject in much greater detail cannot do better than read the New Naturalist book on the subject – *Mountain Flowers* by John Raven and Max Walters.

Conservation

Most mountain plants, particularly the high alpine species, do not take kindly to uprooting or to the English garden. Besides, there are numerous good plant firms which will supply the needs of those keen gardeners who want to grow alpines. Most of these plants are from long established stocks that have been proven in our lowland climates and are therefore the best to grow.

A good deal is said about conservation nowadays and there is a real danger that the subject might be overstressed. It is perfectly true that many species are in danger of extinction, or at least of disappearing in certain areas. Over-collecting, the use of herbicides in the past twenty years and the destruction of habitats have all lead to rarity. National parks and reserves set up in many countries in recent years provide protected areas for some of the more vulnerable plants and animals. There is always the danger, however, that the visitors who turn up in such huge numbers will trample to death that which the park or reserve is intended to preserve. Habitat destruction, however it occurs, leads to a rapid decline in species.

Of course not all the species are equally threatened and it would be absurd to think that every plant in a particular reserve is endangered. However, there are plants, like the beautiful Lady's Slipper Orchid or the mysterious Edelweiss, that have declined alarmingly in the past fifty years purely because they are well known and have been over-picked. Such plants need and deserve careful protection.

Each country has its own laws relating to reserves and parks and many plants are protected by law – to PICK THESE IS AN OFFENCE. Left alone the plants will have a greater chance of survival by being able to set seed and thus reproduce in their natural environment. In this way the beauty of the mountains will be preserved for all to enjoy.

Societies to join

Alpine Garden Society, the Secretary, Lye End Link, St. John's, Woking, Surrey, GU21 1SW. The leading world-wide society for anyone interested in alpine and mountain plants.

Scottish Rock Garden Club, the Honorary Secretary, 48 St. Alban's Road, Edinburgh, EH9 2LU.

Further reading

Alpenflora, by G. Hegi; good illustrations but far from complete.

Alpine Flora, by H. Correvon and P. H. Robert.

Alpine Flowers in Colour, by T. P. Barneby; many colour photographs, but far from complete.

Alpine Plants of Europe, by H. S. Thompson.

Atlas de la Flore Alpine, by H. Correvon.

The English Rock Garden, by R. Farrer; a standard work dealing with the cultivation of alpine, mountain and rock plants.

Europäische und Mediterrane Orchideen, by H. Sundermann; generally good but the nomenclature is muddling.

Fleurs Alpines, by E. & O. Danesch; very incomplete but with superb photographs.

The Wild Flowers of Britain & Northern Europe, by R & A. Fitter & M. Blamey; the companion volume to this book, for Western and Northern Europe, but also for many lowland species outside the scope of this volume.

Flora della Alpi, by L. Fenaroli.

Flora Europaea, edited by T. G. Tutin and others (Cambridge); the standard European Flora, eventually in five volumes – *not* illustrated.

Flora von Mittel-Europa, by G. Hegi; a marvellously detailed work in many volumes.

Flowers of South-West Europe – a field guide, by O. Polunin & B.E. Smythies (Oxford); covers the Pyrenees in part. Designed to be used with *Flowers of Europe*.

Mountain Flowers, by A. Huxley; a very useful guide, mostly illustrated, of many of the alpine and mountain flowers of Europe, including Britain and Scandinavia but not south-east Europe.

Mountain Flowers, by J. E. Raven and M. Walters; although concerned with Great Britain there is a very useful introduction to the subject in Chapter 3.

Orchideen Europas, by E. & O. Danesch; beautifully illustrated guides to European orchids.

Orchideen (Hallwag Taschenbuch) – a pocket guide; excellent small guide with many photographs in colour.

Quarterly Bulletins, Alpine Garden Society; many authors. Invaluable articles on alpine and mountain plants and places visited in Europe. Very useful for those wishing to visit mountain areas in Europe. Many illustrations.

Orchids of Europe, A. Duperrex; one of the best and most concise pocket guides to European Orchids. Poorly illustrated.

Index of English Names

Aaron's Rod, 206
Aconite, Winter, 38
Adenostyles, 248
Aethionema, 68
Agrimony, 96
  Bastard, 96
  Fragrant, 96
Alder, 28
  Green, 28
  Grey, 28
Alkanet, 188
Alpenrose, 164
  Dwarf, 164
  Hairy, 164
Alpine Bells, 172
Alyssoides, 64
Alyssum, Alpine, 304
  Diffuse, 304
  Italian, 304
  Mountain, 64
  Pyrenean, 64
  Wulfen's, 64
Amelanchier, 106
Ancient King, The, 88
Andromeda, Marsh, 164
Anemone, Blue Wood, 42
  Monte Baldo, 42
  Narcissus-flowered, 42
  Wood, 42
  Yellow, 42
Angelica, Pyrenean, 156
  Wild, 158
Aphyllanthes, Blue, 264
Aposeris, 262
Arabis, Blue, 64
Archangel, Balm-leaved, 196
  Yellow, 196
Arnica, 248
Asarabacca, 288
Ash, 310
  Manna, 310
  Mountain, 106
Asparagus, Bath, 272
Aspen, 26
Asphodel, 268
  Scottish, 264
  Tofields, 264
Aster, Alpine, 240
  False, 240
  Pyrenean, 240
  Three-veined, 240
Athamanta, 154
Auricula, 168
Avens, Alpine, 96
  Creeping, 96
  Mountain, 96
  Pyrenean, 96
  Water, 96
Azalea, Creeping, 164

Baldmoney, 154
Balm, Bastard, 198
Balsam, Small, 138
Baneberry, 40

Barberry, 56
Barrenwort, 56
Bartsia, Alpine, 212
  Red, 212
Basil-thyme, 200
Bastard Toadflax, Alpine, 288
  Bavarian, 288
  Branched, 288
  Pyronoan, 288
Bearberry, 164
  Alpine, 164
Bearsear, 168
Bedstraw, Coniferous, 312
  Cushion, 314
  Fen, 312
  Hedge, 312
  Lady's, 324
  Marsh, 312
  Northern, 312
  Reddish, 314
  Swiss, 224
  Wood, 312
Beech, 20
Bellflower, Alpine, 232
  Arctic, 234
  Bearded, 232
  Clustered, 234
  Cottian, 236
  Creeping, 232
  Crimped, 232
  Flax-leaved, 234
  French, 236
  Giant, 234
  Jaubert's, 236
  Large-flowered, 232
  Mt. Cenis, 232
  Nettle-leaved, 232
  Paniculate, 236
  Peach-leaved, 232
  Perforate, 236
  Pyrenean, 232
  Rainer's, 234
  Rampion, 232
  Rock, 234
  Solitary, 236
  Spanish, 236
  Spiked, 234
  Spreading, 232
  Yellow, 234
Betony, 198
  Alpine, 198
  Yellow, 198
Bilberry, 162
Bindweed, Field, 174
Birch, Downy, 28
  Dwarf, 28
  Silver, 28
Birdsnest, Orchid, 286
  Yellow, 160
Bistort, 30
  Alpine, 30
  Yellow, 160
Bittercress, Alpine, 62
  Asarum-leaved, 62

Coral-root, 64
  Drooping, 62
  Ivy-leaved, 64
  Kitaibel's, 62
  Large, 62
  Mignonette-leaved, 64
  Radish-leaved, 62
  Seven-leaflet, 62
  Trifoliate, 62
Bittersweet, 316
Blackberry, 90
Blackcurrant, 78
Blackthorn, 108
Bladder Senna, 112
Bladderwort, Lesser, 220
Blue-eyed Mary, 190
Bogbean, 176
Borage, 184
Box, 308
Box Thorn, 316
Bramble, 90
  Arctic, 90
  Rock, 90
Braya, Alpine, 60
Briar, Field, 94
  Sweet, 94
  White-stemmed, 94
Broadbean, 118
Brooklime, 210
Brookweed, 174
Broom, Ardoin, 108
  Black, 122
  Hairy, 110
  Hedgehog, 112
  Lugano, 108
  Purple, 110
  Pyrenean, 108
  Silvery, 110
  Spanish, 110
Broomrape, Alsace, 222
  Amethyst, 222
  Clover-scented, 222
  Common, 222
  Germander, 222
  Knapweed, 222
  Mugwort, 222
  Sage, 222
  Sand, 222
  Sermountain, 222
  Slender, 222
  Thistle, 222
  Thyme, 222
  Yellow, 222
Bryony, Black, 274
Buckler, Mustard, 72
  Chicory-leaved, 72
  Rosetted, 72
  Scapose, 72
Buckthorn, 308
  Alder, 308
  Alpine, 308
  Dwarf, 308
  Rock, 308
  Sea, 138

Bugle, Common, 192
  Pyramidal, 192
  Tenore's, 192
Bugloss, 188
  Viper's, 186
Bulbocodium, Spring, 264
Bunium, 152
Burnet, Great, 96
  Italian, 96
  Salad, 96
Butterbur, 248
  Alpine, 248
  White, 248
Buttercup, Aconite-leaved, 50
  Alpine, 50
  Amplexicaule, 52
  Bulbous, 50
  Carinthian, 48
  Crenate, 50
  Glacier, 52
  Goldilock's, 50
  Gouan's, 48
  Grenier's, 48
  Hooked, 48
  Hybrid, 50
  Meadow, 48
  Mountain, 48
  Multiflowered, 48
  Parnassus-leaved, 52
  Pygmy, 50
  Pyrenean, 52
  Séquier's, 50
  Thore's, 50
  Wood, 48
  Woolly, 48
Butterwort, Alpine, 220
  Common, 220
  Large-flowered, 220
  Long-leaved, 220
  Southern, 220

Cabbage, Alpine, 70
  Wallflower, 72
Calamint, Alpine, 200
  Large-flowered, 200
  Lesser, 200
  Wood, 200
Callianthemum, 38
Calycocorsus, 260
Camomile, Corn, 242
Campion, Bladder, 32
  Moss, 32
  Red, 335
  Rock, 32
Candytuft, Annual, 70
  Dauphine, 70
  Spoon-leaved, 70
Caraway, 156
Cassiope, 162
  Matted, 162
Catchfly, Alpine, 32
  Eared, 335
  Heart-leaved, 32
  Large-flowered, 32
  Narrow-leaved, 32
  Northern, 32
  Nottingham, 32

  Pyrenean, 32
  Spanish, 32
  Sticky, 30
  Tufted, 32
  Valais, 32
Catmint, Broad-leaved, 196
  Common, 196
  Lesser, 198
Catsear, Giant, 260
  Spotted, 260
Catsfoot, 242
  Alpine, 242
  Carpathian, 242
Celandine, Greater, 58
Centaury, Common, 182
  Lesser, 182
Chaenorhinum, 204
Chaffweed, 174
Cherry, Bird, 108
  Cornelian, 150
  St. Lucie's, 108
  Wild, 108
Chervil, Hairy, 152
Chickweed, Common, 292
Chives, 266
Cinquefoil,
  Alchemilla-leaved, 100
  Alpine, 100
  Carnic, 102
  Creamy, 102
  Creeping, 100
  Dwarf, 100
  Eastern, 100
  Golden, 100
  Grey, 100
  Hoary, 98
  Large-flowered, 100
  Marsh, 98
  Pennsylvanian, 98
  Pink, 102
  Pyrenean, 100
  Rock, 98
  Shrubby, 98
  Snowy, 98
  Tufted, 102
Clary, Meadow, 194
  Whorled, 194
Clematis, Alpine, 46
  Erect, 46
  Simple-leaved, 46
Cloudberry, 90
Clover, Alpine, 124
  Alsike, 124
  Brown, 124
  Cream, 124
  Crimson, 124
  Dutch, 124
  Large Brown, 124
  Mountain, 124
  Pale, 124
  Red, 124
  Sulphur, 126
  White, 124
  Zigzag, 124
Coltsfoot, 248
  Alpine, 248
Columbine, Alpine, 54
  Bertoloni's, 54

  Common, 54
  Dark, 54
  Einsel's, 54
  Pyrenean, 54
Comfrey, Common, 188
Conopodium, 152
Coris, Yellow, 142
Corn Cockle, 30
Cornel, Dwarf, 150
Cornflower, 258
  Mountain, 258
Corydalis, Bulbous, 56
  Solid-tubered, 56
  Yellow, 56
Cotoneaster, Wild, 104
Cotton-thistle, 250
  Stemless, 250
Cowberry, 162
Cowslip, 166
Cow-wheat, Common, 210
  Crested, 210
  Field, 210
  Wood, 210
Crab, Wild, 106
Cranberry, 162
Cranesbill, Ashy, 130
  Bloody, 130
  Dovesfoot, 132
  Dusky, 130
  Knotted, 130
  Marsh, 130
  Meadow, 130
  Pyrenean, 132
  Rock, 130
  Spreading, 130
  Western, 130
  Wood, 130
Creeping Jenny, 174
Cress, Chamois, 68
Crocus, Autumn, 264
  Leafless, 278
  Purple, 278
  Riviera, 278
  White, 278
Crowberry, 162
Crowfoot, Glacier, 52
Cuckoo Flower, 64
Cuckoo Pint, 274
Cudweed, Dwarf, 242
  Highland, 242
  Wood, 242
Currant, Black, 78
  Mountain, 78
  Red, 78
  Rock Red, 78

Daffodil, 276
  Lesser Wild, 276
  Wild, 276
Daisy, 240
  Ox-eye, 244
Dame's Violet, 62
  Alpine, 62
Dandelion, Apennine, 330
  Alpine, 330
  Brenner, 328
  Broad-leaved, 330
  Brownish, 330

Cut-leaved, 330
Dark, 330
Glacier, 328
Hoppe's, 330
Marsh, 330
Shröter's, 328
Deadnettle, Henbit, 196
Red, 196
Spotted, 196
White, 196
Delphinium, 42
Dethawia, 154
Devil's Claw, 228
Dewberry, 90
Diapensia, 162
Dipcadi, 272
Dock, Apennine, 290
Mountain, 290
Northern, 290
Rubble, 290
Snow, 290
Dodder, Common, 174
Greater, 337
Dog-daisy, Carpathian, 242
Mountain, 242
Southern, 242
Dogwood, Common, 150
Dragonhead, Northern, 200
Dragonmouth, 204
Dropwort, 90
Dutchman's Pipe, 160

Echinospartum, 110
Edelweiss, 242
Edraianthus, 230
Elder, Alpine, 226
Common, 226
Dwarf, 226
Red-berried, 226
Endressia, 156
Ephedra, 22
Eryngo, Alpine, 152
Pyrenean, 152
Silver, 152
Everlasting Pea,
Broad-leaved, 120
Eyebright, Alpine, 212
Common, 212
Dwarf, 212
Eastern, 212
Glossy, 212
Irish, 212
Wind, 212

Fairy's Thimble, 236
Felwort, 182
Marsh, 182
Figwort, Alpine, 318
Common, 318
Green, 318
Italian, 318
Pyrenean, 318
Water, 318
Yellow, 318
Fir, Silver, 22
Flax, 337
Perennial, 136
Purging, 136

Pyrenean, 136
Sticky, 136
Yellow, 136
Fleabane, Alpine, 240
Blue, 240
Greek, 240
Neglected, 242
One-flowered, 242
Variable, 240
Fleawort, Field, 326
Forget-me-not,
Alpine Wood, 188
Bur, 190
Changing, 188
Field, 188
Lammotte's, 188
Tufted, 188
Water, 188
Wood, 188
Fox and Cubs, 262
Foxglove, Fairy, 204
Large Yellow, 208
Small Yellow, 208
Fritillary, Tyrolean, 268
Fumitory, Common, 56

Gagea, Hairy, 264
Pyrenean, 264
Yellow, 264
Garland flower, 140
Garlic, Crow, 266
Field, 266
Gean, 108
Genipi, 324
Yellow, 324
Gentian, Bavarian, 180
Bladder, 180
Brown, 176
Clusius's, 178
Cross, 178
Field, 180
German, 182
Great Yellow, 176
Karawanken, 178
Marsh, 176
Prostrate, 178
Purple, 176
Pyrenean, 178
Pyrenean Trumpet, 178
Short-leaved, 180
Slender, 180
Snow, 180
Southern, 178
Spotted, 176
Spring, 180
Styrian, 178
Triglav, 180
Trumpet, 178
Willow-leaved, 176
Gentianella, Apennean, 180
Germander, Mountain, 190
Pyrenean, 190
Wall, 190
Globeflower, 38
Globularia, Apennine, 220
Common, 220
Leafless-stemmed, 220
Matted, 220

Goatsbeard, 262
Goatsbeard Spiraea, 90
Golden Drop, 184
Pyrenean, 184
Swiss, 184
Golden Rod, 240
Gooseberry, Wild, 78
Gorse, 112
Grafia, 154
Grass of Parnassus, 78
Greenweed, Dyer's, 110
German, 110
Southern, 110
Winged, 110
Gromwell, 184
Blue, 184
Corn, 184
Shrubby, 184
Ground Ivy, 200
Ground Pine, 192
Groundsel, Common, 326
Grey Alpine, 328
One-flowered
Alpine, 328
Guelder Rose, 226
Gypsophila, Alpine, 34

Hacquetia, 150
Harebell, 234
Hare's-ear, 156
Long-leaved, 156
Pyrenean, 156
Rock, 156
Sickle, 156
Three-veined, 156
Hawksbeard, 332
Alpine, 332
Golden, 334
Marsh, 332
Mountain, 332
Northern, 332
Pink, 332
Pygmy, 334
Pyrenean, 332
Triglav, 334
Hawkbit, Alpine, 338
Mountain, 262
Pyrenean, 262
Swiss, 338
Hawkweed, Alpine, 262
Dwarf, 262
Mouse-ear, 338
Orange, 262
Woolly, 262
Hawthorn, 106
Hazel, 28
Heath, Cornish, 164
Cross-leaved, 164
Mountain, 164
Spring, 164
Heather, Bell, 164
Heartsease, 146
Hellebore, Green, 38
Stinking, 38
Helleborine,
Black False, 260
Broad-leaved, 284
Dark Red, 284

Marsh, 339
Red, 286
Violet, 339
White, 286
White False, 268
Hemp-nettle, Common, 194
Downy, 192
Hairy, 194
Large-flowered, 192
Pyrenean, 192
Red, 192
Henbane, 316
Hepatica, 44
Herb Bennet, 96
Herb Paris, 274
Robert, 132
Hogweed, 158
Austrian, 158
Holly, 339
Honesty, 64
Perennial, 64
Honeysuckle, Alpine, 224
Black-berried, 224
Blue-berried, 224
Common, 224
Fly, 224
Pyrenean, 224
Honeywort, Lesser, 186
Smooth, 186
Horehound, Black, 196
Shrubby, 196
Hound's Tongue, 190
Houseleek, Catalonian, 335
Cobweb, 74
Common, 74
Dolomitic, 74
Hen-and-chickens, 76
Large-flowered, 74
Limestone, 74
Mountain, 74
Wulfen's, 74
Hyacinth, Grape, 272
Pyrenean, 272
Tassel, 272
Hymenolobus, 68
Hyssop, 202

Iris, English, 278
Flag, 278
Garden, 278
Grassy-leaved, 278
Stinking, 278
Variegated, 278
Yellow, 278
Isopyrum, Rue-leaved, 40
Ivy, Common, 150
Ground, 200

Jacob's Ladder, 174
Juniper, Common, 22
Dwarf, 22
French, 22
Jupiter's Distaff, 194
Jurinea, 250

Kernera, 68
Kidney-vetch, Common, 126
Mountain, 126

King of the Alps, 190
Knapweed, 256
Alpine, 258
Austrian, 256
Bluish, 258
Cardoon, 260
Doubtful, 258
Giant, 260
Paniculate, 258
Plume, 258
Rhaetian, 258
Rock, 256
South-eastern, 256
Wig, 258
Knawel, 292
Knotgrass, 30
Alpine, 30

Laburnum, 108
Lady's Mantle, 104
Alpine, 102
Cut-leaved, 102
Deceptive, 104
Hairless, 104
Hoppe's, 104
Intermediate, 104
Rock, 102
Small, 104
Lady's Smock, 64
Lady's Tresses, Autumn, 286
Creeping, 286
Summer, 286
Larch, 22
Larkspur, Alpine, 42
Mountain, 42
Laurel, Spurge, 140
Lavender, 204
Lavender Cotton, 244
Leek, Alpine, 266
Round-headed, 266
Lentil, Mountain, 112
Wild, 112
Leopardsbane, 337
Austrian, 248
Heart-leaved, 248
Large-flowered, 248
Lepidium, 72
Lettuce, Blue, 260
Edible, 260
Mountain, 260
Purple, 260
Lily, Carnic, 270
Lent, 276
Martagon, 270
May, 272
Orange, 270
Red, 270
Snakeshead, 268
Snowdon, 273
St. Bernard's, 264
St. Bruno, 264
Yellow Turk's-cap, 270
Lily of the Valley, 274
Lime, Large-leaved, 314
Small-leaved, 314
Ling, 164
Liquorice, Wild, 114
Lloydia, 272

Lovage, 158
Alpine, 158
Unbranched, 158
Love-in-a-mist, 38
Lomatogonium, 182
Lord and Ladies, 274
Lousewort, Ascending, 216
Beaked, 216
Beakless Red, 214
Common, 214
Crested, 214
Crimson-tipped, 214
Fern-leaved, 216
Flesh-pink, 216
Lapland, 214
Leafy, 214
Long-beaked Yellow, 214
Marsh, 214
Mt. Cenis, 216
Pink, 214
Pyrenean, 216
Stemless, 214
Tufted, 216
Verticillate, 214
Lucerne, 122
Lungwort, Common, 186
Long-leaved, 186
Mountain, 186
Narrow-leaved, 186
Styrian, 186
Lychnis, Alpine, 30
Purple, 30

Madwort, 188
Mallow, Common, 138
Cut-leaved, 138
Dwarf, 138
Maple, Field, 306
Italian, 306
Norway, 306
Marigold, Marsh, 40
Marjoram, 202
Masterwort, 158
Bavarian, 150
Great, 150
Lesser, 150
Southern, 158
May, 106
Meadow-rue, Alpine, 54
Great, 54
Large-fruited, 54
Lesser, 52
Small, 52
Stinking, 52
Meadowsweet, 90
Medick, Black, 122
Pyrenean, 122
Sprawling, 122
Medlar, False, 106
Melilot, Common, 122
Tall, 122
White, 122
Mercury, Annual, 134
Dog's, 134
Merendera, 264
Mezereon, 140
Alpine, 140
Rock, 140

Michelmas Daisy,
European, 240
Micromeria, 200
Mignonette, Corn, 72
Pyrenean, 72
Milfoil, Andorran, 246
Dwarf, 244
Large-leaved, 246
Musk, 244
Silvery, 246
Simple-leaved, 246
Tansy, 246
Milk-vetch,
Alpine, 112
Austrian, 114
Central-Alps, 114
Foucaud's, 116
Inflated, 116
Meadow, 116
Mountain, 116
Northern, 116
Norwegian, 114
Pallid, 112
Purple, 112
Samnitic, 116
Silky, 118
Southern, 114
Sprawling, 114
Stemless, 114
Stinking, 118
Tyrolean, 114
Woolly, 118
Yellow, 116
Milkwort, Alpine, 136
Bitter, 136
Mountain, 136
Pyrenean, 136
Shrubby, 136
Thyme-leaved, 136
Tufted, 136
Mint, Corn, 202
Horse, 202
Spear, 202
Water, 202
Mistletoe, 288
Monkshood, Common, 40
Variegated, 40
Yellow, 40
Molopospermum, 152
Moltkia, 186
Monk's Rhubarb, 290
Moon Daisy, 244
Alpine, 244
Saw-leaved, 244
Moschatel, 226
Mossy Cyphel, 298
Mouse-ear, Arctic, 294
Alpine, 294
Bell-flowered, 296
Broad-leaved, 294
Carinthian, 296
Common, 296
Dwarf, 296
Field, 294
Glacier, 294
Grey, 296
Italian, 294
Julian, 294

Little, 296
Narrow-leaved, 294
Sea, 296
Slovenian, 296
Starwort, 294
Sticky, 296
Mousetail, 46
Mucizonia, 74
Mudwort, 318
Mullein, Dark, 206
Hoary, 206
Lanate, 206
Orange, 206
Murbeckiella, 60
Myricaria, 142

Narcissus,
Pheasant's-eye, 276
Poets, 276
Rock, 276
Rush-leaved, 276
Nettle, Annual, 288
Common, 288
Nightshade, Alpine, 148
Black, 316
Deadly, 316
Alpine Enchanter's, 148

Oak, Downy, 28
Durmast, 28
Pedunculate, 28
Pyrenean, 28
Sessile, 28
White, 28
Odontites, Sticky, 212
Yellow, 212
Onion, Mountain, 266
Narcissus-flowered, 266
Rock, 266
Strap-leaved, 266
Yellow, 266
Orchid, Bee, 280
Birdsnest, 286
Black Vanilla, 282
Bog, 280
Broad-leaved Marsh, 282
Bug, 280
Burnt, 282
Common Spotted, 282
Coralroot, 286
Early Marsh, 282
Early Purple, 280
Early Spider, 280
Elder-flowered, 282
False, 284
Flecked Marsh, 338
Fly, 338
Fragrant, 284
Frog, 284
Ghost, 286
Greater Butterfly, 286
Green-winged, 280
Heath Spotted, 338
Lady, 282
Lady's Slipper, 280
Late Spider, 280
Lesser Butterfly, 286
Lizard, 284

Loose-flowered, 280
Man, 282
Military, 280
Monkey, 282
Musk, 284
Pale-flowered, 280
Provence, 280
Pyramidal, 284
Rosy Vanilla, 338
Round-headed, 286
Small White, 284
Soldier, 280
Spidor, 280
Toothed, 282
Vanilla, 282
Violet Birdsnest, 284
Orpine, 76
Oxalis Yellow, 128
Oxlip, 166
Ox-eye Daisy, 244
Yellow, 218

Paederota, Bluish, 210
Paeony, 56
Pansy, Alpine, 146
Bertoloni's, 146
Duby's, 146
Horned, 146
Long-spurred, 146
Maritime Alps, 146
Mountain, 146
Mt. Cenis 146
Pasque Flower, Alpine, 44
Common, 44
Small, 44
Spring, 44
White, 44
Pea, Chickling, 120
Yellow, 120
Pear, Cultivated, 106
Southern, 106
Wild, 106
Pearlwort, Alpine, 292
Cushion, 292
Knotted, 292
Procumbent, 292
Pennycress, 70
Alpine, 70
Apennean, 70
Early, 70
Mountain, 70
Round-leaved, 70
Small-flowered, 70
Pennyroyal, 202
Periwinkle, Lesser, 182
Petrocoptis, 30
Pheasant's-eye, 46
Apennine, 46
Pyrenean, 46
Pimpernel, Bog, 174
Yellow, 174
Pine, Arolla, 22
Austrian, 22
Black, 22
Mountain, 22
Scots, 22
Pink, Alpine, 36
Carthusian, 36

369

Cheddar, 36
Common, 36
Deptford, 36
Fringed, 34
Glacier, 36
Maiden, 36
Painted, 34
Pyrenean, 36
Sequier's, 34
Short, 36
Tall, 36
Three-veined, 36
Wood, 36
Plantain, Alpine, 322
Dark, 322
Fleshy, 322
Greater, 322
Hoary, 322
Ribwort, 322
Pleurospermum, 154
Plum, Marmot, 108
Poplar, Black, 26
White, 26
Poppy, Alpine, 58
Arctic, 58
Common, 58
Corn, 58
Long-headed, 58
Prickly, 58
Pyrenean, 58
Rhaetian, 58
Welsh, 58
Potentilla, Cut-leaved, 98
Lax, 100
Thuringian, 100
Primrose, 166
Allioni's, 168
Birdseye, 166
Entire-leaved, 168
Glaucous, 166
Least, 166
Long-flowered, 166
Marginate, 168
Northern, 168
Piedmont, 168
Scottish, 168
Spectacular, 166
Sticky, 166
Villous, 168
Viscid, 168
Wulfen's, 166
Purslane, 000

Queen of the Alps, 152

Ragwort, Alpine, 326
Chamois, 326
Pinnate-leaved, 326
Rock, 326
Southern, 326
Tournefort's, 326
Wood, 328
Ramonda, 218
Rampion, Betony-leaved, 228
Black, 230
Dark, 337
Dwarf, 230
Globe-headed, 230

Horned, 230
Maritime, 230
Rhaetian, 230
Round-headed, 230
Rosette-leaved, 230
Scorzonera-leaved, 228
Spiked, 228
Ramsons, 266
Ragged Robin, 30
Raspberry, 90
Rattle, Apennine, 218
Aristate, 218
Burnat's Yellow, 218
Greater Yellow, 218
Narrow-leaved, 218
Red, 214
Southern Yellow, 218
Yellow, 218
Redcurrant, 78
Rock, 78
Restharrow, Mt. Cenis, 122
Round-leaved, 122
Shrubby, 122
Spiny, 122
Yellow, 122
Rhododendron, Lapland, 162
Rhynchosinapis, 72
Rockcress, Alpine, 302
Annual, 302
Bristol, 302
Cevenne, 302
Compact, 302
Corymbose, 302
Dwarf, 302
Hairy, 302
Scopoli's, 302
Soyer's, 302
Tall, 64
Rocket, Austrian, 60
False London, 60
London, 60
Tansy-leaved, 60
Rock-jasmine, Alpine, 170
Annual, 172
Blunt-leaved, 170
Ciliate, 170
Cylindric, 170
Elongated, 172
Hairy, 170
Mathilda's, 170
Milkwhite, 170
Northern, 172
Pink, 170
Pyrenean, 170
Swiss, 170
Rockrose, Alpine, 140
Apennine, 140
Common, 140
Hoary, 140
Shrubby, 140
White, 140
Rose, Alpine, 92
Apple, 94
Blue-leaved, 92
Blunt-leaved Dog, 94
Burnet, 92
Christmas, 38
Cinnamon, 92

Dog, 94
Downy, 94
Field, 92
Mountain, 92
Provence, 92
Styled, 92
Rosemary, Bog, 164
Roseroot, 78
Rowan, 106

Sage, Sticky, 194
Salvia, Wind, 194
Sainfoin, Alpine, 128
Mountain, 128
Rock, 128
Silvery, 128
Small, 128
White, 128
Sandwort, Apennean, 298
Austrian, 298
Bergamasque, 298
Cadi, 300
Carnic, 300
Ciliate-leaved, 300
Creeping, 300
Imbricate, 300
Large-flowered, 300
Mossy, 300
Narrow-leaved, 300
Northern, 298
Norwegian, 300
Pink, 300
Rock, 298
Sickle-leaved, 298
Two-flowered, 300
Vernal, 298
Sanicle, 150
Mountain, 150
Sarcocapnos, 56
Saussurea, Alpine, 250
Dwarf, 250
Heart-leaved, 250
Savin, 335
Sawwort, 260
Single-flowered, 260
Saxifrage, Awl-leaved, 82
Arctic, 336
Bergamasque, 82
Biennial, 80
Blue, 84
Bulbous, 84
Burnet, 154
Cobweb, 80
Columnar, 84
Drooping, 84
Eastern, 82
Encrusted, 86
Fragile, 80
French, 80
Geranium-like, 82
Greater Burnet, 154
Hairless mossy, 82
Hairy, 84
Hawkweed, 80
Host's, 86
Livelong, 88
Marsh, 80
Meadow, 84

Mossy, 336
Musky, 82
Neglected, 82
One-flowered, 84
Orange, 88
Paniculate, 88
Piedmont, 82
Purple, 84
Pyramidal, 86
Pyrenean, 88
Reddish, 86
Retuse-leaved, 84
Rough, 80
Round-leaved, 80
Rue-leaved, 80
Scented-leaved, 82
Scree, 82
Slender, 326
Spoon-leaved, 80
Starry, 80
Thick-leaved, 86
Two-flowered, 84
Water, 82
White Musky, 82
Wood, 80
Yellow, 84
Yellow Mountain, 82
Scabious, Alpine, 238
Devil's Bit, 238
Pyrenean, 238
Shining, 238
Tyrolean, 238
Wood, 238
Scopolia, 316
Scurvy-grass, Alpine, 68
Common, 68
Self-heal, 200
Cut-leaved, 200
Large, 200
Senna, Bladder, 112
False, 126
Serapias, 282
Sermountain, 158
Sesamoides, 72
Seseli, Pyrenean, 154
Sheepsbit, Dwarf, 228
Mountain, 228
Shepherd's Purse, 304
Sibbaldia, 102
Sideritis, 192
Silverweed, 98
Skullcap, Alpine, 190
Sloe, 108
Snakeshead, Pyrenean, 268
Slender-leaved, 268
Three-bracted, 268
Tyrolean, 268
Snapdragon, Common, 204
Creeping, 204
Rock, 204
Soft, 204
Sneezewort, 246
Alpine, 244
Cream-flowered, 246
Dark-stemmed, 244
Pyrenean, 246
Snowbell, Alpine, 172
Austrian, 172

Dwarf, 172
Hungarian, 172
Least, 172
Mountain, 172
Pyrenean, 172
Snowdrop, 276
Snowflake, Spring, 276
Summer, 276
Snow in Summer, 294
Soapwort, Dwarf, 34
Rock, 34
Spoon-leaved, 34
Tufted, 34
Yellow, 34
Solenanthus, 190
Solomon's Seal, Common, 274
Scented, 274
Whorled, 274
Sorrel, Common, 290
French, 290
Mountain, 30
Sheep's, 290
Sowbread, 172
Sow-thistle, Alpine, 260
Spearwort, Creeping, 52
Greater, 52
Lesser, 52
Speedwell, Alpine, 208
Common, 320
Field, 320
Germander, 210
Heath, 320
Ivy-leaved, 320
Large, 208
Leafless-stemmed, 210
Long-leaved, 320
Marsh, 320
Pyrenean, 208
Rock, 208
Spanish, 208
Spiked, 210
Spring, 320
Stemless, 320
Thyme-leaved, 208
Violet, 208
Wall, 320
Water, 320
Wood, 320
Spignel, 154
Spikenard, Celtic, 226
Spindle-tree, 304
Alpine, 304
Spiraea, Elm-leaved, 90
Hairy, 90
Spruce, Norway, 22
Spurge, Blue, 134
Carnian, 134
Cypress, 134
Glaucous, 134
Irish, 134
Pyrenean, 134
Rock, 134
Squill, Alpine, 272
Pyrenean, 272
Spring, 272
St. John's Wort,
Alpine, 142

Imperforate, 142
Mountain, 142
Perforate, 142
Trailing, 142
Western, 142
Star of Bethlehem, 272
Starwort, Abortive, 310
Water, 310
Stitchwort, Bog, 292
Greater, 292
Lesser, 292
Long-leaved, 292
Wood, 292
Stock, Sad, 62
Stonecrop, Annual, 78
Biting, 76
Chickweed, 76
Creamish, 76
Dark, 78
English, 76
Hairy, 78
Pink, 78
Reddish, 76
Rock, 76
Thick-leaved, 76
White, 76
Whorled-leaved, 76
Storksbill, Alpine, 132
Common, 132
Large Purple, 132
Rock, 132
Strawberry, Hautbois, 102
Wild, 102
Streptopus, 274
Sundew, Common, 74
Long-leaved, 74
Swallow-wort, 184
Sweet Cicely, 152
Sweet William, 34
Sycamore, 306

Tare, Hairy, 118
Thistle, Acanthus-leaved
Carline, 250
Carnic, 256
Alpine, 252
Apennean, 252
Brook, 254
Carline, 250
Carline-leaved, 252
Carnic, 256
Common Carline, 250
Corymbose, 254
Cotton, 250
Great Marsh, 252
Melancholy, 254
Musk, 252
Pale Yellow, 256
Pyrenean, 254
Southeastern, 252
Spiniest, 256
Stemless, 254
Stemless Carline, 250
Stemless Cotton, 250
Waldstein's, 254
Welted, 252
Woolly, 254
Yellow Melancholy, 256

Thrift, Haller's, 174
  Mountain, 174
  Plantain-leaved, 174
Thyme, Glabrescent, 202
  Hairy, 202
  Larger Wild, 202
  Wild, 202
Thymelaea, Annual, 138
  Hairy, 138
  Twisted, 138
Toadflax, Alpine, 204
  Bergamasque, 204
  Common, 204
  Ivy-leaved, 204
  Pyrenean, 204
  Striped, 204
Tolpis, 262
Toothwort, 218
Tormentil, 100
Touch-me-not, 138
Tozzia, 210
Tragacanth, Mountain, 114
Traveller's Joy, 46
Treacle-Mustard,
  Decumbent, 60
  Hawkweed-leaved, 60
  Hoary, 60
  Wood, 60
Trefoil,
  Alpine Birdsfoot, 126
  Birdsfoot, 126
  Hungarian, 126
  Red, 126
Trochiscanthes, 154
Tulip, Wild, 268
Tunic Flower, 34
Twayblade, Common, 286
  Lesser, 286
Twin Flower, 224

Valerian, Common, 226
  Dwarf, 228
  Elongated, 228
  Entire-leaved, 226
  Globularia-leaved, 226
  Marsh, 226
  Narrow-leaved, 228
  Pyrenean, 226
  Rock, 228
  Three-leaved, 226
Veronica, Yellow, 210
Vervain, 190
Vetch, Bitter, 120
  Bush, 118
  Brown, 120
  Cirrhose, 120
  Common, 118
  False, 116
  Felted, 120
  Horseshoe, 126
  Pale, 118
  Purple, 112
  Pyrenean, 118
  Scorpion, 126
  Silvery, 118
  Slender, 120
  Small Scorpion, 126
  Tufted, 118

Wood, 118
Vetchling, Meadow, 120
  Spring, 120
  Yellow, 120
Violet, Austrian, 144
  Bog, 144
  Common Dog, 144
  Dog's Tooth, 270
  Early Dog, 144
  Finger-leaved, 146
  Hairy, 144
  Heath Dog, 144
  Hill, 144
  Pyrenean, 144
  Sweet, 144
  Teesdale, 144
  Water, 172
  White, 144
  Yellow Wood, 144
Vipergrass, Austrian, 338
  Bearded, 262
Vitaliana, 168

Waldsteinia, 98
Wallpepper, 76
Water Lily, Fringed, 176
Wayfaring Tree, 226
Weld, 72
Whitebeam, 106
Whitlow-grass, Austrian, 66
  Bald White, 66
  Carinthian, 66
  Engadine, 66
  Pyrenean, 68
  Sauter's, 66
  Spring, 68
  Starry, 66
  Twisted, 66
  Wall, 66
  Woolly-fruited, 66
  Yellow, 66
Whortleberry, 162
  Bog, 162
Willow, Alpine, 24
  Apuan, 24
  Austrian, 24
  Blue-leaved, 26
  Finely-toothed, 24
  Hairless, 24
  Lagger's 26
  Lapland, 26
  Large-stipuled, 26
  Least, 24
  Mountain, 26
  Netted, 24
  Polar, 24
  Pyrenean, 24
  Retuse-leaved, 24
  Silky, 24
  Swiss, 26
  Woolly, 26
Willowherb, Alpine, 148
  Chickweed, 148
  Greater, 148
  Mountain, 148
  Nodding, 148
  Pimpernel-leaved, 148
  Rosebay, 148

  Western, 148
  Whorled-leaved, 148
Windflower, Snowdrop, 42
Wintergreen, Chickweed, 174
  Intermediate, 160
  Lesser, 160
  Nodding, 160
  One-flowered, 160
  Pale-green, 160
  Round-leaved, 160
  Umbellate, 160
Woad, 326
  Alpine, 60
Wolfsbane, 40
  Northern, 40
Woodruff, 312
  Blue, 224
  Pyrenean, 224
  Six-leaved, 224
  Southern, 224
Wood-sorrel, 128
Wormwood, Arctic, 324
  Dark Alpine, 324
  Digitate-leaved, 324
  Field, 324
  Glacier, 324
  Narrow-leaved, 324
  Norwegian, 324
  Pyrenean, 324
  Sea, 324
Woundwort, Alpine, 198
  Downy, 198
  Hedge, 198
  Marsh, 198
  Yellow, 198
Wulfenia, 208

Xatardia, 154

Yam, Pyrenean, 274
Yellow-wort, 182
Yew, 22

Index of Scientific Names

Trinominal names indicate subspecies, for example *Achillea oxyloba mucronulata* is *Achillea oxyloba* subspecies *mucronulata* and written throughout the book *Achillea o. mucronulata* or *A.o. mucronulata*.

Abies alba, 22
Acer campestre, 306
  lobelii, 306
  opalus, 306
  platanoides, 306
  pseudoplatanus, 306
Aceras anthropophorum, 282
Achillea atrata, 244
  barrelieri, 244
  chamaemelifolia, 246
  clavennae, 246
  collina, 246
  erba-rotta, 246
  e. rupestris, 246
  macrophylla, 246
  millefolium, 246
  moschata, 244
  nana, 244
  odorata, 246
  oxyloba, 244
  o. mucronulata, 244
  pannonica, 246
  ptarmica, 246
  pyrenaica, 246
  tanacetifolia, 246
  t. distans, 246
Acinos alpinus, 200
  arvensis, 200
Aconitum anthora, 40
  compactum, 40
  lamarckii, 40
  napellus, 40
  paniculatum, 40
  septentrionale, 40
  tauricum, 40
  variegatum, 40
  vulparia, 40
Adenostyles alliariae, 248
  a. hybrida, 248
  alpina, 248
  leucophylla, 248
Actaea spicata, 40
Adonis aestivalis, 46
  annua, 46
  distorta, 46
  flammea, 46
  pyrenaica, 46
  vernalis, 46
Adoxa moschatellina, 226
Aethionema saxatile, 68
Agrimonia eupatoria, 96
  procera, 96
Agrostemma githago, 30
Ajuga chamaepitys, 192
  genevensis, 192
  pyramidalis, 192
  reptans, 192
  tenorii, 192
Alchemilla alpina, 102
  basaltica, 102

  conjuncta, 104
  fallax, 104
  fissa, 104
  glaucescens, 104
  hoppeana, 104
  pentaphyllea, 102
  plicatula, 104
  saxatilis, 102
  splendens, 104
  subsericea, 102
  vulgaris, 104
Allium carinatum, 266
  flavum, 266
  montanum, 266
  narcissiflorum, 266
  ochroleucum, 266
  oleraceum, 266
  saxatile, 266
  schoenoprasum, 266
  sphaerocephalum, 266
  strictum, 266
  suaveolens, 266
  ursinum, 266
  victorialis, 266
  vineale, 266
Alnus glutinosa, 28
  incana, 28
  viridis, 28
Alyssoides utriculata, 64
Alyssum alpestre, 304
  argenteum, 304
  cuneifolium, 304
  diffusum, 304
  montanum, 64
  ovirense, 64
  serpyllifolium, 304
  wulfenianum, 64
Amelanchia ovalis, 106
Anacamptis pyramidalis, 284
Anagallis minima, 174
  tenella, 174
Anchusa arvensis, 188
  barrelieri, 198
  officinalis, 188
Andromeda polifolia, 164
Androsace alpina, 170
  brevis, 337
  carnea, 170
  c. brigantiaca, 337
  c. laggeri, 337
  c. rosea, 337
  chaixii, 172
  chamaejasme, 170
  ciliata, 170
  cylindrica, 170
  elongata, 172
  hausmannii, 170
  helvetica, 170
  imbricata, 170
  lactea, 170

  mathildae, 170
  maxima, 172
  obtusifolia, 170
  pubescens, 170
  pyrenaica, 170
  septentrionalis, 172
  vandellii, 170
  villosa, 170
  wulfeniana, 170
Anemone appennina, 42
  baldensis, 42
  hepatica, 44
  narcissiflora, 42
  nemorosa, 42
  ranunculoides, 42
  sylvestris, 42
  trifolia, 42
Angelica archangelica, 158
  razulii, 158
  sylvestris, 158
Antennaria alpina, 242
  carpatica, 242
  dioica, 242
Anthemis arvensis, 242
  carpatica styriaca, 242
  cretica, 242
  c. alpina, 242
  triumfetti, 242
Anthericum liliago, 264
  ramosum, 264
Anthyllis montana, 126
  m. jacquinii, 126
  vulneraria, 126
  v. alpestris, 126
  v. carpatica, 126
  v. forondae, 126
  v. pyrenaica, 126
  v. vulneraroides, 126
Antirrhinum majus, 204
  molle, 204
  sempervirens, 204
Aphyllanthes monspeliensis, 264
Aposeris foetida, 262
Aquilegia alpina, 54
  aragonensis, 54
  atrata, 54
  bertolonii, 54
  einseleana, 54
  nigricans, 54
  pyrenaica, 54
  thalictrifolia, 54
  vulgaris, 54
Arabis allionii, 302
  alpina, 302
  caerulea, 64
  cebennensis, 302
  corymbiflora, 302
  hirsuta, 302
  pedemontana, 302

373

pumila, 302
recta, 302
scopoliana, 302
serpyllifolia, 302
soyeri, 302
s. jacquinii, 302
stricta, 302
turrita, 302
vochinensis, 302
Arctostaphylos alpinus, 164
uva-ursi, 164
Aremonia agrimonioides, 96
Arenaria aggregata, 300
a. erinacea, 300
biflora, 300
ciliata, 300
grandiflora, 300
humifusa, 300
huteri, 300
ligericina, 300
norvegica, 300
purpurascens, 300
tetraqueta, 300
Armeria alliacea, 174
maritima alpina, 174
m. halleri, 174
ruscinonensis, 337
Arnica alpina, 248
montana, 248
Artemisia atrata, 324
campestris, 324
c. alpina, 324
c. borealis, 324
chamaemelifolia, 324
genipi, 324
glacialis, 324
herba-alba, 324
insipida, 324
maritima vallesiaca, 324
mutellina, 324
nitida, 324
norvegica, 324
petrosa, 324
Arum maculatum, 274
Aruncus dioicus, 90
Asarina procumbens, 204
Asarum europaeum, 288
Asperugo procumbens, 188
Asperula arvensis, 224
cynanchica, 337
hexaphylla, 224
hirta, 337
pyrenaica, 224
taurina, 224
Asphodelus albus, 268
Aster alpinus, 240
amellus, 240
bellidiastrum, 240
pyrenaeus, 240
sedifolius trinervis, 240
Astragalus alpinus, 112
australis, 114
austriacus, 114
centralpinus, 114
cicer, 112
danicus, 112
depressus, 114
exscapus, 114

frigidus, 112
glycyphyllos, 114
leontinus, 114
monspessulanus, 116
norvegicus, 114
penduliflorus, 112
purpureus, 112
sempervirens, 114
vesicarius, 116
v. pastellianus, 116
Astrantia bavarica, 150
carniolica, 337
major, 150
m. carinthiaca, 337
minor, 150
Athamanta cortiana, 154
cretensis, 154
Atropa belladonna, 316

Ballota frutescens, 196
nigra, 196
Bartsia alpina, 212
spicata, 212
Bellis perennis, 240
Berardia subacaulis, 250
Berberis vulgaris, 56
Betula nana, 28
pendula, 28
pubescens carpatica, 28
Biscutella brevifolia, 72
cichoriifolia, 72
flexuosa, 72
intermedia, 72
laevigata, 72
scaposa, 72
Blackstonia perfoliata, 182
Borago officinalis, 184
Brassica gravinae, 70
repanda, 70
Braya alpina, 60
linearis, 335
Buglossoides arvensis, 184
gastonii, 184
purpurocaerulea, 184
Bulbocodium vernum, 264
Bunium alpinum petraeum, 152
Buphthalmum salicifolium, 248
Bupleurum angulosum, 156
falcatum, 156
f. cernuum, 156
longifolium, 156
petraeum, 156
ranunculoides, 156
r. gramineum, 156
stellatum, 156
Buxus sempervirens, 308
alpina, 200

Calamintha alpina, 200
grandiflora, 200
nepeta, 200
sylvatica, 200
Callianthemum
anemonoides, 38
coriandrifolium, 38
kerneranum, 38

Callitriche lophocarpa, 310
hamulata, 310
obtusangula, 310
palustris, 310
platycarpa, 310
stagnalis, 310
Calluna vulgaris, 164
Caltha palustris, 40
Calycocorsus stipitatus, 260
Campanula
affinis bolosii, 232
allionii, 232
alpestris, 232
alpina, 232
apennina, 236
barbata, 232
beckiana, 236
bertolae, 236
bononiensis, 234
carnica, 234
cenisia, 232
cespitosa, 236
cochleariifolia, 236
excisa, 236
foliosa, 234
fritschii, 236
glomerata, 234
hispanica catalanica, 236
jaubertiana, 236
latifolia, 234
morettiana, 234
patula, 232
p. costae, 232
persicifolia, 232
petraea, 234
precatoria, 236
pseudostenocodon, 236
pulla, 236
pusilla, 236
raineri, 234
rapunculus, 232
rapunculoides, 232
recta, 236
rhomboidalis, 234
rotundifolia, 234
scheuchzeri, 234
speciosa, 232
s. corbariensis, 232
s. oliveri, 232
spicata, 234
stenocodon, 236
thyrsoides, 234
trachelium, 234
uniflora, 234
witasekiana, 236
zoysii, 232
Capsella bursa-pastoris, 304
rubella, 304
Cardamine amara, 62
asarifolia, 62
bellidifolia, 64
b. alpina, 64
bulbifera, 64
crassifolia, 64
enneaphyllos, 62
heptaphylla, 62
kitaibelii, 62
opizii, 62

pentaphyllos, 62
plumieri, 64
pratensis, 64
raphanifolia, 62
resedifolia, 64
trifolia, 62
Cardaminopsis arenosa, 64
halleri, 64
petraea, 64
Carduus acanthoides, 252
affinis, 254
carduelis, 252
carlinifolius, 252
carlinoides, 254
chrysacanthus, 252
crispus, 252
defloratus, 252
litigiosus, 252
nutans, 252
personatus, 252
Carlina acanthifolia, 250
a. cynara, 250
acaulis, 250
a. simplex, 250
vulgaris, 250
Carum carvi, 156
rigidulum, 156
Cassiope hypnoides, 162
tetragona, 162
Centaurea alpestris, 256
alpina, 258
badensis, 256
carniolica, 338
cyanus, 258
dichroantha, 256
grinensis, 256
leucophaea, 258
maculosa, 258
montana, 258
nigrescens, 258
paniculata, 258
phrygia, 258
procumbens, 258
rhaetica, 258
rupestris, 256
spinabadia, 258
thuillieri, 258
transalpina, 338
triumfetti, 258
uniflora, 258
Centaurium erythraea, 182
pulcellum, 182
suffruticosum, 182
Centranthus angustifolius,
228
lecoqii, 228
Cephalanthera damasonium,
286
longifolia, 286
rubra, 286
Cephalaria alpina, 238
Cerastium alpinum,
294
a. lanatum, 294
arcticum, 294
arvense, 294
brachypetalum, 294
carinthiacum, 296

c. austroalpinum, 296
cerastoides, 294
diffusum, 296
fontanum, 296
glomeratum, 296
julicum, 294
latifolium, 294
lineare, 294
pedunculatum, 296
pumilum, 296
pyrenaicum, 294
scaranii, 294
semidecandrum, 296
subtriflorum, 296
tomentosum, 294
uniflorum, 294
Cerinthe glabra, 186
minor, 186
m. auriculata, 186
Chaenorhinum origanifolium,
204
Chaerophyllum elegans, 152
hirsutum, 152
villarsii, 152
Chamaecytisus hirsutus, 110
polytrichus, 110
purpureus, 110
Chamaepericlymenum
suecicum, 162
Chamaespartium sagittale,
110
Chamorchis alpina, 284
Chelidonium majus, 58
Chimaphila umbellata, 160
Cicerbita alpina, 260
plumieri, 260
Circaea alpina, 148
Cirsium acaulon, 254
carniolicum, 266
erisithales, 256
eriophorum, 254
glabrum, 256
helenioides, 254
montanum, 254
morisianum, 254
richteranum, 254
rivulare, 254
spathulatum, 254
spinosissimum, 256
waldsteinii, 254
Clematis alpina, 46
integrifolia, 46
recta, 46
vitalba, 46
Cochlearia officinalis, 68
pyrenaica, 68
Coeloglossum viride, 284
Colchicum alpinum, 264
autumnale, 264
Colutea arborescens, 112
Conopodium pyrenaicum,
152
Convallaria majalis, 274
Convolvulus arvensis, 174
Corallorhiza trifida, 286
Cornus mas, 150
sanguinea, 150
suecica, 150

Coronilla emerus, 126
vaginalis, 126
Cortusa matthiola, 172
Corydalis acaulis, 56
bulbosa, 56
intermedia, 56
lutea, 56
pumila, 56
solida, 56
Corylus avellana, 28
Cotoneaster integerrimus,
104
nebrodensis, 104
Crataegus laevigatus, 106
macrocarpa, 106
monogyna, 106
Crepis albida, 332
alpestris, 332
aurea, 334
a. glabrescens, 334
conyzaefolia, 332
lampsanoides, 332
mollis, 332
paludosa, 332
pontana, 332
praemorsa, 332
pygmaea, 334
pyrenaica, 332
rhaetica, 334
terglouensis, 334
Crocus albiflorus, 278
caeruleus, 278
napolitanus, 278
nudiflorus, 278
purpureus, 278
vernus, 278
versicolor, 278
Cuscuta epithymum, 174
europaea, 337
Cyclamen europaeum, 172
hederifolium, 172
neapolitanum, 172
purpurascens, 172
Cymbalaria muralis, 204
pallida, 337
Cynoglossum magellense,
190
nebrodense, 190
officinale, 190
Cypripedium calceolus, 280
Cytisus ardoini, 108
decumbens, 108
emeriflorus, 108
hirsutus, 110
purgans, 108
purpureus, 110
sauzeanus, 108
sessilifolius, 337

Dactylorhiza cruenta, 338
fuchsii, 282
incarnata, 282
maculata, 338
majalis, 282
m. alpestris, 338
sambucina, 282
traunsteineri, 338

Daphne alpina, 140
  cneorum, 140
  laureola, 140
  l. philippi, 140
  mezereum, 140
  petraea, 140
  striata, 140
Delphinium dubium, 42
  elatum, 42
  e. austriacum, 42
  e. helveticum, 42
  montanum, 42
Dethawia tenuifolia, 154
Dianthus alpinus, 36
  armeria, 36
  barbatus, 34
  b. compactus, 335
  carthusianorum, 36
  deltoides, 36
  furcatus, 34
  giganteus, 36
  glacialis, 36
  gratianopolitanus, 36
  monspessulanus, 34
  m. marsicus, 34
  m. sternbergii, 34
  neglectus, 36
  pavonius, 36
  plumarius, 36
  pontederae, 36
  pungens, 36
  pyrenaicus, 36
  p. catalaunicus, 335
  seguieri, 34
  s. italicus, 34
  serotinus, 36
  subacaulis, 30
  superbus, 34
  sylvestris, 36
Diapensia lapponica, 162
Digitalis grandiflora, 208
  lutea, 208
Dioscorea pyrenaica, 274
Dipcadi serotina, 272
Doronicum austriacum, 248
  carpetanum, 338
  cataractarum, 337
  clusii, 338
  columnae, 248
  grandiflorum, 248
  pardalianches, 337
  plantagineum, 338
Draba aizoides, 66
  alpina, 66
  aspera, 66
  carinthiaca, 66
  dubia, 66
  fladnizensis, 66
  hoppeana, 66
  incana, 66
  kotschyi, 66
  ladina, 66
  lasiocarpa, 66
  muralis, 66
  nemorosa, 66
  nivalis, 66
  sauteri, 66
  stellata, 66

  tomentosa, 66
Dracocephalum austriacum, 200
  ruyschiana, 200
Drosera anglica, 74
  intermedia, 74
  rotundifolia, 74
Dryas octopetala, 96

Echinospartum horridum, 110
Echium vulgare, 186
Edraianthus graminifolius, 230
Empetrum nigrum, 162
  n. hermaphroditum, 162
Endressia pyrenaica, 156
Ephedra distachya, 22
  d. helvetica, 22
Epilobium alpestre, 148
  alsinifolium, 148
  anagallidifolium, 148
  angustifolium, 148
  collinum, 148
  dodonaei, 148
  duriaei, 148
  fleisheri, 148
  hirsutum, 148
  lanceolatum, 148
  montanum, 148
  nutans, 148
  palustre, 148
  parviflorum, 148
  roseum, 148
  tetragonum, 148
Epimedium alpinum, 56
Epipactis atrorubens, 284
  helleborine, 284
  leptochila, 284
  microphylla, 284
  muelleri, 284
  palustris, 339
  purpurata, 339
Epipogium aphyllum, 286
Eranthis hyemalis, 38
Erica carnea, 164
  cinerea, 164
  herbacea, 164
  tetralix, 164
  vagrans, 164
Erigeron acer, 240
  alpinus, 240
  atticus, 240
  epiroticus, 240
  gaudinii, 240
  humilis, 242
  neglectus, 242
  polymorphus, 240
  uniflorus, 242
Erinacea anthyllis, 112
Erinus alpinus, 204
Eritrichium nanum, 190
Erodium alpinum, 132
  cicutarium, 132
  manescavi, 132
  petraeum, 132
  p. crispum, 132
  p. glandulosum, 132

  p. lucidum, 132
Erophila verna, 68
Eryngium alpinum, 152
  bourgatii, 152
  spinalba, 152
Erysimum decumbens, 60
  helveticum, 60
  hieracifolium, 60
  incanum, 60
  pumilum, 60
  sylvestre, 60
  virgatum, 60
Erythronium dens-canis, 270
Euonymus europaeus, 304
  latifolius, 304
  verrucosus, 304
Euphorbia carniolica, 134
  chamaebuxus, 134
  cyparissias, 134
  hyberna, 134
  h. canuti, 134
  kerneri, 134
  myrsinites, 134
  saxatilis, 134
  valliniana, 134
Euphrasia alpina, 212
  christii, 212
  drosocalyx, 212
  hirtella, 212
  micrantha, 212
  minima, 212
  nemorosa, 212
  rectinata, 212
  picta, 212
  portae, 212
  rostkoviana, 212
  salisburgensis, 212
  stricta, 212
  tricuspidata, 212

Fagus sylvatica, 28
Filipendula ulmaria, 90
  vulgaris, 90
Fragaria moschata, 102
  vesca, 102
  viridis, 102
Frangula alnus, 308
  rupestris, 308
Fraxinus excelsior, 310
  ornus, 310
Fritillaria delphinensis, 268
  involucrata, 268
  meleagris, 268
  pyrenaica, 268
  tenella, 268
  tubiformis, 268
  t. moggridgei, 268
Fumaria officinalis, 56
  schleicheri, 56
  vaillantii, 56

Gagea fistulosa, 264
  lutea, 264
  minima, 264
  pratensis, 264
  soleirolii, 264
  villosa, 264
Galanthus nivalis, 276

Galeopsis angustifolia, 192
  bifida, 194
  ladanum, 192
  pubescens, 194
  pyrenaicum, 192
  reuteri, 192
  segetum, 192
  speciosa, 192
  tetrahit, 194
Galium album, 312
  aniosophyllum, 314
  austriacum, 314
  boreale, 312
  caespitosum, 314
  cometerhizon, 314
  helveticum, 224
  laevigatum, 312
  lucidum, 312
  mollugo, 312
  obliquum, 314
  odoratum, 312
  palaeoitalicum, 314
  palustre, 312
  pyrenaicum, 314
  rubrum, 314
  saxosum, 314
  sylvaticum, 312
  trifidum, 312
  triflorum, 312
  uliginosum, 312
  verum, 312
Genista cinerea, 110
  germanica, 110
  hispanica, 110
  h. occidentalis, 110
  horrida, 110
  pilosa, 110
  purgans, 108
  radiata, 110
  sagittalis, 110
  sericea, 110
  tinctoria, 110
Gentiana acaulis, 178
  alpina, 178
  angustifolia, 178
  asclepiadea, 176
  bavarica, 180
  brachyphylla, 180
  b. favratii, 180
  burseri, 176
  clusii, 178
  cruciata, 178
  frigida, 178
  froelichii, 178
  kochiana, 178
  ligustica, 178
  lutea, 176
  l. symphyandra, 176
  nivalis, 180
  occidentalis, 178
  pannonica, 176
  pneumonanthe, 176
  prostrata, 178
  pumila, 180
  p. delphinensis, 180
  punctata, 176
  purpurea, 176
  pyrenaica, 178

  rostanii, 180
  tergestina, 180
  terglouensis, 180
  t. schleicheri, 180
  utriculosa, 180
  verna, 180
  v. tergestina, 180
Gentianella amarella, 182
  campestris, 180
  columnae, 180
  germanica, 182
  hypericifolia, 180
  nana, 180
  tenella, 180
Geranium, argenteum, 130
  cinereum, 130
  c. subcaulescens, 130
  divaricatum, 130
  endressii, 130
  macrorrhizum, 130
  molle, 132
  nodosum, 130
  palustre, 130
  phaeum, 130
  pratense, 130
  purpureum, 132
  pusillum, 132
  pyrenaicum, 132
  robertianum, 132
  rotundifolium, 132
  sanguineum, 130
  sylvaticum, 130
  s. rivulare, 130
Geum montanum, 96
  pyrenaicum, 96
  reptans, 96
  rivale, 96
  urbanum, 96
Glechoma hederacea, 200
Globularia cordifolia, 220
  gracilis, 220
  incanescens, 220
  meridionalis, 220
  nudicaulis, 220
  punctata, 220
  repens, 220
Gnaphalium (= Omalotheca)
  hoppeanum, 242
  norvegicum, 242
  supinum, 242
  sylvaticum, 242
Goodyera repens, 286
Gratia golaka, 154
Gymnadenia albida, 284
  conopsea, 284
  odoratissima, 284
Gypsophila repens, 34

Hacquetia epipactis, 150
Hedera helix, 150
Hedysarum boutignyanum,
  128
  hedysaroides, 128
  h. exaltatum, 128
Helianthemum apenninum,
  140
  canum, 140
  c. piloselloides, 140

  lunulatum, 140
  nummularium, 140
  n. berterianum, 140
  n. glabrum, 140
  n. grandiflorum, 140
  n. ovatum, 140
  n. pyrenaicum, 140
  n. semiglabrum, 140
  oelandicum alpestre, 140
Helleborus dumetorum, 38
  foetidus, 38
  niger, 38
  viridis, 38
Hepatica nobilis, 44
  triloba, 44
Heracleum austriacum, 158
  minimum, 158
  sphondylium, 158
Herminium monorchis, 284
Hesperis inodora, 62
  matronalis, 62
  m. candida, 62
Hieracium alpinum, 338
  arolae, 338
  aurantiacum, 262
  humile, 262
  lactucella, 338
  lanatum, 262
  pilosella, 338
Himantoglossum hircinum,
  284
Hippocrepis comosa, 126
Hippophae rhamnoides, 138
Homogyne alpina, 248
  discolor, 337
  sylvestris, 337
Horminum pyrenaicum, 204
Hottonia palustris, 172
Hugueninia tanacetifolia, 60
  t. suffruticosa, 60
Hutchinsia alpina, 68
  a. brevicaulis, 68
Hyacinthus amethystinus,
  272
Hymenolobus pauoiflorus, 68
Hyoscyamus niger, 316
Hypericum coris, 142
  hirsutum, 142
  humifusum, 142
  maculatum, 142
  montanum, 142
  nummularium, 142
  perforatum, 142
  richeri, 142
  r. burseri, 142
Hypochoeris maculata, 260
  radicata, 260
  uniflora, 260
Hyssopus officinalis, 202

Iberis aurosica, 70
  spathulata, 70
  s. nana, 70
  stricta, 70
  s. leptophylla, 70
Ilex aquifolium, 330
Impatiens noli-tangere, 138
  parviflora, 138

Iris aphylla, 278
  foetidissima, 278
  germanica, 278
  graminea, 278
  pseudacorus, 278
  variegata, 278
  xiphioides, 278
Isatis allionii, 60
  tinctoria, 335
Isopyrum thalictroides, 40

Jasione crispa, 228
  laevis, 337
  montana, 228
Jovibarba allionii, 76
  arenaria, 76
  hirta, 76
  sobolifera, 76
Juniperus communis, 22
  c. hemisphaerica, 22
  c. nana, 22
  sabina, 335
  thurifera, 22
Jurinea humilis, 250
  mollis, 250

Kernera alpina, 68
  saxatilis, 68
Knautia sylvatica, 238
Koenigia islandica, 30

Laburnum alpinum, 108
  anagyroides, 108
Lactuca perennis, 260
  quercina, 260
  sativa, 260
Lamiastrum galeobdolon,
  196
  g. flavidum, 196
  g. montanum, 196
Lamium album, 196
  amplexicaule, 196
  hybridum, 196
  maculatum, 196
  orvala, 196
  purpureum, 196
Lappula deflexa, 190
  squarrosa, 190
Larix decidua, 22
Laserpitium halleri, 158
  krapfii, 158
  latifolium, 158
  nestleri, 158
  nitidum, 158
  peucedanoides, 158
  siler, 158
Lathraea squamaria, 218
Lathyrus aphaca, 120
  bauhinii, 120
  cirrhosus, 120
  filiformis, 120
  heterophyllus, 120
  laevigatus, 120
  l. occidentalis, 120
  latifolius, 120
  montanus, 120
  pannonicus, 120
  pratensis, 120

  sativus, 120
  setifolius, 120
  sylvestris, 120
  tuberosus, 120
  venetus, 120
  vernus, 120
Lavandula angustifolia, 204
  a. pyrenaica, 204
Lembotropis nigricans, 110
Leontodon alpinus, 262
  autumnalis, 338
  croceus, 338
  helveticus, 338
  hispidus alpinus, 338
  montaniformis, 338
  montanus, 262
  pyrenaicus helveticus, 338
Leontopodium alpinum, 242
  nivale, 242
Lepidium villarsii, 72
Leucanthemopsis alpina, 244
Leucanthemum alpinum, 244
  atratum, 244
  vulgare, 244
Leucoium aestivum, 276
  vernum, 276
Leuzea centauroides, 260
  rhapontica, 260
  r. bicknellii, 260
  r. heleniifolia, 260
Levisticum officinale, 158
Ligusticum ferulaceum, 158
  lucidum, 158
  mutellina, 158
  mutellinoides, 158
Lilium bulbiferum, 270
  b. croceum, 270
  tonzigii, 204
  vulgaris, 204
Linnaea borealis, 224
Linum austriacum, 337
  cartharticum, 136
  flavum, 136
  perenne, 136
  suffruticosum salsoloides,
  136
  carniolicum, 270
  martagon, 270
  pomponium, 270
  pyrenaicum, 270
Limodorum abortivum, 284
Limosella aquatica, 318
Linaria alpina, 204
  angustissima, 204
  repens, 204
  supina, 204
  usitatissimum, 337
  viscosum, 136
Listera cordata, 286
  ovata, 286
Lithodora oleifolia, 184
Lithospermum officinale, 184
Lloydia serotina, 272
Loiseleuria procumbens, 164
Lomatogonium
  carinthiacum, 182
Lonicera alpigena, 224
  caerulea, 224

  c. pallasii, 337
  nigra, 224
  periclymenum, 224
  pyrenaica, 224
  xylosteum, 224
Lotus alpinus, 126
  corniculatus, 126
Lunaria annua, 64
  rediviva, 64
Lychnis alpina, 30
  flos-cuculi, 30
  flos-jovis, 30
  viscaria, 30
Lycium barbarum, 316
Lysimachia nemorum, 174
  nummularia, 174

Maianthemum bifolium, 272
Malus sylvestris, 106
Malva alcea, 138
  moschata, 138
  neglecta, 138
  sylvestris, 138
Matthiola fruticulosa, 62
Meconopsis cambrica, 58
Medicago hybrida, 122
  lupulina, 122
  sativa, 122
  suffruticosa, 122
Melampyrum arvense, 210
  cristatum, 210
  italicum, 210
  nemorosum, 210
  pratense, 210
  subalpinum, 210
  sylvaticum, 210
  vaudense, 210
  velebeticum, 210
Melilotus alba, 122
  altissima, 122
  officinalis, 122
Melittis melissophyllum, 198
Mentha acquatica, 202
  arvensis, 202
  longifolia, 202
  pulegium, 202
  spicata, 202
Menyanthes trifoliata, 176
Mercurialis annua, 134
  ovata, 134
  perennis, 134
Merendera montana, 264
  pyrenaica, 264
Meum athamanticum, 154
Micromeria marginata, 200
Minuartia austriaca, 298
  biflora, 298
  capillacea, 298
  cerastiifolia, 298
  cherlerioides, 298
  graminifolia, 298
  g. clandestina, 298
  grignensis, 298
  lanceolata, 298
  laricifolia, 298
  l. kitaibelii, 298
  recurva, 298
  rupestris, 298

aedoidea, 298
verna, 298
v. collina, 298
villarsii, 298
Moehringia bavarica, 300
b. insubrica, 300
ciliata, 300
dielsiana, 300
diversifolia, 300
glaucovirens, 300
muscosa, 300
tommasinii, 300
Molopospermum
peloponnesiacum, 152
Moltkia suffruticosa, 186
Moneses uniflora, 160
Monotropa hypopitys, 160
Mucizonia sedoides, 74
Murbeckiella pinnatifida, 60
zanonii, 60
Muscari atlanticum, 272
botryoides, 272
comosum, 272
racemosum, 272
Myosotis alpestris, 188
alpina, 188
arvensis, 188
caespitosa, 188
decumbens, 188
discolor, 188
lamottiana, 188
laxa, 188
ramoisissima, 188
scorpioides, 188
speluncicola, 188
stenophylla, 188
stricta, 188
sylvatica, 188
Myosurus minima, 46
Myricaria germanica, 142
Myrrhis odorata, 152

Narcissus bicolor, 276
juncifolius, 276
minor, 276
poeticus, 276
p. radiiflorus, 276
pseudonarcissus, 276
p. abscissus, 276
p. alpestris, 276
p. moschatus, 276
p. nobilis, 276
p. pallidiflorus, 276
requienii, 276
rupicola, 276
Neottia nidus-avis, 286
Nepeta cataria, 196
latifolia, 196
nepetella, 198
Nigella arvensis, 38
Nigritella miniata, 282
nigra, 282
rubra, 338
Nymphoides peltata, 176

Odontites lutea, 212
verna, 212
viscosa, 212

Omalotheca carpatica, 242
dioica, 242
hoppeana, 242
norvegica, 242
supina, 242
sylvatica, 242
Omphalodes scorpioides,
190
verna, 190
Onobrychis arenaria, 128
a. taurerica, 128
argentea hispanica, 128
montana, 128
pyrenaica, 128
saxatilis, 128
Ononis aragonensis, 122
cenisia, 122
cristata, 122
fruticosa, 122
natrix, 122
ropono, 122
rotundifolia, 122
spinosa, 122
striata, 122
Onopordum acanthium, 250
a. gautieri, 250
acaulon, 250
rotundifolium, 250
Onosma arenaria, 184
austriaca, 184
bubanii, 184
helvetica, 184
vaudensis, 184
Ophrys apifera, 280
fuciflora, 280
holosericea, 280
insectifera, 338
muscifera, 338
sphegodes, 280
Orchis coriophora, 280
laxiflora, 280
mascula, 280
militaris, 280
morio, 280
pallens, 280
palustris, 280
provincialis, 280
purpurea, 282
simia, 282
tridentata, 282
ustulata, 282
Origanum vulgare, 202
Ornithogalum
pyrenaicum, 272
umbellatum, 272
Orobanche alba, 222
alsatica, 222
amethystea, 222
arenaria, 222
caryophyllacea, 222
elatior, 222
flava, 222
gracilis, 222
laserpitii-sileris, 222
loricata, 222
lucorum, 222
lutea, 222
minor, 222

purpurea, 222
reticulata, 222
salviae, 222
teucrii, 222
Orthilia secunda, 160
Oxalis acetosella, 128
corniculata, 128
Oxyria digyna, 30
Oxytropis amethystea, 116
campestris, 116
c. tiroliensis, 116
fetida, 118
foucaudii, 116
gaudinii, 116
halleri, 118
h. velutina, 118
jacquinii, 116
lapponica, 116
pilosa, 118
pyrenaica, 116
triflora, 116

Paederota bonarota, 210
lutea, 210
Paeonia officinalis, 56
Papaver argemone, 58
burseri, 58
dubium, 58
hybridum, 58
kerneri, 58
radicatum, 58
rhaeticum, 58
rhoeas, 58
sendtneri, 58
suaveolens, 58
s. endressii, 58
Paradisea liliastrum, 264
Paris quadrifolia, 274
Parnassia palustris, 78
Pedicularis aoaulis, 214
ascendens, 216
asparagoides, 214
asplenifolia, 216
cenisia, 216
comosa, 214
elegans, 216
elongata, 214
flammea, 214
foliosa, 216
gyroflexa, 216
g. praetutiana, 216
hacquetii, 214
hirsuta, 214
julica, 214
kerneri, 214
lapponica, 214
mixta, 216
oederi, 214
palustris, 214
portenschlagii, 216
pyrenaica, 216
recutita, 214
rosea, 214
r. allionii, 214
rostratoapitata, 216
rostratospicata, 216
sylvatica, 214

tuberosa, 214
verticillata, 214
Petasites albus, 248
frigidus, 248
hybridus, 248
paradoxus, 248
Petrocallis pyrenaica, 68
Petrocoptis crassifolia, 335
hispanica, 30
pyrenaica, 30
Petrorhagia prolifera, 335
saxifraga, 34
Peucedanum ostruthium, 158
venetum, 158
verticillare, 158
Phyllodoce caerulea, 164
Physoplexis comosa, 228
Phyteuma balbisii, 230
betonicifolium, 228
charmelii, 230
comosum, 228
globulariifolium, 230
g. pedemontana, 230
hallleri, 337
hedraianthifolium, 230
hemisphaericum, 230
humile, 230
michelii, 337
nanum, 230
nigrum, 230
orbiculare, 230
ovatum, 337
pyrenaicum, 337
rupicola, 230
scheuchzeri, 230
scorzonerifolium, 228
sieberi, 230
spicatum, 228
zahlbruckneri, 337
Picea abies, 22
a. alpestris, 335
Pimpinella major, 154
saxifraga, 154
siifolia, 154
Pinguicula alpina, 220
grandiflora, 220
leptoceras, 220
longifolia, 220
l. reichenbachiana, 220
villosa, 220
vulgaris, 220
Pinus cembra, 22
mugo, 22
nigra, 22
n. salzmannii, 22
sylvestris, 22
uncinata, 335
Plantago alpina, 322
argentea, 322
atrata, 322
holosteum, 322
lanceolata, 322
major, 322
maritima serpentina, 322
media, 322
monosperma, 322
Platanthera bifolia, 286
chlorantha, 286

Pleurospermum austriacum,
154
Polemonium caeruleum, 174
Polygala alpestris, 136
alpina, 136
amara, 136
amarella, 337
carueliana, 337
chamaebuxus, 136
comosa, 136
nicaeensis, 136
serpyllifolia, 136
vayredae, 136
Polygonatum multiflorum,
274
officinale, 274
verticillatum, 274
Polygonum alpinum, 30
aviculare, 30
bistorta, 30
viviparum, 30
Populus alba, 26
nigra, 26
tremula, 26
Potentilla alchimilloides, 100
apennina, 102
anserina, 98
argentea, 98
aurea, 100
brauniana, 100
calabra, 98
carniolica, 102
caulescens, 100
cineria, 100
clusiana, 100
crantzii, 100
delphinensis, 100
erecta, 100
frigida, 100
fruticosa, 98
gammopetala, 102
grandiflora, 100
multifida, 98
neglecta, 98
nitida, 102
nivalis, 000
nivea, 98
palustris, 98
pennsylvanica, 98
pyrenaica, 100
reptans, 100
rupestris, 98
saxifraga, 102
tabernaemontani, 100
thuringiaca, 100
valderia, 100
Prenanthes purepurea, 260
Primula allionii, 168
apennina, 168
auricula, 168
clusiana, 166
daonensis, 168
elatior, 166
e. intricata, 166
farinosa, 166
glaucescens, 166
glutinosa, 166
halleri, 166

hirsuta, 168
integrifolia, 168
latifolia, 168
longiflora, 166
marginata, 168
minima, 166
pedemontana, 168
scandinavica, 168
scotica, 168
spectabilis, 166
stricta, 168
tyrolensis, 168
veris, 166
villosa, 168
viscosa, 168
vulgaris, 166
wulfeniana, 166
Prunella grandiflora, 200
laciniata, 200
vulgaris, 200
Prunus avium, 108
brigantina, 108
mahaleb, 108
padus, 108
spinosa, 108
Ptilotrichum lapeyrousianum,
64
pyrenaicum, 64
Pulmonaria angustifolia, 186
kerneri, 186
longifolia, 186
montana, 186
obscura, 186
officinalis, 186
stiriaca, 186
visianii, 186
Pulsatilla alba, 44
alpina, 44
a. apiifolia, 44
halleri, 44
h. styriaca, 44
montana, 44
pratensis, 44
rubrum, 44
vernalis, 44
vulgaris, 44
Pyrola chlorantha, 160
media, 160
minor, 160
norvegica, 160
rotundifolia, 160
Pyrus amygdaliformis, 106
austriaca, 106
communis, 106
nivalis, 106
pyraster, 106

Quercus petraea, 28
pubescens, 28
p. palensis, 28
pyrenaica, 28
robur, 28

Ramonda myconi, 218
Ranunculus aconitifolius, 50
acris, 48
aduncus, 48
alpestris, 50

amplexicaulis, 52
auricomus, 50
bilobus, 50
brevifolius, 50
bulbosus, 50
carinthiacus, 48
crenatus, 50
flammula, 52
glacialis, 52
gouanii, 48
grenieranus, 48
hybridus, 50
hyperboreus, 50
lanuginosa, 48
lingua, 52
montanus, 48
nemorosus, 48
n. polyanthemophyllus, 48
n. serpens, 48
nivalis, 50
oreophilus, 40
parnassifolius, 52
platanifolius, 50
polyanthemos, 48
p. polyanthemoides, 40
pygmaeus, 52
pyrenaeus, 52
repens, 48
reptans, 52
ruscinonensis, 48
sardous, 50
seguieri, 50
thora, 50
traunfellneri, 50
venetus, 48
Reseda glauca, 72
lutea, 72
luteola, 72
phyteuma, 72
Rhamnus alpinus, 308
catharticus, 308
pumilus, 308
saxatilis, 308
Rhinanthus alectorolophus, 218
alpinus, 218
angustifolius, 218
antiquus, 218
aristatus, 218
borbasii, 218
burnatii, 218
carinthiacus, 218
minor, 218
ovifugus, 218
songeonii, 218
wettsteinii, 218
Rhizobotrya alpina, 68
Rhodiola rosea, 78
Rhododendron ferrugineum, 164
hirsutum, 164
lapponicum, 162
Rhodothamnus chamaecistus, 164
Rhynchosinapsis cheiranthos, 72
richeri, 72
Ribes alpinum, 78
nigrum, 78

petraeum, 78
rubrum, 78
uva-crispa, 78
Rosa
agrestis, 94
alpina, 92
andegavensis, 92
arvensis, 92
canina, 94
elliptica, 94
gallica, 92
glauca, 92
jundzillii, 92
majalis, 92
micrantha, 94
montana, 92
obtusifolia, 94
pendulina, 92
pimpinellifolia, 92
rubiginosa, 94
rubrifolia, 92
stylosa, 92
tomentosa, 94
villosa, 94
vosagiaca, 94
Rubus arcticus, 90
caesius, 90
chamaemorus, 90
fruticosus, 90
idaeus, 90
saxatilis, 90
Rumex acetosa, 290
acetosella, 290
alpinus, 290
amplexicaulis, 290
arifolius, 290
gussonii, 290
longifolius, 290
nivalis, 290
scutatus, 290

Sagina apetala, 292
caespitosa, 292
glabra, 292
intermedia, 292
nodosa, 292
procumbens, 292
saginoides, 292
Salix alpina, 24
appendiculata, 26
arbuscula, 26
bicolor, 24
breviserrata, 24
caesia, 26
crataegifolia, 24
glabra, 26
glandulifera, 26
glaucosericea, 24
hastata, 24
hegetschweileri, 24
helvetica, 26
herbacea, 24
laggeri, 26
lanata, 26
lapponum, 26
mielichhoferi, 24
myrsinites, 24
phylicifolia, 24
polaris, 24

pyrenaica, 24
reticulata, 24
retusa, 24
serpyllifolia, 24
waldsteiniana, 26
Salvia glutinosa, 194
nemorosa, 194
pratensis, 194
verticillata, 194
Sambucus ebulus, 226
nigra, 226
racemosa, 226
Samolus valerandi, 174
Sanguisorba dodecandra, 96
minor, 96
officinalis, 96
Sanicula europaea, 150
Santolina chamaecyparissus, 244
tomentosa, 244
Saponaria bellidifolia, 34
caespitosa, 34
lutea, 34
ocymoides, 34
pumilio, 34
Sarcocapnos enneaphylla, 56
Saussurea alpina, 250
a. depressa, 338
a. macrophylla, 338
discolor, 250
pygmaea, 250
Saxifraga adscendens, 80
a. parnassica, 336
aizoides, 82
aizoon, 88
androsacea, 82
aphylla, 336
aquatica, 82
arachnoidea, 80
aretioides, 84
aspera, 80
biflora, 84
bryoides, 336
bulbifera, 84
burseriana, 84
caesia, 84
callosa, 86
cebennensis, 83, 336
cernua, 84
cespitosa, 84
clusii, 80
cochlearis, 86
corbariensis, 336
cotyledon, 86
crustata, 86
cuneata, 336
cuneifolia, 80
depressa, 336
diapensioides, 84
exarata, 82
facchinii, 336
florulenta, 88
foliolosa, 336
geranioides, 82
× geum, 336
glabella, 336
granulata, 84
hariotii, 336
hieracifolia, 80

hirculus, 80
hirsuta, 336
h. paucicrenata, 336
hostii, 86
ssp. rhaetica, 86
hypnoides, 336
lingulata, 86
longifolia, 88
marginata, 336
media, 86
moschata, 82
muscoides, 336
mutata, 88
nervosa, 82
nivalis, 336
oppositifolia, 84
o. blepharophylla, 84
o. latina, 84
o. murithiana, 84
o. rudolphiana, 84
o. speciosa, 84
paniculata, 88
paradoxa, 80
pedemontana, 82
p. prostii, 336
pentadactylis, 82
petraea, 336
porophylla, 86
praetermissa, 82
presolanensis, 82
pubescens, 84
p. iratiana, 84
retusa, 84
r. augustana, 336
rivularis, 336
rotundifolia, 80
sedoides, 82
seguieri, 336
squarrosa, 337
stellaris, 80
s. alpigena, 336
tenella, 82
tenuis, 336
tombeanensis, 84
tridactylites, 80
umbrosa, 80
valdensis, 86
vandellii, 336
Scabiosa cinerea, 238
c. hladnikiana, 238
graminifolia, 238
lucida, 238
stricta, 238
vestina, 238
Scilla bifolia, 272
liliohyacinthus, 272
verna, 272
Scleranthus perennis, 292
p. polycnemoides, 292
Scopolia carniolica, 316
Scorzonera aristata, 262
austriaca, 338
humilis, 338
purpurea, 338
rosea, 338
Scrophularia alpestris, 318
auriculata, 318
canina, 318

c. hoppii, 318
nodosa, 318
pyrenaica, 318
scopolii, 318
umbrosa, 318
vernalis, 318
Scutellaria alpina, 190
Sedum acre, 76
album, 76
alpestre, 336
alsinefolium, 76
anacampseros, 76
anglicum, 76
annuum, 78
anopetalum, 76
atratum, 78
a. carinthiacum, 78
cepaea, 78
dasyphyllum, 76
hispanicum, 78
magellense, 336
monregalense, 76
ochroleucum, 76
reflexum, 76
rubens, 76
sartorianum hildebrandtii, 76
sexangulare, 76
telephium, 76
villosum, 76
Selinum pyrenaeum, 156
Sempervivum andreanum, 335
arachnoideum, 74
a. tomentosum, 335
calcareum, 74
cantabricum, 74
dolomiticum, 74
grandiflorum, 74
montanum, 74
m. burnatii, 335
m. stiriacum, 336
pittonii, 335
tectorum, 74
wulfenii, 74
Senecio abrotanifolius, 326
adonidifolius, 326
alpinus, 326
balbisianus, 326
cacaliaster, 328
doronicum, 326
helenitis, 326
incanus, 328
i. carniolicus, 328
i. persoonii, 328
integrifolius, 326
i. aurantiacus, 326
i. capitatus, 326
lanatus, 326
leucophyllus, 328
nemorensis, 328
n. fuchsii, 328
ovirensis, 326
rivularis, 326
squalidus, 326
subalpinus, 326
tournefortii, 326
uniflorus, 328

viscosus, 328
vulgaris, 328
Serapias vomeracea, 282
Serratula lycopifolia, 260
tinctoria, 260
Sesamoides pygmaea, 72
Sesili nanum, 154
Sibbaldia procumbens, 102
Sideritis endressii, 192
hyssopifolia, 192
Silene acaulis, 32
a. longiscapa, 32
alpestris, 32
auriculata, 335
borderi, 32
campanula, 32
cordifolia, 32
dioica, 335
elizabetha, 32
furcata, 335
italica, 335
nutans, 32
otites, 32
pusilla, 335
rupestris, 32
saxifraga, 32
vallesia, 32
veselskyi, 335
vulgaris, 32
v. glareosa, 335
v. prostrata, 335
wahlbergella, 32
Sisymbrium austriaca, 60
a. chrysanthum, 60
irio, 60
loeselii, 60
Solanum dulcumara, 316
nigrum, 316
Soldanella alpina, 172
austriaca, 172
hungarica major, 172
minima, 172
m. samnitica, 172
montana, 172
pusilla, 172
villosa, 172
Solenanthus apenninus, 190
Solidago virgaurea, 240
Sorbus aria, 106
aucuparia, 106
austriaca, 106
chamaemespilus, 106
mougeotii, 106
torminalis, 106
Spiraea chamaedryfolia, 90
decumbens tomentosa, 90
Spiranthes aestivalis, 286
spiralis, 286
Stachys alopecuros, 198
alpina, 198
annua, 198
densiflora, 198
germanica, 198
monieri, 198
officinalis, 198
palustris, 198
recta, 198
sylvatica, 198

Stellaria alsine, 292
  calycantha, 292
  crassipes, 292
  graminea, 292
  holostea, 292
  longifolia, 292
  media, 292
  neglecta, 292
  nemorum, 292
  pallida, 292
Streptopus amplexifolius, 274
Succisa pratensis, 238
Swertia perennis, 182
Symphytum officinale, 188
  tuberosum, 188

Tamus communis, 274
Taraxacum apenninum, 330
  ceratophorum, 330
  cucullatum, 330
  dissectum, 330
  fontanum, 330
  glaciale, 328
  hoppeanum, 330
  nigricans, 330
  pacheri, 328
  palustre, 330
  phymatocarpum, 328
  schroterianum, 328
Taxus baccata, 22
Teucrium chamaedrys, 190
  lucidum, 190
  montanum, 190
  pyrenaicum, 190
Thalictrum alpinum, 54
  aquilegifolium, 54
  foetidum, 52
  macrocarpum, 54
  minus, 52
  simplex, 52
  tuberosum, 54
Thesium alpinum, 288
  bavarum, 288
  divaricatum, 288
  linophyllon, 288
  pyrenaicum, 288
  p. alpestre, 288
Thlaspi alpestre, 70
  alpinum, 70
  brachypetalum, 70
  goesingense, 70
  kerneri, 70
  montanum, 70
  praecox, 70
  rotundifolium, 70
  r. cepaefolium, 70
  stylosum, 70
Thymelaea calycina, 138
  dioica, 138
  passerina, 138
  pubescens, 138
  tinctoria, 138
Thymus alpestris, 202
  glabrescens, 202
  g. decipiens, 202
  nervosus, 202
  praecox polytrichus, 202

pulegioides, 202
  serpyllum, 202
Tilia cordata, 314
  platyphyllos, 314
  × vulgaris, 314
Tofieldia calyculata, 264
  pusilla, 264
Tolpis staticifolia, 262
Tozzia alpina, 210
Tragopogon dubium, 262
  pratensis, 262
Traunsteinera globosa, 286
Trientalis europaea, 174
Trifolium alpestre, 124
  alpinum, 124
  badium, 124
  hybridum, 124
  incarnatum, 124
  medium, 124
  montanum, 124
  noricum, 124
  ochroleucon, 126
  pallescens, 124
  pannonicum, 126
  pratense, 124
  repens, 124
  rubens, 126
  saxatile, 124
  spadiceum, 124
  thalii, 124
Trochiscanthes nodiflora, 154
Trollius europaeus, 38
Tulipa australis, 268
  didieri, 268
  sylvestris australis, 286
Tussilago farfara, 248

Ulex europaeus, 112
  minor, 112
Urtica dioica, 288
  kioviensis, 288
  urens, 288
Utricularia minor, 220

Vaccinium microcarpum, 162
  myrtillus, 162
  oxycoccus, 162
  uliginosum, 162
  virtis-idaea, 162
Valeriana celtica, 226
  dioica, 226
  elongata, 228
  globulariifolia, 226
  montana, 226
  officinalis, 226
  o. collina, 226
  pyrenaica, 226
  saliunca, 226
  saxatilis, 228
  supina, 228
  tripteris, 226
Veratrum album, 268
  nigrum, 268
Verbascum argenteum, 206
  chaixii, 206
  densiflorum, 206
  lanatum, 206

lychnitis, 206
  nigrum, 206
  phlomoides, 206
  pulverulentum, 206
  thapsus, 206
  t. crassifolium, 206
Verbena officinalis, 190
Veronica agrestis, 320
  allionii, 320
  alpina, 208
  anagallis-aquatica, 320
  aphylla, 210
  aragonensis, 208
  arvensis, 320
  austriaca, 208
  a. teucrium, 208
  beccabunga, 210
  bellidioides, 208
  b. lilacina, 208
  catenata, 320
  chamaedrys, 210
  fruticans, 208
  fruticulosa, 208
  hederifolia, 320
  longifolia, 320
  montana, 320
  nummularia, 208
  officinalis, 320
  opaca, 320
  polita, 320
  ponae, 208
  prostrata, 208
  scutellata, 320
  serpyllifolia, 208
  s. humifusa, 208
  spicata, 210
  verna, 320
Viburnum lantana, 226
  opulus, 226
Vicia argentea, 118
  cracca, 118
  faba, 118
  hirsuta, 118
  incana, 118
  onobrychioides, 118
  oroboides, 118
  pyrenaica, 118
  sativa, 118
  sepium, 118
  sylvatica, 118
  tenuifolia, 118
  tetrasperma, 118
Vinca minor, 182
Vincetoxicum hirundinaria, 184
Viola alba scotophylla, 144
  alpina, 146
  ambigua, 144
  bertolonii, 146
  biflora, 144
  bubanii, 146
  calcarata, 146
  c. villarsiana, 146
  c. zoysii, 146
  canina, 144
  conisia, 146
  collina, 144
  comollia, 146

cornuta, 146
diversifolia, 146
dubyana, 146
eugeniae, 146
hirta, 144
lutea, 146
magellensis, 146
mirabilis, 144
nummulariifolia, 146
odorata, 144

palustris, 144
pinnata, 146
pyrenaica, 144
reichenbachiana, 144
riviniana, 144
rupestris, 144
suavis, 144
thomasiana, 144
tricolor, 146
t. subalpina, 146

valderia, 146
Viscum album, 288
a. abietis, 288
a. austriacum, 288
Vitaliana, primuliflora, 168

Waldsteinia ternata, 98
Wulfenia carinthiaca, 208

Xatardia scabra, 154

METRES/ FEET CONVERSION

0	500	1000	2000	3000	4000	5000 METRES
0	1640	3280	6560	9840	13120	16400 FEET